Rehav Rubin
Portraying the Land

Rehav Rubin

Portraying the Land

Hebrew Maps of the Land of Israel from Rashi
to the Early 20th Century

DE GRUYTER   MAGNES

ISBN 978-3-11-068320-2
e-ISBN (PDF) 978-3-11-057065-6
e-ISBN (EPUB) 978-3-11-056893-6

Library of Congress Control Number: 2018934508

Bibliographic information published by the Deutsche Nationalbibliothek
The Deutsche Nationalbibliothek lists this publication in the Deutsche Nationalbibliografie; detailed bibliographic data are available in the Internet at http://dnb.dnb.de.

© 2019 Walter de Gruyter GmbH, Berlin/Boston
This volume is text- and page-identical with the hardback published in 2018.
& Hebrew University Magnes Press, Jerusalem
Published in Hebrew as T*surat Ha-aretz. Eretz Yisrael ba-mappa ha-Ivrit mi-Rashi ve-ad reshit ha-meah ha-esrim* © 2014 Yad Izhak Ben-Zvi, Jerusalem.
Cover image: A map of Eretz Yisrael in the Oxford manuscript of Be'er mayim ḥayim, opp. 195, 234b, Bodleian Libray
Printing and binding: CPI books GmbH, Leck

www.degruyter.com
www.magnespress.co.il

Contents

Preface —— VII

List of the Maps Presented in the Book —— IX

Introduction —— XV

1   Rashi and His Maps —— 1

2   Following in Rashi's Footsteps —— 23

3   Between Jewish and Christian Maps: The Sixteenth-Century Map of the Israelites' Peregrinations from Mantua —— 80

4   Following in the Footsteps of Christian Cartographers —— 102

5   The Debut of the Tribal Allotments on the Traditional Hebrew Map —— 173

6   Cartographic Tableaux of the Holy Places —— 204

7   From Tradition to Modernity—the Hebrew Map between the Nineteenth Century and Early Twentieth Century —— 221

Conclusion —— 283

Cartographic Material —— 293

Bibliography —— 297

Index —— 317

Preface

The work before you is dedicated to maps and graphic descriptions of the Land of Israel, of the Israelites' peregrinations through the wilderness, the allotments of the tribes, and places deemed to be sacred in the Hebrew literature from the days of Rashi until the early twentieth century. This book is rooted in the study of ancient maps of Jerusalem and the Holy Land—a field that I have been immersed in for quite some time. However, until now, my focus has been on maps that were drafted by Christians in Europe. The provenance of this study is an invitation that I received in the fall of 2006 to lecture on the development and meaning of Jewish maps at a conference in honor of Virginia Garrett at the University of Texas-Arlington. While preparing for this paper, the broad horizons of this corpus opened up before my eyes, as I discovered a bevy of maps that I was hitherto unfamiliar with. At the time, I did not fathom just how multi-faceted this project was destined to become or the rich details that were latent in these works. Only upon delving into this topic did I realize the formidable challenge that laid ahead: A map is a work that brings together content and form, namely the respective message and means by which the former is presented and transmitted. With this in mind, I endeavored to sort the maps into types and groups, to glean their sources and the key processes that inform their development, and to decipher the connections between the various types and groups. Above all, I sought to take stock of the perpetual dialogue between the jug and the wine, the synergy between content and form that creates the message.

Alongside Hebrew maps, I also encountered a small handful in Yiddish and Ladino as well as multilingual maps that were printed in Hebrew and European languages, all of which have been included in this book. My access to the myriad maps housed in collections the world over, not least the National Library of Israel in Jerusalem, and the sabbatical year that I spent at the University of Pennsylvania enabled me to expand this project on the fly. In this fashion, the research background, the topic itself, the sources of the maps, and the opportunity all fell into place.

As the book's author, I owe a debt of gratitude to the large number of organizations and individuals without whom this project would not have come to fruition: to the staff and management of the National Library of Israel in Jerusalem, particularly Ayelet Rubin, the director of the Eran Laor Cartographic Collection, from which the lion's share of the work's maps were taken; to the librarians at the Center for Advanced Judaic Studies in Philadelphia; to the Israel Museum in Jerusalem and Ariel Tishby; to the National Maritime Museum in Haifa; to the Yad Ben-Zvi library and Dr. Dov Cohen; to the Zurich Central Library; the Bavarian State Library in Munich; the Berlin State Library; Oxford University's Bodleian Library; the British Library in London; the Bibliothèque nationale de France in Paris; the Vatican Apostolic Library; the Biblioteca Ambrosiana in Milan; and the Jewish Museum in New York. I would also like to thank those private collectors— Mr. René Braginsky, Mr. William

Gross, the late Dr. Alfred Moldovan, Mr. Amnon Dor, and those who preferred to remain anonymous —who were kind enough to place their treasures at my disposal.

This book was originally published in Hebrew in 2014 by Yad Izhak Ben-Zvi in Jerusalem; my gratitude goes to the dedicated and professional staff at the Yad Ben-Zvi press for the production of the Hebrew edition. I also wish to thank the James Amzalak Fund for Research in Historical Geography for supporting the English translation.

Furthermore, I would like to thank those colleagues who read parts of the manuscript and enlightened me on various matters: Professor Haim Goren, Professor Mayer Gruber, Professor Mordechai (Motti) Zalkin, Dr. Zur Shalev, Dr. Aviad Stollman and Rabbi Menachem Silber. I am also indebted to Professor Gad Freudenthal, Professor Aharon Maman, Professor Shalom Sabar, Mr. Reuven Salomons, and Prof. Doron Bar who helped me with specific research questions; to Professor Dennis P. Reinhartz and Professor Gerald Saxon from the University of Texas-Arlington; to Dorit Ayalon, Shai Cohen and Ilana Har-Tal, my students and research assistants at The Hebrew University. Special thanks to Avi Aronsky who was much more than a translator and whose kin eyes improved every page in this book.

I am especially grateful to my wife Dr. Milka Levy-Rubin who shared her wisdom, advice, and extensive knowledge. As always, she was my first reader and critic.

All of these talented people lent a hand and were indeed responsible for the few virtues this book has to offer. I bear full responsibility, though, for the shortcomings and mistakes that taint this work.

Last but certainly not least, this book is dedicated to the loving memory of my son Yoav, who assisted me at several phases of this study. A precocious and prodigious erudite with insatiable curiosity, who mastered both the wisdom of Israel and general knowledge, Yoav left us before his time. Insufficient as it may be, it is my hope that this book shall honor his memory, the goodness of his heart, and his abundant wisdom that I so sorely miss.

Rehav (Buni) Rubin
Jerusalem 2017

List of the Maps Presented in the Book

Chapter 1

Fig. 1 Rashi's map for Numbers 33, the Munich manuscript, Bayerische Staatsbibliothek München (Cod. Hebr. 5(1), fol. 139v).
Fig. 2 Rashi's map for Numbers 33, the Paris manuscript, Bibliothèque Nationale de France (Hébreu 154, fol. 35v).
Fig. 3 Rashi's map for Numbers 34 in the Munich manuscript, Bayerische Staatsbibliothek München (cod. Hebr. 5(1), fol. 140r).
Fig. 4 Rashi's map for Numbers 34 in the Paris manuscript, Bibliothèque Nationale de France (Hébreu 155, fol. 179v).
Fig. 5 Map of the Land of Canaan from Rashi's commentary on the Pentateuch, manuscript from the National Library of Israel, Jerusalem (copied by Yekutiel ben Shlomo), pp. 246–247.
Fig. 6 Rashi's diagram of the Shechem [Nablus] road, Judges 21:19, the Oxford manuscript, Bodleian Library (opp. fol. 125a).
Fig. 7 Rashi's maps for Ezekiel 45 and 48, Paris manuscript, Bibliothèque Nationale de France (Hébreu, 154, fol. 135v and fol. 139r).
Fig. 8 A Latin map for Ezekiel 58 that draws on Rashi's map, the Paris manuscript, Bibliothèque Nationale de France (Latin 3438, fol. 81).

Chapter 2

Fig. 9 A map in a manuscript of the Ḥazzekuni commentary on Mas'ei, the Oxford manuscript, The Bodleian Library, University of Oxford, ms. Mich. 568.
Fig. 10 The map in a manuscript of R. Joseph Kara's commentary on the Latter Prophets, Paris, Bibliothèque Nationale de France, Hébreu, 162, fol. 174r.
Fig. 11 Map of the Land of Israel in a manuscript of Rashi's commentary, the British Library, Add. MS 27128, fol. 83..
Fig. 12 Map in the Berlin manuscript, Berlin State Library, Ms. or. fol. 121, page 134b.
Fig. 13 Map in the Vatican manuscript, Biblioteca Apostolica Vaticana. Biblioteca Apostolica Vaticana. MS Borg. Ebr. 4
Fig. 14 Map from the manuscript of Kaftor va'ferech, Veneranda Biblioteca Ambrosiana, Milano, De Agostini Picture Library.
Fig. 15 Elijah Mizrachi, Commentary on Rashi on the Torah (Venice: Daniel Bomberg, 1545), p. 282a, 170 x 140 mm.
Fig. 16 Mizrachi's map as rendered in Thomas Shaw, Travels, or Geographical, Physical and Miscellaneous Observations Relating to Several Parts of Barbary and the Levant, 150 x 145 mm.
Fig. 17 Map of Eretz Yisrael's borders in Simeon Halevi Aschaffenburg, Sefer Devek Tov page 113a, 142 x 91 mm.
Fig. 18 Illustrations in Sefer devek tov: Jacob's ladder on page 28b; diagram of the Red Sea on page 53a.
Fig. 19 Map of Natan ben Shimshon Shapira and the cover of his book Bai'urim al ha'eshel ha'gadol, p. 150, 165 x 143 mm.
Fig. 20 Illustration of Jacob's ladder and the cluster of grapes, in Natan ben Shimshon Shapira, ibid. pp. 28b and 53, respectively.
Fig. 21 The map of Mordecai Yoffe in Levush ha'orah, page 82b, 185 x 145 mm.

Fig. 22	A map of Eretz Yisrael in the Oxford manuscript of Be'er mayim ḥayim, opp. 195, 234b.
Fig. 23	Map of Yosef ben Yissachar Miklish, Sefer yosef da'at, p. 128a, 180 x 130 mm.
Fig. 24	The Map as per Mizrachi (150 x 93 mm), p. 59b; and the map as per Yoffe (169 x 142 mm), Four Commentaries on Rashi, part 4: Numbers, p. 1a (Warsaw: Isaac Rathauer Print, 1864) [Hebrew].
Fig. 25	Map comprised of print characters, Me'am lo'ez, Book of Numbers (Constantinople: 1764), p. 158, Yad Ben-Zvi Library, 185 x 300 mm.
Fig. 26	The lined map of the Livorno edition of the Me'am lo'ez (1822–1823), p. 143b, Amnon Dor collection, 155 x 230 mm.
Fig. 27	Map in the Izmir edition of the Me'am lo'ez (1867), the Dov Cohen Collection p. 158a, 230 x 160.
Fig. 28	Pair of illustrations from Me'am lo'ez, Book of Numbers (1715), p. 65a.
Fig. 29	Pair of illustrations from Me'am lo'ez, Book of Numbers (Livorno), p.59a.
Fig. 30	Pair of illustrations from Me'am lo'ez, Book of Numbers (Izmir), p. 63a.
Fig. 31	The layout of the tribal camp, Me'am lo'ez (Livorno edition), between pages 8 and 9, Amnon Dor collection, 230 x 315 mm.
Fig. 32	Map of Eretz Yisrael's borders by Haim Isaiah Hacohen, Sefer tshu'ot ḥayim, p. 76, 156 x 112 mm.
Fig. 33	Rashbam and Rashi's maps, Five Books of the Torah […], volume 4, Book of Numbers, page 152b, 165 x 95 mm and 205 x 125 mm, respectively.
Fig. 34	Eidels' diagrams, Sefer ḥiddushei halakhot, chapter I, p. 123b, 36 x 60 mm; 34 x 52 mm.
Fig. 35	Diagram of Acre and its environs in Meir ben Gedalia, Sefer meir einai ḥakhamim (1709), p. 115b, 70 x 40 mm.
Fig. 36	Diagram of Acre and its environs in Yisrael Lipschitz, Tiferet Yisrael, p. 163a (i.e., 325), 40 X 30 mm.
Fig. 37	Map of shalosh artzot le'shevi'it in Shlomo Adeni, Malekhet shlomo, p. 162b, 75 x 75 mm.
Fig. 38	Diagram in Ḥiddushei halakhot
Fig. 39	Two diagrams in Áron Chorin, "Dirat aharon" (Áron's Apartment), Emek ha'shaveh: the first as per the MaHaRShA (p. 68b, circa 70 x 70 mm); and the second as per Chorin (p. 69a, 120 x 90 mm).

Chapter 3

Fig. 40	The Mantua Map, the Central Library of Zurich. 377 x 509 mm
Fig. 41	Jerusalem and the Temple in the Mantua Map, and the Venice Haggadah.
Fig. 42	Illustration of the spies in the Mantua Map and in Natan Shapira, Venice, 1593.
Fig. 43	Text of the Mantua Map.
Fig. 44	Cover page of the Mantua Haggadah and the biographical details of Bassan, Seder haggadah shel pesaskh.
Fig. 45	A comparison between the candelabrums in the Mantua Map and Shapira's supercommentary.
Fig. 46	The map of The Geneva Bible, Richard Harrison edition (London: 1562), 154 x 210 mm.
Fig. 47	Details from The Geneva Bible map.

Chapter 4

Fig. 48	Christian van Adrichom, "Situs Terrae Promissionis," Theatrum Terrae Sanctae, 1010 x 354 mm.
Fig. 49	Yaacov ben Avraham Zaddik, "A Drawing of the Situation of the Lands of Canaan," Bibliothèque Nationale Française, 1626 x 523 mm.
Fig. 50	The portrait and cartouche in Zaddik's map.
Fig. 51	The Dead Sea in Adrichom and Zaddik's maps.
Fig. 52	The Sea of Galilee in Adrichom and Zaddik's maps.
Fig. 53	A ship with the Star of David on Zaddik's map.
Fig. 54	Abraham bar Yaacov's map in the Amsterdam Haggadah, 480 x 262 mm.
Fig. 55	The Hondius-Mercator map, 494 x 370 mm; Gerardus Mercator and Jodocus Hondius, "Situs Terrae Promissionis."
Fig. 56	Figures of Moses and Aaron on the Amsterdam Haggadah's cover and in the Hondius-Mercator map.
Fig. 57	Hondius-Mercator as the source of bar Yaacov's map: Jonah's boat in the two versions.
Fig. 58	The personification of Egypt in both the bar Yaacov and Plancius map; Petrus Plancius, "Orbis Terrarum Typus de Integro Multis in Locis Emendatus."
Fig. 59	The Temple in the Amsterdam Haggadah and its source: Matthaeus Merian, Figures de la Bible.
Fig. 60	Map in the manuscript of the Herlingen Haggadah, 1730, 505 x 292 mm. Braginsky Collection, Zurich. Photography by Ardon Bar-Hama, Ra'anana, Israel.
Fig. 61	The cover page of Ḥug ha'areṣ in the manuscript at the National Library of Israel.
Fig. 62	The tents of Kedar in the Adrichom map.
Fig. 63	Shlomo of Chelm's map of Judah, Simeon, Benjamin, and Dan.
Fig. 64	The plain of Jericho and the location of Lydda and Hadid in the Adrichom and Chelm maps.
Fig. 65	Matthaeus Seutter, "Palaestina seu Terra a Mose et Iosua occupata et inter Iudaeos distributa per XII Tribus vulgo Sancta adpellata," 570 x 493 mm.
Fig. 66	Details in the Seutter map: Acre, Be'er Sheva, Rhinocorurah and Lake Bardawil.
Fig. 67	Shlomo of Chelm's map of the Twelve Tribes
Fig. 68	The map of Yehonatan ben Yaacov, 300 x 410 mm, in Ya'ari, "Miscellaneous Bibliographical Notes 41.
Fig. 69	Is Heidmann the source of Yehonatan ben Yaacov's map? Christoffer Heidmann, "Totius Terrae Sanctae Delineatio," 320 x 268 mm.
Fig. 70	Yaacov Auspitz's the sons of Noah map, 158 x 138 mm, from Be'er ha'luḥot.
Fig. 71	Yaacov Auspitz, "These are the Israelites' Journeys" map, 158 x 205 mm.
Fig. 72	Augustin Calmet's map of the Exodus from Egypt, 430 x 310 mm.
Fig. 73	The circumnavigation of the Red Sea in Franciscus Quaresmius' map, "Imago transitvs filiorvm Israel per Mare Rvbrvm," 150 x 118 mm.
Fig. 74	Yaacov Auspitz's map of the Israelites' encampment, 191 x 140 mm.
Fig. 75	Yaacov Auspitz's map of the Holy Land, 445 x 326.
Fig. 76	Tobias Conrad Lotter, Terra Sancta sive Palaestina, 480 X 570 mm.
Fig. 77	The figure of Joshua ben Nun in the Lotter, Auspitz, and Aharon ben Haim maps.
Fig. 78	Yaacov Auspitz, "The Division of Eretz Yisrael in the Future (according to Ezekiel)" Map, 180 x 106 mm.
Fig. 79	The map of Dov Baer Yozpa, 388 x 235 mm.
Fig. 80	Source and adaptation: details from the Bar Yaacov and Yozpa maps.
Fig. 81	Warsaw edition of the Yozpa map, 388 x 235 mm
Fig. 82	Moshe Danzigerkron's edition of the Yozpa map, 380 x 300 mm.

Fig. 83   The map in Aharon ben Haim, Sefer moreh derekh, 370 x 282 mm.
Fig. 84   The crossing of the Red Sea in maps of bar Yaacov, Auspitz, and ben Haim.
Fig. 85   Binder's map, 317 x 404, in "Charte Worauf das Iüdische Land vornehmlich, wie es zur Christi und der Apostel Zeiten gewesen ist, vorgestellet wird. Nach dem Entwurff des Herrn W. A. Bachiene von neuem gezeichnet von M. Binder aus Schæssburg in Siebenbürgen, Sebast."
Fig. 86   The Meir Isaac Hirsch edition of the Aharon ben Haim map, 375 x 285 mm.
Fig. 87   The Galgor edition of the Aharon ben Haim map, 360 x 280 mm.
Fig. 88   Gershon Chanoch (Henich) Leiner, The "This shall be Your Land as Defined by its Borders All Around" Map, 224 x 360 mm. The map is located at the end of his book, Sidrei tohorot, vol. 1, tractate Keilim.
Fig. 89   Benjamin Lichtenstaedter's "The Lands of the Orient" Map, 395 x 251 mm, in Amtaḥat Binyamin.
Fig. 90   The Visscher family, "Die Gegend des iridischen Paradieses und des Landes Canaan," 505 x 378 mm.

Chapter 5

Fig. 91   Joshua Feivel ben Israel (Teomim), map of Eretz Yisrael, 310 x 200 mm.
Fig. 92   Teomim's map of the tribal allotments as per the Book of Ezekiel, 200 x 310 mm.
Fig. 93   Alexander Ziskind, "Drawing of the Borders of the P[ortion] of Mas'ei," 145 x 170 mm, in Yesod ve'shoresh ha'avodah, 1795, p. 102b.
Fig. 94   Alexander Ziskind, "Drawing of Joshua's Borders," in Yesod ve'shoresh ha'avodah, 135 x 165 mm, p. 102a.
Fig. 95   Yehiel Hillel Altschuler, "Map of the Apportionment of Eretz Yisrael and its Borders," 174 x 212 mm.
Fig. 96   Altschuler and Auspitz's maps of Eretz Yisrael's future division as per Ezekiel.
Fig. 97   Calmet's map of the division of the Land according to Ezekiel, 186 x 355 mm.
Fig. 98   Two versions of Elijah ben Shlomo Zalman's "Layout of the Land" map in Ṣurat ha'areṣ le'gvuloteha: MS Shklov 1802, 209 x 209; and the 1905 Jerusalem edition, 210 x 215 mm.
Fig. 99   "Eretz Yisrael's Apportionment for All its Borders," 470 x 380 mm.
Fig. 100   Aryeh ben Isaac's map in the second edition of Kiṣvai areṣ, 378 x 326 mm.
Fig. 101   Isaac ben Phinehas Berman, the "Here Before You are the Borders of EY" map, 180 x 205 mm.
Fig. 102   Jacob Emden's map of Eretz Yisrael, 260 x 180 mm.
Fig. 103   Jacob Emden, the map in Sefer leḥem shamayim, ca. 115 x 115 mm.

Chapter 6

Fig. 104   Moshe Ganbash, "This Very Drawing Encompasses All of Eretz Yisrael" map, Constantinople, 1839, 1041 x 864 mm.
Fig. 105   Joshua Alter and Mosheh ben Pinhas Fainkind, "Layout of the Holy Land for All its Borders" map, 520 x 400 mm.
Fig. 106   "The Layout of the Holy Land for All its Borders" map, Vienna, 600 x 480 mm.
Fig. 107   "The Layout of the Holy Land for All its Borders" cloth map, 620 x 474 mm.
Fig. 108   "The Layout of the Holy Land for All its Borders" map, ca. 1900.

Chapter 7

Fig. 109 Jacob Kaplan, "Map of the Land of Ancient Times," 240 x 345.
Fig. 110 Menachem Mendel Breslauer (Breza), "Map of Eretz Yisrael," 272 x 300 mm.
Fig. 111 Menachem Mendel Breslauer, map of the Israelites' peregrinations, 220 x 180 mm.
Fig. 112 Yehoseph Schwarz, "The Holy Land and its Borders" map, 500 x 645 mm.
Fig. 113 Johann Baptist Homann, "Judea that is Palestine" map, 560 x 484 mm.
Fig. 114 The figures of Moses and Aaron on the Schwarz and Homann maps.
Fig. 115 Yehoseph Schwarz, map of the Land of Israel in the English edition of Tvu'ot ha'areṣ, 290 x 490 mm.
Fig. 116 Yehoseph Schwarz, map of Eretz Yisrael's borders in the English edition of Tvu'ot ha'areṣ, 140 x 170 mm.
Fig. 117 Yehoseph Schwarz, map of the Land of Israel in the German edition of Tvu'ot ha'areṣ, 290 x 484 mm.
Fig. 118 The Tvu'ot ha'areṣ map (English version), 380 x 426 mm; and Kiepert-Robinson "Map of Palestine, Chiefly from the Itineraries and Measurements of E. Robinson and E. Smith [...]," 376 x 426 mm,
Fig. 119 Salomon Munk, "The Land of Israel and its Division between the Twelve Tribes," 105 x 172 mm, in idem, Palestine: Description géographique, historique, et archéologique, tab. 69, facing p. 224
Fig. 120 Salomon Munk, "The Route of the Hebrews' Journey upon Traversing the Desert," 105 x 163 mm, in idem, Palestine: Description géographique, historique, et archéologique, facing p. 122.
Fig. 121 Israel Joseph Benjamin, "And Israel Wrote the Destinations of his Travels, and These [Constitute] the Itinerary of his Travels for All its Destinations" map, in idem, Sefer masei yisrael.
Fig. 122 Jonas Spitz, "Palestine in the Work Eretz Ṣvi" map, 175 x 215 mm.
Fig. 123 Ephraim Israel Blücher, map of Canaan, 69 x 99 mm.
Fig. 124 Ephraim Israel Blücher, "Map of Eretz Yisrael," 420 x 515 mm.
Fig. 125 "The Land of Israel Following the Land's Division into the Twelve Tribes by M. Löwi," circa 1870, 755 x 938 mm.
Fig. 126 Juda Funkenstein, "The Land of Canaan" map, 1003 x 1320 mm.
Fig. 127 Yitskhok Yoyel Linetski, "Die Carta Eretz Yisrael" map.
Fig. 128 Jacob Goldzweig, "A New and Original Biblical Map of the Holy Land."
Fig. 129 Avigdor Malkov, "The Map of the True Path," 790 x 530 mm.
Fig. 130 Shabtai ben Yaakov Matskevitsh, "Map of the Holy Land," 203 x 146 mm.
Fig. 131 Ephraim Michael Grover, "The Well-Researched and Precisioned Map of Eretz Yisrael," 405 x 334 mm.
Fig. 132 Nahum Sokolow, "Map of the Holy Land[,] 'the Precious Land,' Bearbeitet von H. Kiepert, Berlin 1885," 300 x 433 mm.
Fig. 133 Eliyahu Sapir and Ephraim Krauze, "Map of Eretz Yisrael (Palestine)," 1370 x 1920 mm.
Fig. 134 D. Ben Menachem, "Map of the Land-of-Israel Republic," 413 x 600 mm.
Fig. 135 Jabotinsky and Perlman, "Land of Israel" map, 1925, 240 x 420 mm.

Introduction

This book is devoted to "the Hebrew map." Along with a few works in Ladino, Yiddish, and European languages, this term refers to maps in Hebrew that were drafted and printed by Jews and cover Jewish topics, such as the Land of Israel and its borders, the exodus from Egypt and the Israelites' peregrinations through the wilderness, the tribal allotments, holy sites, and other Biblical and Talmudic topics. These maps were incorporated into unequivocally Jewish works: exegetical volumes of the Scriptures, Passover Haggadahs, *ketubot* (wedding contracts), descriptions of sacred places, responsa literature, rabbinic and Enlightenment works, *inter alia*. The majority of the maps indeed came out in Hebrew; a few were bi- and even multilingual, while several were in other languages. For this reason, the terms "Hebrew maps" and "Jewish maps" will appear side by side, without any sharp distinction between the two.

The existing literature on Hebrew maps is so threadbare that *The Hebrew Encyclopedia*'s entry on the maps of Eretz Yisrael (Land of Israel),[1] which was subsequently translated into English and published without substantial changes in *The Encyclopaedia Judaica*, makes no reference whatsoever to Hebrew maps or those printed by Jews.[2] Isaac Schattner (the author of the above-cited entry) also made no mention of these works in his book on the maps of Eretz Yisrael.[3] Similarly, when the Library of Congress published a book on its treasure trove of Judaica, not one Hebrew map was included in the chapter on the mapping of the Land of Israel.[4] While featuring chapters on Muslim, Chinese, and Indian cartography, Harley and Woodward's *The History of Cartography*, a central depository of knowledge in this field, lacks a chapter on Jewish Maps.[5] What is more, the only Hebrew map that was included (albeit as an illustration) in this thick, multi-volume work is a map penned in the Rashi script (see chapter 2), which is regrettably aligned upside down.[6] There are two books on the Hebrew maps of Eretz Yisrael, but both suffice with a partial, album-like, and unsystematic presentation.[7] Moreover, these works do not provide a basis for under-

---

[1] To this day, Jews refer to the Land of Israel using the Hebrew term *eretz yisrael*, regardless of the primary language they happen to be speaking. Henceforth, the terms "Land of Israel" and "Eretz Yisrael" will be employed in a synonymous and interchangeable manner.
[2] Schattner, "The Maps of Eretz Yisrael" [Hebrew]; idem, "Maps of Erez Israel,"*Encyclopaedia Judaica*.
[3] Schattner, *The Map of Eretz Yisrael and Its Annals* [Hebrew].
[4] Karp (ed.), *From the Ends of the Earth, Judaic Treasures of the Library of Congress*.
[5] Kedar thrust similar criticism on Woodard and Harley for their lack of attention to the Hebrew map; see Kedar, "Rashi's Map of the Land of Canaan, ca. 1100, and Its Cartographic Background."
[6] Morse, "The Role of Maps in Later Medieval Society: Twelfth to Fourteenth Century;" Harley and Woodward, *The History of Cartography*, vol. 3, book 1, p. 42.
[7] Vilnay, *The Hebrew Maps of the Holy Land: A Research in Hebrew Cartography* [Hebrew]; Wajntraub and Wajntraub, *Hebrew Maps of the Holy Land*.

standing the Hebrew maps, as there is no discussion about the types of maps, their sources, the topics they tend to, or the path of their development.

The present study undertakes to fill this research void in several ways: first, it depicts the wide range of maps that I have come across in various collections; second, the maps will be sorted into groups on the basis of content and form, the background behind their creation, and their makers' approach; third, over the course of the discussion, I will analyze the principal topics that surface in these works; and lastly, the book will strive to explain different phenomena in the evolution of the Hebrew map and tie each one to its source. In parallel, a basic distinction will be drawn between maps that originated within the Jewish world and those copied from outside sources, namely from Christian European authors.

I do not profess to include every Hebrew map that was ever compiled in this book. Some of the works that were examined during the research phase have been left out for various reasons. In addition, there are bound to be maps that I failed to locate. There are also maps that were alluded to in different texts, but in all likelihood are no longer exist.[8] As is to be expected, most of the maps that reached my hands were printed, while relatively few have survived in manuscript form.

Chronologically speaking, the book opens with Rashi, an eleventh-century Jewish exegete (see chapter 1). It stands to reason that Rashi was the pioneer of the Hebrew map and was certainly the first whose maps influenced generations of Jewish cartographers. His works became a foundation and the cornerstone of what I refer to as "traditional" Hebrew cartography. The outer limits of this work are the early twentieth century, when modern geographic and cartographic knowledge, along with the rise of Zionism, yielded a cornucopia of modern maps of Eretz Yisrael. During this period, the curtain fell on the traditional Hebrew map. The focus of my attention will be on the development of the Hebrew map from the traditional phase to modernity. In consequence, I will not expand on either the modern Hebrew map from the early 1900s onwards or the use that the Zionist movement made of this graphic device for the purposes of education, nation building, and propaganda. As important as this topic may be, it falls under the purview of another study—one that requires different tools than those that I have availed myself of in researching this book. That said, I will touch upon these issues in the final chapter.

Before diving into the heart of the matter, I would like to discuss the terminology that will serve us throughout the length of this book and the connection between religion, maps, and cartography.

---

[8] For example, Joseph Shalit ben Eliezer Riqueti apparently refers to a map that has subsequently been lost to posterity in his book *Sefer Ḥokhmat ha-Mishkan* (the Book of the Tabernacle's Wisdom): "I took the trouble to create a form [i.e., a map] in the shape of the Land of Israel that has been in the making for what is now a couple of years in the holy community of Amsterdam may God protect them and I am indeed making it ornate and fetching with several drawings."

What's in a Map? On Maps in General and Antique Maps in Particular

Cartography is a discipline that combines art, craftsmanship, and knowledge for the sake of formulating maps. As such, the discourse on this field often turns to the following subjects: the history of technology; the rise of geographic knowledge; the development of map projection and coordinates; and the process of discovering the world in the run-up to more precise maps. More recently, the discussion has embraced terms like satellite navigation systems and computerized measurement and map production. These scientific breakthroughs notwithstanding, maps have always served and continue to be used as platforms for expressing ideas and promoting ideologies and ideals. In other words, they have been used to explicate political and religious ideas as well as to advance the goals of individuals, organizations, and states.[9] This has been pursued with a wide variety of maps: old and contemporary, fictitious and precise, schematic and complex, rudimentary and artistic-cum-ornate and those designed with a scientific veneer.

Two different approaches to maps have arisen in the literature on cartography and its history. On the face of things, these approaches are contradictory, but in practice they complement one another: the first views the map in terms of geometry, geodesy, map projection, and accurate surveying; the second portrays the map as a tool for advancing ideas and outlooks. Toeing the line of the first camp, Naftali Kadmon defines the map as "a schematic, planar, miniaturized, oriented, and quantifiable account of the planet."[10] Although this approach is considered among historians of cartography as outdated and even non-relevant anymore, it is commonly accepted among the non-professional users of maps which presumed that maps really represent the reality of the world.

In contrast, Robinson and Petchenik assert that a map is a graphic "representation of the milieu."[11] In choosing the French term "milieu" over, say, "surrounding" or "environment," they not only emphasize the map's physical dimensions, but its cultural and moral freight as well. Tony Campbell toned down this rather sweeping definition by proposing that a document can be considered a map if it meets two basic criteria: the provision of graphic information about the real world; it refers, even if in a schematic, imprecise manner, to the relative position and distance between objects in an expanse.[12] Similarly, Brian Harley and David Woodward describe maps as "graphic representations that facilitate a spatial understanding of things, concepts, conditions, processes or events in the human world."[13] Building on this definition in a series of articles, Harley claims that maps are tools with a propensity

---

9 Cosgrove, Mapping New Worlds, pp. 65–89.
10 Kadmon, "Cartography," *The Encylopaedia Hebraica* [Hebrew].
11 Robinson and Petchenik, *The Nature of Maps*, p. 16
12 Campbell, *The Earliest Printed Maps, 1472–1500*, p. 17.
13 Harley and Woodward, *The History of Cartography*, vol. I, p. xiv.

for conveying principled messages in ideological disputes.¹⁴ Dennis Wood even goes so far as to call maps "weapons" in debates of this sort.¹⁵ A more balanced approach is taken by Anne Godlewska. By their very nature, she writes, maps "attempt to be all things to all people. Maps may simultaneously—and more or less obviously and deliberately—seduce, confuse, and obfuscate [reality]."¹⁶ Delano-Smith and Kain illuminated this discussion when they characterized maps as "the representation of features (places, people, phenomena, real or imagined) in their relative or actual spatial location",¹⁷ thus combining together the real and ideal elements of features that are depicted on maps.The ensuing discussion will indeed adopt the outlook whereby cartographers draft maps for the purposes of disseminating principles and ideas, no less than providing concrete and exacting geographic information. These objectives are achieved in a variety of ways: by choosing a style, graphic elements, and colors, the mapmaker stresses certain components, while downplaying or concealing others; by interspersing symbols and evocative images, the significance of places on a map are ratcheted up or down; and by incorporating meaningful images, the author imparts value to sites of his or her choosing. A drawing, for example, of imaginary islands opposite the shoreline of Eretz Yisrael represents the *"nisin* she-bayam" (islands in the sea), a concept from the literature of the Sages. Likewise, depicting Moses on Mount Sinai or Jonah opposite the Jaffa coast portrays the Biblical narrative. In addition, rendering the first Jewish colonies in the form of a modern house in a late nineteenth-century map places them in a positive light.

Mapmakers, past and present, are wont to choose names and concepts that favor their own views over other accepted terms. A case in point is European cities with both German and Polish names, such as Breslau/Wroclaw and Danzig/Gdańsk. These discrepancies also turn up in modern-day atlases of Ireland, where the use of Irish or English names depends on the cartographer's allegiances. Similar differences inform Greek and Turkish maps of Cyprus as well as Israeli and Palestinian cartographic accounts of their shared, or rather contested, territories. The authors of the maps in question reinforced their own message by highlighting elements from the Biblical age and its glossary of toponyms. These cartographers, both Christians and Jews alike, tended to ignore the contemporaneous reality, while depicting settlements and events that are noted in the Bible, the tribal allotments, and the cities of refuge.

Most of the works in our corpus, and certainly the vast majority of maps that predate the nineteenth century, ignored the geographical reality at the time of their

---

14 Harley, "Deconstructing the Map;" idem, "Silences and Secrecy: the Hidden Agenda of Cartography in Early Modern Europe;" idem, "Maps, Knowledge and Power." Harley's critical papers also appear in the following anthology: idem, *The New Nature of Maps: Essays in the History of Cartography.*
15 Dennis Wood, *The Power of Maps*; idem, "How Maps Work."
16 Anne Godlewska, "The Idea of the Map," p. 20.
17 Catherine Delano-Smith and Roger J. P. Kain, *English Maps: A History*, London: British Library, 1999

drafting that is the surveyable alignment, distances, and altitudinal disparities between the sites they cover. Put differently, eschewing measurement or map projection tools, the authors refrained from drawing their maps to scale. In fact, many of these cartographers never stepped foot in the Near East. On the other hand, though, they were well versed in the Bible and presented the Land of Israel according to their own interpretation of the canon. From the authors' perspective, the maps were intended to help them disseminate their ideas. In turn, the readers used these works as learning aids for comprehending the Bible and its commentaries.

With respect to the Hebrew maps, the main topics were the borders of the Promised Land, the exodus from Egypt, and later the tribal allotments. Additionally, there were maps that dealt with the boundaries of *shalosh artzot le'shevi'it* (literally three lands for the seventh, or zones for the sabbatical year), and Eretz Yisrael's borders in the vicinity of Acre, which were relevant to halachic questions concerning the fallow year and *gittin* (bills of divorce). Most of the traditional Hebrew maps are quite simple from a graphic standpoint, as they constitute schematic outlines without artistic drawings. Their purpose was to convey religious and exegetical ideas to an audience that was interested in the above-mentioned halachas (Jewish law). In contrast, those Jewish maps that were drafted in response to Christian works offered a more comprehensive picture of the Land of Israel, and the same can be said for those of the Enlightenment period. However, as we shall see, only the later maps were predicated on empirical knowledge and described the geographical structure of the land as it "actually" is.

The present study will examine the graphic conventions that emerged in this corpus as well as the content and objectives of its individual maps.

What are Hebrew or Jewish Maps?

For the purposes of this book, a "Hebrew" or "Jewish" map is one that was compiled by a Jewish author, deals with topics pertaining to the Jewish world of knowledge and content, and was printed in a Jewish context.[18] While most of these maps are in Hebrew, several are in Yiddish and Ladino, others are bilingual (i.e., in Hebrew and European languages), and a few are exclusively in a European tongue. These works were included in Bibles, commentaries on the Jewish Bible and Halacha, tractates of the Mishna and Talmud, Passover Haggadahs, *ketubot*, and other Jewish works. On rare occasions, they were printed as stand-alone sheets.

Mappah, the Hebrew term for map, is relatively new. As we shall see, it was preceded by the following terms, all of which were commonplace in the exegetical liter-

---

18 For the most part, this study ignores maps that were produced within the context of general knowledge and make no reference to Jewish sources or the Land of Israel, even those that were compiled or printed by Jewish cartographers, publishers, or printers.

ature: *tzura* (literally: form or shape), *tzurat ha-aretz* (the shape of the Land), plain *tziyur* (picture or drawing), *tziyur tkhumai eretz yisrael* (a drawing of the Land of Israel's boundaries), *tmunat mas'ai bnai yisrael* (a picture of the Israelites' voyages), and *bai'ur al eretz yisrael le'gvuloteha* (an explanation of the Land of Israel within its borders). *Mappah* only entered the mainstream lexicon at the start of the nineteenth century, though there is documented evidence of its usage from as early as the eighteenth century.[19] In parallel, some Jewish mapmakers adopted the terms *carta* (or *land-kart* in Yiddish), under the influence of the European terminology; and the same can naturally be said for the maps that came out in European languages. Be that as it may, the general term that will be used throughout this work will be map.

Over the course of this book, I will introduce these works, analyze them, and track after their sources. The maps will be classified into groups according to their content, objectives, and sources. All these steps will bolster our effort to shed light on the Jewish map's development, from the earliest known versions, which were drafted by Rashi in the eleventh century, until the *fin-de-siècle*. By the early twentieth century, the Hebrew cartographers were influenced by the modern mapping of the Land and sought to promote the nascent Zionist ideology.

Above all, this study grapples with the constant tension and dialogue between the form and content of the Hebrew map. Maps convey their maker's ideas via textual and graphic elements, both of which can be either conservative or innovative. With this in mind, the following questions will be posed: How is the expanse presented in "Jewish maps?" What were their principal topics and graphic elements? How did the spatial outlooks that are represented in these works evolve? And how were they transmitted? What is the provenance of these views and what principles did they serve? These questions will lead us to the crux of this study: What were the root causes behind the development of the Hebrew map? And to what extent can those causes from within the Jewish world be distinguished from outside influences, that is the Christian world in general and European cartography in particular?

Before proceeding to the main discussion, I would like to raise one more question. The notion that the Land of Israel is sanctified prevails throughout the Jewish literature, from the Bible to the present. Among the most salient examples are God's promise to bequeath Eretz Yisrael to Abraham's progeny as well as the numerous commandments and halachas that are dependent on the Land. Furthermore, this view supports the halachic discussions on the borders of Eretz Yisrael. This discourse clearly shows that as early as the days of the Pentateuch and certainly from the Second Temple period onwards, there were well-conceived opinions on geographical is-

---

19 The *Ben Yehuda Dictionary* refers to Israel Ha-Levy, *Ozar Nekhmad* as the earliest source in which the word *mappah* appears; Elieser Ben Yehuda, *A Complete Dictionary* (Tel-Aviv: 1949) vol. vi, p. 3212 [Hebrew].

sues within the Jewish tradition.[20] Even the historiographical chapters of the Hebrew Bible that do not directly pertain to Eretz Yisrael's holiness broach geographical topics, like the tribal allotments, the boundaries of the kingdom, and the distribution of the cities of refuge. In addition, the literal descriptions of the Land of Israel can be viewed as quasi-maps, including the following passages: the account of Joshua dispatching scouts to survey the Land before apportioning it to the tribes (Joshua 18:4); and the verse from Ezekiel "Take a brick…and incise on it a city, Jerusalem" (Ezekiel 4:1). What is more, the Book of Jubilees perhaps alludes to a map of the world that was included therein.[21] Be that as it may, there is no existing Hebrew map of the Land of Israel, any outline of its borders, or even a credible hint as to a lost work of this sort that predates those of Rashi.

This absence is all the more glaring when compared to the rich cartographic world that existed in the Ancient East. Maps were indeed used in Mesopotamia and Egypt. Moreover, they reached a high degree of sophistication under the Greeks and Romans, the early Christians, and Arabs as well. On the basis of Ptolemy's works, the maps deriving from the Roman world,[22] the Madaba Map of the Land of Israel,[23] those used by early medieval church representatives,[24] and the output of Muslim geographers,[25] it is quite evident that the map was well-known in these cultures. Therefore, it stands to reason that there was a full-fledged geographical outlook in the Jewish tradition as well, for Jewish communities could be found within or in proximity to most of these societies. On the other hand, though, there is barely a clue as to the existence of a Hebrew map before those of Rashi. In light of the above, the question that begs asking is thus: Why are pre-eleventh century Jewish works entirely devoid of maps? I am afraid that by the end of the present study, this riddle will remain unsolved.[26]

---

[20] Weinfeld, "The Extent of the Promised Land: Two Points of View [Hebrew];" Wazana, *All the Boundaries of the Land: the Promised Land in Biblical Thought in Light of the Ancient Near East* [Hebrew]; Shilhav, "Interpretations and Misinterpretations of Jewish Territoriality" [Hebrew]. For an in-depth discussion of the Jewish outlook on the borders and the research thereof, see Sussmann, "The 'Boundaries of Eretz-Israel'" [Hebrew].

[21] Alexander, "Notes on the 'Imago Mundi' of the Book of Jubilees;" Scott, *Geography in Early Judaism and Christianity.*

[22] Dilke, *Greek and Roman Maps*; Salway, "The Nature and Genesis of the Peutinger Map;" Albu, "Imperial Geography and the Medieval Peutinger Map."

[23] Piccirillo and Alliata, *The Madaba Map Centenary*; Michael Avi-Yonah, *The Madaba Mosaic Map.*

[24] Edson, *Mapping Time and Space*; Thomas O'Loughlin, "Map as Text: a Mid-Ninth Century Map for the Book of Joshua;" idem, "Map Awareness in the Mid-Seventh Century: Jonas' Vita Columbani."

[25] Sezgin, *The Contribution of the Arabic-Islamic Geographers to the Formation of the World Map*; idem, *Mathematical Geography and Cartography in Islam and Their Contribution in the Occident*; Harley and Woodward, *The History of Cartography*, vol. 2, book 1.

[26] Of course, there is a possibility that there were maps of this sort, but none have been preserved and there is no textual evidence of their existence. Therefore, in my estimation, this scenario is far-fetched.

Cartography, Maps and Religions

Maps depict the world, its continents and lands, mountains, rivers, coasts, settlements, and other sites, as well as the spatial relations between all these elements. At one and the same time, they pass on values and ideas. In consequence, religions embraced the map, along with literature, painting, and other art forms, as a means for communicating their take on and insights concerning the Earth and the places therein. Cartographers in the service of religion presented the sacred sites that are mentioned in their holy texts and interpreted related geographical topics. On numerous occasions, ecclesiastics drew up maps that served as a scriptural-cum-exegetic learning aid or as a platform for discussing eschatological topics.[27] In addition, there were maps that were devoted to holy places and pilgrimage routes.

While Christianity was still in its nascence, Eusebius of Caesarea wrote *The Onomasticon*. This book originally contained a map of the places that are mentioned in the Bible; however, the latter has been lost to posterity. Thanks to Jerome's translation of *The Onomasticon* from Greek to Latin, the work had a marked impact on the Christian world.[28] The Madaba Map, which apparently depicted most of the Holy Land, was also strongly linked to Eusebius' book. In addition, maps were incorporated into several later manuscripts of Jerome's works. Muslims oriented their maps southward, according to the *qibla* (their direction of prayer), thereby demonstrating the importance with which they held their custom of praying towards Mecca.[29]

During the early Middle Ages, maps were sprinkled into Christian manuscripts, and quite a few of them have indeed survived. One of these maps, a representation of the tribal allotments, was aptly affixed to a ninth-century exegesis on the Book of Joshua.[30] Other maps placed the world in a round structure that is divided into three continents. With Jerusalem at the center and Paradise to the east, these maps espoused a religious outlook concerning the layout of the world.[31] As in earlier centuries, some Renaissance and Modern Era maps were designed to help audiences understand Biblical commentaries.[32] For instance, both the supporters and opponents of the Protestant Reformation formulated maps for this purpose. Beginning in the 1500s, maps were printed in Bibles. Protestants were the first to adopt this

---

27 Delano-Smith, "Maps and Religion in Medieval and Early Modern Europe;" Shalev, *Sacred Words and Worlds: Geography, Religion and Scholarship, 1550–1700*.
28 Eusebius, *Das Onomastikon, der Biblischen Ortsnamen*.
29 Sezgin, *The Contribution of the Arabic-Islamic Geographers*.
   For the most part, medieval Muslim maps were produced in the Mediterranean Basin and are thus oriented southward. Needless to say, works of this sort that were compiled in other regions were in all likelihood oriented towards Mecca.
30 O'Loughlin, "Map as Text: a Mid-Ninth Century Map for the Book of Joshua."
31 Edson, *Mapping Time and Space*; Harvey, *Medieval Maps*.
32 Ingram, "Maps as Readers' Aids: Maps and Plans in Geneva Bibles;" Delano-Smith, "Maps as Art and Science: Maps in Sixteenth Century Bibles."

practice, especially in what is known as "the Geneva Bible."[33] Towards the end of the century, Catholic exegetes, like Arias Montano and Christian van Adrichom (also known as Adrichem), also started to include maps in their commentaries.[34] All told, this trend proliferated in the centuries to come.[35] Fiorani suggests that the difference between the Catholic and Protestant maps is tied to the place of Rome in each of the denominations' religio-historical worldview.[36] Over the years, these Christian Bibles developed a nigh permanent compendium of maps. The chapters in the Old Testament section featured maps of Paradise, the peregrinations of the Israelites, and the division of the Promised Land into tribes, whereas the New Testament section included maps of Jesus' travels in the Holy Land and those of the Apostles throughout the Mediterranean Basin. Some of these Bibles also contained maps of Jerusalem during the reign of Solomon along with a panoply of non-cartographic drawings.

One of the major functions of devotional maps was to encourage and facilitate virtual pilgrimages. Given the hardships that overseas travel entailed, many believers sufficed with a spiritual voyage to the holy places from the comfort and safety of their homes. Cartographic renderings of the sacred sites indeed played a central role in this sort of pilgrimage.[37] What is more, Christian groups that were involved in cultivating Holy Land consciousness, such as the Franciscan Custodia Terra Santa and the Greek-Orthodox Patriarchate of Jerusalem, produced and distributed works of this kind. In so doing, these organizations also bolstered their own power. Apart from maps, numerous illustrations of the holy sites were included in the writing of pilgrims and other works that promoted this kind of travel.[38]

In the eyes of their makers, the Hebrew maps constituted an integral part of the Biblical or Talmudic text, as they helped the devotee read and interpret the Jewish canon. These works simultaneously conveyed ideological ideas, such as the Promised Land, and eschatological ideas (e.g., the future division of the Land as per Ezekiel's prophecy). These cartographic representations of sacred sites as well as the Israelites' wanderings through the wilderness and their life in the Holy Land certainly enabled Jewish readers to venture forth on an emotive virtual tour of the Sinai Desert

---

[33] Delano-Smith and Ingram, *Maps in Bibles, 1500–1600: an Illustrated Catalogue*; Lloyd (ed.), *The Geneva Bible*.
[34] Shalev, "Sacred Geography, Antiquarianism and Visual Erudition: Benito Arias Monatno and the Maps of the Antwerp Polyglot Bible;" also see Adrichom, *Theatrum Terrae Sanctae*.
[35] For example, see Calmet, *Commentaire litteral sur tous les livres de l'Ancien et du Nouveau Testament, Les trois premiers livres des Rois*.
[36] Fiorani, *The Marvel of Maps, Art, Cartography and Politics in Renaissance Italy*.
[37] Connolly, "Imagined Pilgrimage in the Itinerary Maps of Matthew Paris;" Delano-Smith, "The Intelligent Pilgrim: Maps and Medieval Pilgrimage to the Holy Land;" Rudy, *Virtual Pilgrimages in the Convent: Imagining Jerusalem in the Late Middle Ages*.
[38] Rubin, "One City, Different Views: a Comparative Study of Three Pilgrimage Maps of Jerusalem;" idem and Levy-Rubin, "An Italian Version of a Greek-Orthodox Proskynetarion;" Rubin, "Iconography as Cartography: Two Cartographic Icons of the Holy City and its Environs."

and Eretz Yisrael. This function stands out in, say, the Mantua Map (chapter 3) that graces Solomon Ben Moses of Chelm's book (chapter 4) and the tables of the sacred places (chapter 6), but it informs all the maps under review.

Structure of the Book

With respect to both their graphic design and content, the maps that comprise our corpus were produced on two separate tracks. To begin with, there are the traditional maps. This category consists of Rashi's maps and countless replications and adaptations thereof. Featuring a schematic, usually square or rectangular structure, these maps focus on Jewish Scriptures and their commentaries, especially in all that concerns the borders of the Promised Land. The second group consists of those maps that were created for the purpose of contending with the Christian maps of the Holy Land. With respect to content and artistic design, this category is more elaborate than its traditional counterpart. For instance, these maps depict various sites and events from the Holy Scriptures with the help of miniature drawings. This subgenre includes maps that copied their layout from Christian works, while adjusting the content to the Jewish tradition. Other maps that fall under this heading merely respond to those of Christians; although these works took certain elements from the latter, they are not identical to them.

There are complex mutual relations between the two defined tracks. Of course, not all the maps in either group were produced within the same timeframe, and some authors created maps in each of the styles. Lastly, some of the works in our corpus belong to neither group, whereas others integrate elements from both.

The book's chapters revolve around these two central axles. The first and second chapters are dedicated to traditional maps—the unadorned format that Rashi inaugurated. Chapter 3 hones in on a map that belongs to neither the traditional nor simulative stream. In my estimation, this work was a reaction to the maps that appeared in printed Christian Bibles. The fourth chapter explores maps that were influenced by the European Christian mapping of the Holy Land. These works contend with this rival Christian corpus, while adopting its graphical conventions. More specifically, they follow in the footsteps of the Christian maps by using artistic means, namely miniature illustrations, for presenting the events that are mentioned in the Holy Scriptures.[39] In the fifth chapter, the two main tracks cross paths. More specifically, ideas that derive from the Christian world, especially the depiction of the tribal allotments, penetrate the maps that took form within the internal, more conservative Jewish framework. The category that is examined in the sixth chapter does not imbibe

---

[39] As noted, there are several instances of an author producing maps that fall under two different categories or of a certain topic that was depicted in more than one style. For these works, I deviated from the above-mentioned structure, pushing forward or delaying a discussion on a certain map or phenomenon from one chapter to the next.

directly from either of the two aforementioned traditions, but emulates the earlier, nineteenth-century tables of holy places. In the final chapter, I touch upon the synergy between tradition and modernity that informs the maps from the Enlightenment era and the dawn of Zionism. At this phase, scientific techniques came to dominate the crafting of Hebrew maps of Eretz Yisrael, thereby development subjecting the tension between reality and its idyllic representation to quantitative measurements and standards of accuracy. At this juncture, the book reaches the end of its line.

1 Rashi and His Maps

Widely known as Rashi,[1] Rabbi Shlomo Yitzhaki (1041–1105) is considered one of the greatest exegetes in Jewish history and perhaps the most important one of them all.[2] Over the course of his lifetime, Rashi authored comprehensive commentaries on the entire Hebrew Bible and Babylonian Talmud. The scope of his enterprise—its depth and breadth—has long been admired by Jewish and Christian scholars alike.[3] Rashi's works were copied time and again; and with the advent of the printing press, print versions of his works began to spread. Be that as it may, there are no extant manuscripts by Rashi himself. The earliest and most trustworthy manuscripts were produced by his close peers and students. A few of these texts contain one or more of the following cartographic works: schematic maps of the Promised Land's borders as delineated in the Torah portion[4] of *Mas'ei* (Numbers 33:1–36:13); maps of the future tribal allotments as enumerated in the prophecy of Ezekiel; a geographic sketch pertaining to the final chapter of the Book of Judges; and a geographic diagram of the strip of land between Acre and Achzib, within the framework of a Talmudic discussion on a related issue. These items constitute the earliest known Hebrew maps. Against this backdrop, it is only fitting that we open this history of the Hebrew map with a survey of Rashi's cartographic diagrams.

The Authenticity of Rashi's Maps

Since we lack so much as a single manuscript that Rashi penned on his own, researchers have grappled with the following question: Did Rashi actually draw up the maps that are attributed to him, or at the very least similar versions thereof? And how close are the existing maps, which were produced by copyists, to the originals?[5] Scholars have convincingly argued that these maps are indeed authentic replications of Rashi's output.[6] Over the next few paragraphs, I will survey the main reasons for this conclusion:

---

[1] *RaShI* is the Hebrew acronym of Rabbi Shlomo Yitzhaki's title and name.
[2] "Rashi," *Encyclopedia Judaica*; Grossman, *Rashi* [Hebrew]. There are researchers who have proposed an earlier year of birth (1030), but we will defer to Grossman's estimate (1041).
[3] Grossman and Japhet, *Rashi: The Man and His Work* [Hebrew]. See ibid for a disquisition on the wide array of fields that Rashi covers in his oeuvre.
[4] The Pentateuch is divided into 54 *parashot* (portions), which are read on a cyclical basis by Jewish congregations over the course of each year.
[5] Ofer, "The Maps of the Land of Israel in Rashi's Commentary on the Torah and the Status of MS Leipzig 1" [Hebrew].
[6] Gruber, "What Happened to Rashi's Pictures?"; idem, "The Sources of Rashi's Cartography"; idem, "Notes on the Diagrams in Rashi's Commentary to the Book of Kings;" idem, "Light on Rashi's Diagrams from the Asher Library of Spertus College of Judaica."

1. In one of the manuscripts, R. Samuel ben Meir (Rashi's grandson, who is known by the Hebrew acronym the *RaShBaM*) explicitly noted, alongside the map to Numbers 34, that "Our rabbi my grandfather [i.e., Rashi] interpreted and drew borders" of Eretz Yisrael.[7]
2. In all likelihood, the Leipzig manuscript was copied from that of Shemaiah of Soissons, Rashi's close disciple, who also bears witness to the maps' authenticity.[8]
3. The maps for Numbers 33 and 34, which are the most pertinent to this study, turn up in the most credible manuscripts, including the Oxford, Leipzig, Paris, and Munich manuscripts.[9] Moreover, the various versions of these maps are highly uniform with respect to their placement in the text and consistent from the standpoint of form and content. For example, the map to Numbers 33 is nearly always embedded into the text itself, whereas its bigger and more detailed counterpart (Numbers 34) usually extends across the width of the page. This uniformity in all that concerns alignment, shape, and content testifies to the fact that the copyists viewed these diagrams to be integral parts of the original texts.
4. While a couple of the manuscripts, such as the Munich version, feature rich illustrations, their maps are schematic and simple. This implies that the maps were not perceived as decorative elements that merit embellishment, but as being part and parcel of the commentary itself.[10]
5. Two of the Oxford manuscripts have an empty space that is reserved, according to an adjacent note, for "the form [i.e., map] of Eretz Yisrael and its borders." Although the diagram itself was not included for various reasons, this insert also bears witness to the fact that the copyist deemed the map to be a full-fledged part of the commentary.[11]
6. Rashi was indeed known to employ diagrams on a regular basis for the sake of clarifying his arguments. For instance, Penkower identifies a map of the Jewish Temple according to the prophecy of Ezekiel in the St. Petersburg manuscript.[12] Another diagram of this sort was most likely integrated into Rashi's discussion on the concept of *mavoi akum* ("crooked alley") in his commentary on the Babylonian Talmud (Eruvin 6a). More specifically, the text includes the phrase *ka'zeh* ("like this"), which apparently refers to the said diagram. Illustrations

---

[7] Grossman, *Rashi*, p. 180; Grossman and Kedar, "Rashi's Maps of the Land of Israel and their Historical Meaning" [Hebrew].
[8] Grossman, "Marginal Notes and Addenda of R. Shemaiah and the Text of Rashi's Biblical Commentary" [Hebrew].
[9] Sed-Rajna, "Some Further Data on Rashi's Diagrams to his Commentary of the Bible."
[10] Metzger, "Le manuscript enluminé Cod. Hebr. 5 de la Bibliothèque d'Etat à Munich;" Sed-Rajna, ibid.
[11] Gruber, "What Happened," esp. pp. 116–117.
[12] Penkower, "A Replica of a Diagram of the Temple and its Courts that Rashi Sent to R. Samuel of Tzavyareh," p. 322 [Hebrew].

of a *mavoi akum* also surface in incunabula of this commentary and later print editions.

In light of the above, these maps can be regarded as the fruit of Rashi's labor, or at least geographic elements that were formulated within the savant's inner circle, with his knowledge, and in accordance to his worldview. Therefore, it seems that "Rashi's maps"—the prevalent term in the research—is a valid heading.

From a cartographic vantage point, all the diagrams that will be explored in this chapter are schematic maps. These items warrant our attention because the content is indicative of Rashi's outlook on and interest in topics that pertain to the Land of Israel, not least the question of its borders.[13] To the best of our knowledge, these are the earliest maps that were produced in the Jewish world. Moreover, they were copied on numerous occasions and served as the basis and inspiration for countless other maps, both in manuscript and print form. As a result, they are to be viewed as formative documents of the Hebrew cartography.[14]

Rashi's Maps for Numbers 33 and 34

Mas'ei (Travels), the final portion of the Book of Numbers (33–36), is the textual source for numerous maps of Eretz Yisrael, from the time of Rashi onwards. In Numbers 33, there is a detailed account of the forty-two stops that the Israelites made on their voyage from Raamses to the River Jordan. The opening section of chapter 34 describes the Promised Land's borders west of the Jordan. Maps focusing on both of these topics were integrated into Rashi's commentary on *Mas'ei*. While these items are quite uniform throughout most of the manuscripts, there are differences with respect to the quality and drafting of the maps as well as other ancillary discrepancies.[15]

The first map can be found in all the manuscripts that I examined, adjacent to chapter 33. Comprised of simple lines, this general account takes up relatively little space. While Ofer has dubbed it "the Map of the Bordering Lands,"[16] in my estimation the principal topics are the exodus and the Israelites' peregrinations through the

---

13 Grossman, *Rashi: Religious Beliefs and Social Views* [Hebrew]; idem, " Eretz Israel in the Thought of Rashi," [Hebrew].
14 Narkiss, "Rashi's Maps" [Hebrew].
15 For the purposes of this work, the key manuscripts are MS Munich, Bayerische Staatsbibliothek, Munich, Cod. Heb. 5, fol. 139v-140r; Leipzig Universitätsbibliotek, B. H. 1, fol. 160r-160v; Paris, Bibliothèque Nationale de France, Hébreu 154, fol. 35v, 36r; Hébreu 155, fol. 178v-179v; Hébreu 156, fol. 158r-158v; Oxford, Bodlean, opp. 34 fol. 97v; Parma Palatina, De Rossi, 543, 99a. For an in-depth look at some of these manuscripts, see Sed-Rajna, *Les Manuscrits Hébreux Enluminés des Bibliothèques de France*.
16 Grossman, *Rashi*, p. 180; Ofer, "The Maps of the Land of Israel."

wilderness. Content-wise, this map is directly tied to Rashi's commentary on Numbers 33, which assays the geographic relation between Egypt, Edom, and Moab in their capacity as the southern neighbors of the Land of Canaan. The diagram thus constitutes a learning aid for understanding the Israelites' progression through the wilderness.

This rectangular map faces north and depicts Eretz Yisrael *vis-à-vis* its neighbors to the south and east: Egypt, Edom, Moab, and the Land of Sihon and Og. The map's most salient feature is a series of interconnected rectangular compartments. Atop the upper red line, there is no content save for the word "north." On the lefthand side, to the "west," lies "the Great Sea," namely the Mediterranean. In the Leipzig (B.H. 1) and Paris (155) manuscripts, the box denoting the Great Sea also consists of smaller text reading "Philistines."[17] The Nile and Egypt, which are rendered as two parallel rectangles, appear in the lower portion. Below Egypt is a compartment spanning the width of the map. It contains a passage that Rashi cites in his commentary for Numbers 34:3: "Then they went along through the wilderness, and compassed the Land of Edom, and the Land of Moab, and came by the east side of the Land of Moab" (Judges 11:18).[18] Kedar aptly suggests that the unfolding of this verse mimics the route that the Israelites took: after eschewing a direct course to Canaan through "the path of the Land of the Philistines," they headed east until the south of Edom.[19] In so doing, the physical alignment of this verse replaces the schematic approach of signifying a route with a line or double line. Upon reaching Edom, Rashi posited, the Israelites "turned their faces to the north until they had circuited its entire eastern boundary along its breadth" (Rashi for Numbers 34:3, statement beginning "From the Wilderness of Zin"). Put differently, they turned north at Trans-Jordan and reached the Promised Land from the east. The right side of the map is comprised of two columns: the inner one is divided into three boxes representing the Jordan River, the Dead Sea, and Edom; the outer frame houses Moab, the plains of Moab, and the Land of Sihon. As per Rashi's interpretation, the rectangle denoting Edom in the Leipzig (B.H. 1), Oxford (Bodleian 34), and Paris (154) manuscripts is diverted to the left, so that this land is located due south of Canaan.

Somewhat larger and more detailed than the first map, the next one depicts the borders of Eretz Yisrael in close adherence to Numbers 34:2–12.[20] Here too, the basic

---

17 MS Paris 155 appears to be the addition of a proofreader. Therefore, I will not expound on the discrepancies between this text and the other versions.
18 In some of the manuscripts, the verse is written backwards; in others, the opening words face upwards. The significance of this sort of inverted writing is discussed in the next chapter.
19 Kedar, "Rashi's Map of the Land of Canaan and Its Cartographic Background."
20 Many of the maps that will be discussed below are based on this passage. Therefore, it is brought here *in toto:* "Command the Israelites and say to them: 'When you enter Canaan, the land that will be allotted to you as an inheritance is to have these boundaries: Your southern side will include some of the Wilderness of Zin along the border of Edom. Your southern boundary will start in the east from the southern end of the Dead Sea, cross south of the Ascent of Akrabbim, continue on to Zin and go south of Kadesh Barnea. Then it will go to Hazar Addar and over to Azmon, where it will turn, join the

Fig. 1: Rashi's map for Numbers 33, the Munich manuscript,
Bayerische Staatsbibliothek München (Cod. Hebr. 5(1), fol. 139v).

Wadi of Egypt and end at the Mediterranean Sea. Your western boundary will be the coast of the Mediterranean Sea. This will be your boundary on the west. For your northern boundary, run a line from

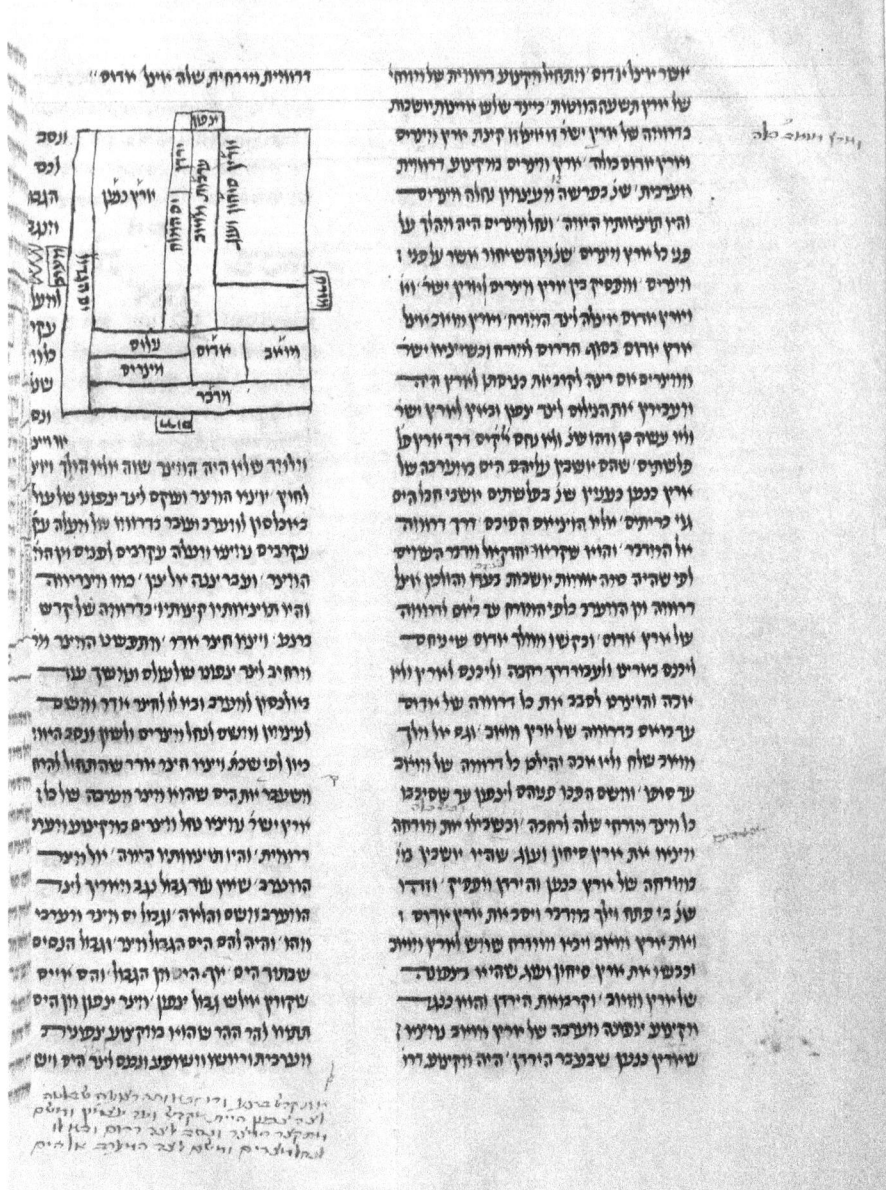

Fig. 2: Rashi's map for Numbers 33, the Paris manuscript, Bibliothèque Nationale de France (Hébreu 154, fol. 35v).

the Mediterranean Sea to Mount Hor and from Mount Hor to the entrance of Hamath. Then the boundary will go to Zedad, continue to Ziphron and end at Hazar Enan. This will be your boundary on the north. For your eastern boundary, run a line from Hazar Enan to Shepham. The boundary will go down from Shepham to Riblah on the east side of Ain and continue along the slopes east of the

contours are rectangular,[21] but the map faces east. In the middle of the diagram is a long passage explaining that the Mount Hor at hand is not the identically-named burial site of Aaron (see Numbers 20), but a mountain on the northwestern tip of the Land of Israel, which is mentioned in tractate Gittin.[22] On the bottom portion of this diagram sits the Great Sea (the Mediterranean). The left column is broken down into a number of rectangles, each of which denotes a toponym from the list of sites on the northern border: Mount Hor, which is also identified as Tor Amnon, the entrance of Hamath, Zedad, Ziphron, and Hazar Enan.[23] The row of boxes along the map's upper edge constitutes the eastern border. One compartment reads, here floweth the Jordan. The other toponyms lining this side are Shepham, or what the Leipzig manuscript refers to as Pamyas;[24] the Riblah, to which the Leipzig version adds the site's Biblical elaboration—"to the east of Ain;" the Sea of Galilee; the Jordan; and the Dead Sea. The right wing offers a representation of the southern border, starting with the tip of the Dead Sea to the east. Across the outer edge is a running and segmented citation from Numbers 34:3–4, with the passage's toponyms tucked in compartments.[25] Each name in the Munich manuscript is surrounded by a rectangular frame, whereas a line twists between the sites in the Leipzig version. The boxes containing the Ascent of Akrabbim (scorpions) and Kadesh Barnea are usually on the Canaanian side of the border. With respect to the double line, the map corresponds with Rashi's written commentary (Numbers 34:4, statement beginning "And your"), as the Ascent of Akrabbim is to the north of the border. Next to each of the four lines delineating the borders of the Promised Land is the phrase "boundary thread" (ḥut metzer) as well as the attendant direction (eastern, western, etc.). On the right side is the following note: "Within these four threads is Eretz Yisrael." The use of the term "boundary thread" in this context alludes to Rashi's commentary for Numbers 34:4 and tractate Gittin 1a.

Rashi hints to the importance that he places on the study of the maps and thus the borders of Eretz Yisrael in his commentary on Numbers 34:2 (statement beginning "This is the Land"): "Since many commandments have to be practiced in the Land of Israel and are not to be practiced outside the Land, it [the Scripture] feels compelled to delimit the boundaries that encompass it, so as to tell you what are the borders

---

Sea of Galilee. Then the boundary will go down along the Jordan and end at the Dead Sea. This will be your land, with its boundaries on every side.'"

21 Of all the manuscripts that I perused, the exception is Paris 156. To begin with, the map is drawn on the lower margins of the page. Moreover, it is condensed and rife with glaring errors. For instance, the words "eastern boundary thread" appear on the west end. See Sed-Rajna, *Les Manuscrits*, p. 232.
22 In MS Paris 155, the text is written backwards.
23 In MS Paris 155, the northern border is terraced, rather than straight. Additionally, the west side is diverted leftwards (i.e., southwards) *vis-à-vis* the east side.
24 The equivalent of Pamyas in the language of the Sages is Paneas—a place in northern Eretz Yisrael presently known as Banyas.
25 For a detailed analysis of the map in MS Leipzig, see Kedar, "Rashi's Map of the Land of Canaan, ca. 1100, and Its Cartographic Background," pp. 159–164.

Fig. 3: Rashi's map for Numbers 34 in the Munich manuscript, Bayerische Staatsbibliothek München (cod. Hebr. 5(1), fol. 140r).

within which the commandments are to be practiced." At any rate, the map that graces this passage is a schematic diagram which helps the reader make sense of the Biblical account and the author's exegesis. Neither Rashi's map nor his commentary attempt to reconcile the border sites with eleventh-century toponyms, for both the exegete and his audience did not live in the Land and were unfamiliar with its geography. Even those few places that are incorporated, such as Pamyas, merely tie the Biblical names to those in the Talmud, rather than contemporaneous toponyms. The schematic and minimalistic character of Rashi's maps should come as neither a surprise nor a disappointment, for they are strictly intended as learning aids for the commentary. These works are certainly not meant to depict the expanse's geometric shape or the mathematic alignment between the places therein.[26] It also bears noting that we will encounter this minimalistic style in chapters 2 and 5.

The similarities in design and content between the two maps for *Mas'ei* render them an inseparable pair.[27] In fact, copyists continued to replicate these works for

---

26 Delano-Smith, "The Exegetical Jerusalem: Maps and Plans for Ezekiel Chapters 40–48."
27 Delano Smith and Gruber, "Rashi's Legacy: Maps of the Holy Land."

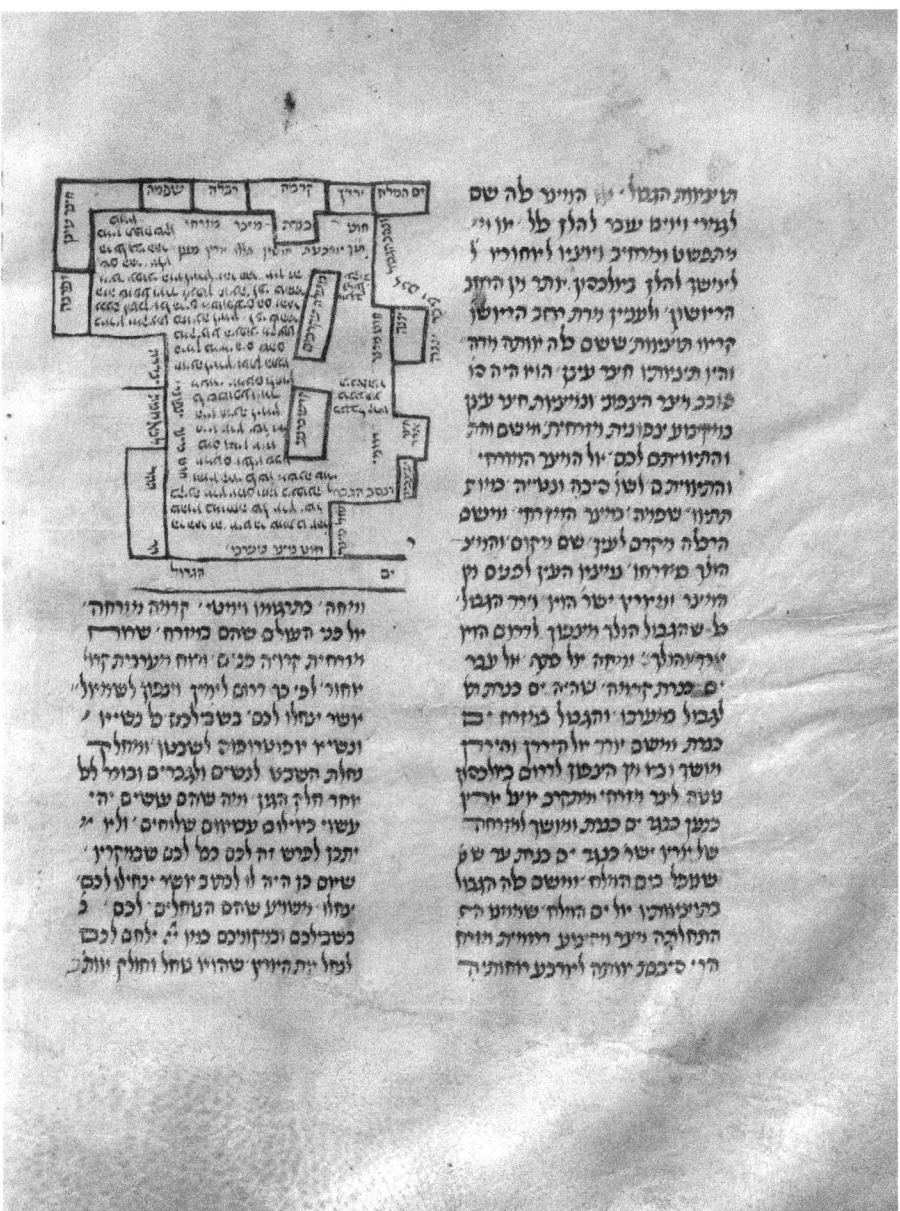

Fig. 4: Rashi's map for Numbers 34 in the Paris manuscript, Bibliothèque Nationale de France (Hébreu 155, fol. 179v).

several centuries to come, even after the introduction of the printing press. For example, the National Library of Israel holds a manuscript from the mid-1400s containing maps that follow in Rashi's footsteps.

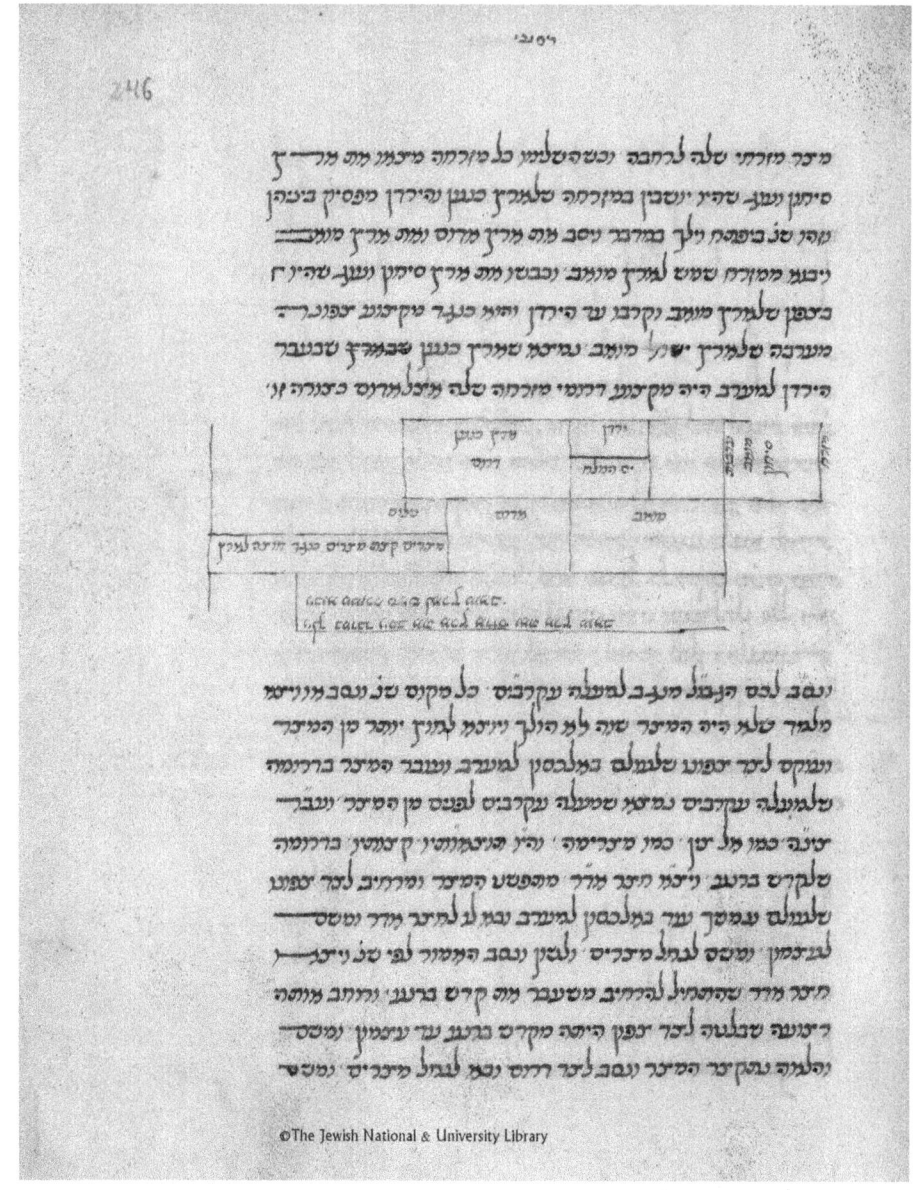

Fig. 5: Map of the Land of Canaan from Rashi's commentary on the Pentateuch, manuscript from the National Library of Israel, Jerusalem (copied by Yekutiel ben Shlomo), pp. 246–247.

In fashioning these diagrams, Rashi transformed the portion of *Mas'ei* into a cornerstone of not only the exegetical discourse on the Promised Land's borders, but of Hebrew cartography as well. With the dawn of the printing age, R. Elijah Mizrachi

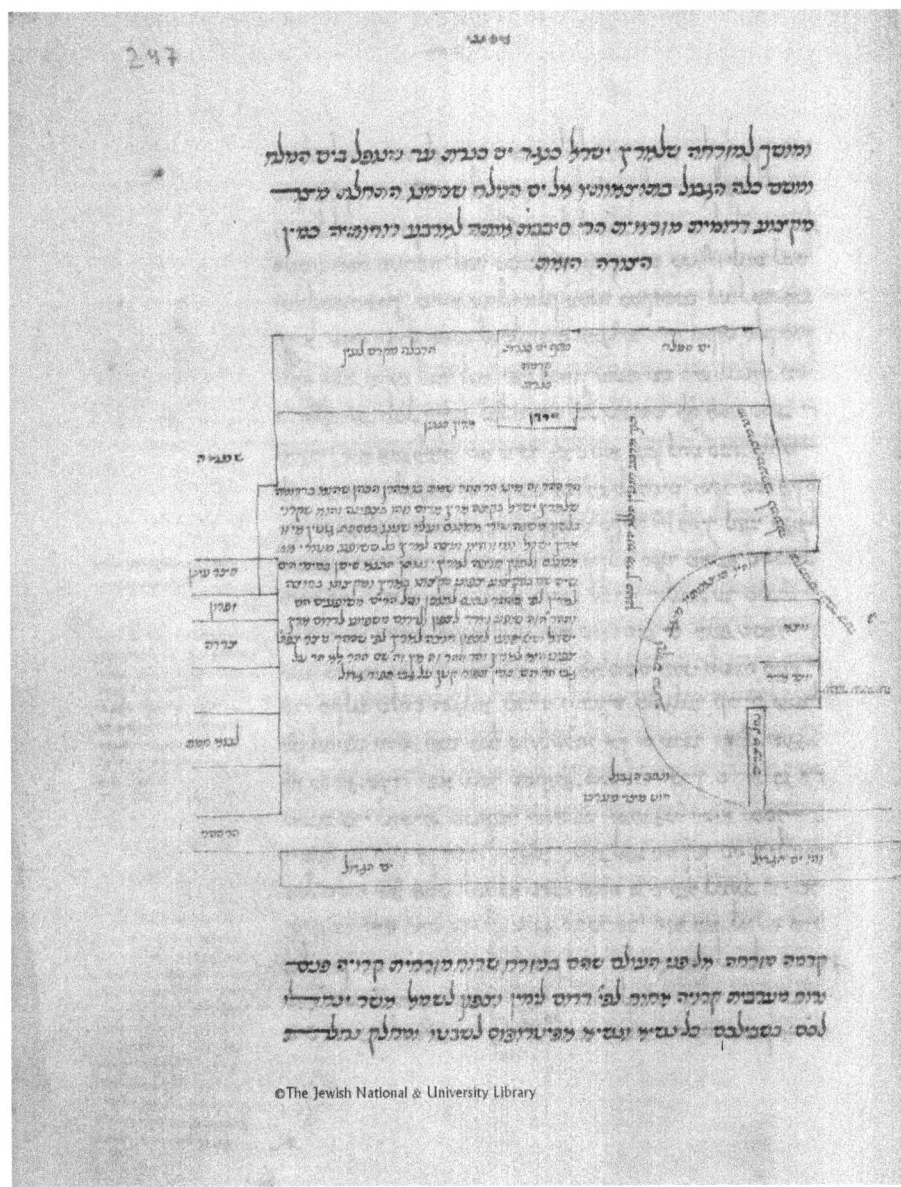

©The Jewish National & University Library

would unite these two maps into one in his commentary on Rashi (see discussion in the next chapter), thereby setting the standard for many later authors.

While Rashi's map depicts the borders of Eretz Yisrael, it only hints to the Israelites' exodus and peregrinations, even though both events are discussed at length in the same portion. In contrast, several of Rashi's successors added a dotted line marking the Israelite's route through the wilderness. This very topic would indeed become a focal point of Christian cartography at the beginning of the modern era. As elucidated in chapters 3 and 4, the journey to the Promised Land would also penetrate the Jewish maps—by way of replication and emulation—at a later stage in their development.

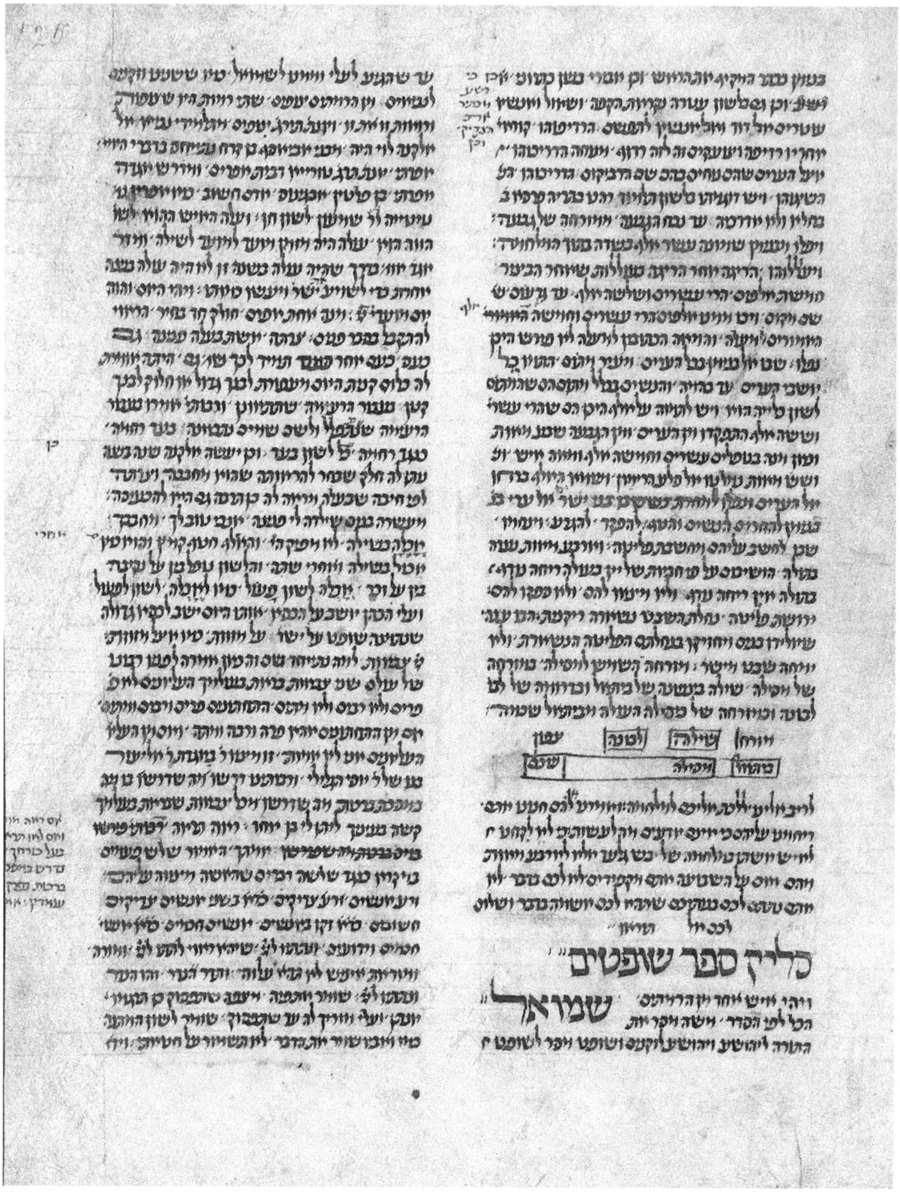

Fig. 6: Rashi's diagram of the Shechem [Nablus] road, Judges 21:19, the Oxford manuscript, Bodleian Library (opp. fol. 125a).

Rashi's Diagram for Judges 21

The third map is a schematic diagram of the road from Jerusalem to Shechem (Nablus). In several manuscripts, this exceedingly modest chart appears several lines before the end of Rashi's commentary on the Book of Judges. Its purpose is indeed to explain the adjacent verse: "… the highway that goeth up from Bethel to Shechem" (21:19).[28] Put differently, it exhibits the relative position of the Shechem, Shiloh, Bethel, and the road between them. In most of the manuscripts, the diagram is incorporated into the actual text or on its margins. It appears to be a didactic tool for comprehending the Biblical text, which Rashi assumed was less than clear for readers unfamiliar with the layout of the Land.[29] To begin with, this unpretentious sketch is important to our discussion because it bears witness to Rashi's appreciation for the expository value of cartographic diagrams and to his propensity for implementing these devices in his writing. *Ex post facto*, it demonstrates that later generations repeatedly copied this diagram and integrated it into Rashi's commentary on the Talmud, thereby attesting to the importance of the exegete's maps among Jewish scholars well after his passing.

The Maps for Ezekiel 45 and 48

The next two maps, which surface in a number of manuscripts, are concerned with Ezekiel 45 and 48.[30] The prophecies in these chapters pertain to the "end of days."[31] More specifically, chapter 45 describes the Jewish Temple and the estates of the priests, Levites, and the *nasi* (political head) in the greater Jerusalem area upon the coming of the messiah. It is on this basis that Rashi drafted his attendant map of the Temple, Jerusalem, and the vicinity. Chapter 48 represents the tribal allotments during the epoch under review. Jerusalem is positioned at the center of the map, with the twelve allotments running parallel to either side of the city: seven to the north; and five to the south. Needless to say, this layout is arranged according to the description in Ezekiel and completely differs from the historical account in the Book of Joshua and other sources. Rashi himself apparently referred to this particular

---

[28] Gruber, "The Sources of Rashi's Cartography;" idem, "Rashi's Map Illustrating His Commentary on Judges 21:19."
[29] See, for example, MS Oxford, the Bodleian Library, opp. 2, fol. 36; opp. 34, fol. 125r; and MS Parma, the Palatine Library, De Rossi # 387, p. 16.
[30] MS Oxford, Bodleian, opp. 34, fol. 178–179; opp. 2, fol. 204, fol. 207; MS Parma, Palatina, De Rossi # 387, pp. 99a, 102a; MS Berlin, Staatbibliotek, or. fol. 122, fol. 68v, 72r; MS Paris, Bibliothèque Nationale de France, Hebreu 154, fol. 135v, 139r.
[31] Sed-Rajna, "Some Further Data," esp. p. 153; Brodsky, "Ezekiel's Map of Restoration."

map in one of his letters, so that scholars would be hard-pressed to deny that he authored the original version.[32]

Diagram of Acre and its Environs in Rashi's Commentary on Tractate Gittin

The sixth map is a schematic diagram that Rashi included in his commentary on tractate Gittin of the Babylonian Talmud. The question of Eretz Yisrael's southern and northern borders along the Mediterranean coast riveted the attention of Jewish exegetes. Above all, they were interested in whether Acre fell within the Land's boundaries.[33] The provenance of this question is a debate between R. Judah and R. Meir in Mishna Gittin (1a) over the city's standing. The case involves a bill of divorce that a messenger delivered from the "province of the sea" (i.e., anywhere except the Land of Israel or Babylon). This issue also came up in other Mishnaic and Talmudic forums. Since Acre was an important town and port of entry to Eretz Yisrael, and given the Halakhic doubts as to whether it fell within the Land's boundaries, the discourse on this matter garnered a prominent spot on the Halakhic agenda. In fact, both Rashi and Maimonides contended with this geographical issue by, among other things, formulating diagrams to bolster their arguments.[34] It bears noting, albeit with utmost humility, that these two spiritual titans did not fully understand the geographical reality in the Holy Land. For instance, their quasi-maps of the Acre region—the primary function of which was to elucidate their respective interpretations—situate both the city and Achzib (a coastal settlement nine miles north of Acre) in the northeast, rather than their true location in the northwest. Although these diagrams were also copied and printed in commentaries on the Babylonian Talmud, their impact on the continued development of the Hebrew maps was meager. Consequently, I will not expound on either document.

Why did Rashi Embrace Maps?

The study of Rashi's maps leads us to the question of why he embraced these devices in the first place. At any rate, the revered scholar broke new ground in this field as well. To the best of our knowledge, he was indeed the first Jewish sage to draw maps as a means for advancing scholarly and didactic pursuits. Given the complexity and scope of this question, I have divided it into several parts:

---

32 Sed-Rajna, ibid., esp. p. 154.
33 For a general discussion on the question of Acre's status, see Ne'eman, "Is Acre Part of Eretz Israel?" [Hebrew].
34 Warhaftig, "The Borders of Eretz Yisrael according to Maimonides" [Hebrew].

יום קנה יונית היערך לוז שעשרים וחמשה אלף. וידות כנגד הזרה שעזרה בה על חמש ורוחב שלהי
הבית עזרין אלפים הלל ותרומה קדש מן הארץ היאהתרומה הנית. וכהנים משרתי המקדש יהיה הע
העזר. על חוות ויאות שלהי הבית. שני עשר אלף. ומאתיס וחמשה. לזרח. וכנגדו לוחרב והמקדש
ביונ ערב. ויתכן אלפים ושכע אות וחמשים יעמון וכנגדן לדרום ויה הם מקום ולכתס העזר הוה
עשב כ הוזצא ולקדש ולקדש ויחט על חמש ורוחב הדרונ צפ יה ווקדשים וע
הקדש וחמש ועשרים אלף. קנה יורך ועשרים אלפים רוחב תרויהו רעה אחרת יול ווז ויעה
של וזעט הרוא כן ומפרש כן. הספר שהי בעפן. ואחוזת העיר הקן. העיר והווות וישב
הווזלה ובעת בה ישל בתים תהנו חורשת ולפים רוחב בנפתה של שעירה ואורך טרדר ש
היועות ערמות כל התרומה ורחכ ער על בה לפים על בה אלפים תרומה וויעורת הקדש מלטותה
אורך ידעות ותרומת הקדש לא בית ישל יהיה אותה רעה שלישית. הדי וישב לדרום ולפוני
וגזרמה יתדומת הקדש ואחוזת העיר בסון הספר הזא חולך את ירן ישר ומידחה והעברה
ולש עשרה ידעות שתי עשרה ולמייז השבטים ולא ואחת עשרם ואחת אלן קנה חכב וירן
כאורך יטל וידעה אחת ותרומה יורכה ין גבל וחרח וער גבל ועבר נבו ורחבה בה אלפים קדש
כשיעור כל חוק חלך. וחוותו רעה בה הרם בזרועית שום רעות הלל ואטריות ומעלה ש
בה אלפים ובעשך עליה יהודה ער סון גבל הירן וכן וחרב יהיה ונשיאי והוה ומחה
ולעערב יא פני תרות הקדש ואל פני אוחוזת העיר כנגד ט רחב שלוט ט רעות הוזברה דך
ותרומה הקדש של רעות הבהנים והליים ואחוזות העיר ופניעיי יה מוערב תרות העא
והעיר עד מערב ועד הגבול מחינ קדמה קדישה ותרחתה התרומה לודיעד וזדרח ...

מזרח

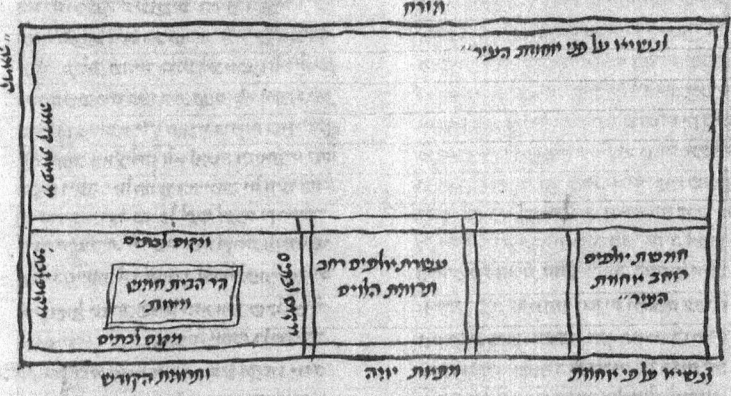

ועשייו על פני אחוזות העיר

ועשוות אחד החלקים של שנטים המפרט בסון. הספר שהן הגבל שלים ארץ ישר עד גבל קדש
וירן ויהי לואחוזה הגל יועל דין מרעה הגי לדגי לדבא לוחספר ולא אומעה והוא טעולים בלות
הריה עושותם פלקיות שיות הגרטים וות ענין וטולתם וימה וחרב. ובש דך ואחד חלן
וכן אחד להם וו עשיריות. וחוור העור. ונשיאת. אמעור החוור בכנת. וכן עשירות חויד יבטו

Fig. 7: Rashi's maps for Ezekiel 45 and 48, Paris manuscript, Bibliothèque Nationale de France (Hébreu, 154, fol. 135v and fol. 139r).

In the first place, can we retrace the source of Rashi's motivation for drawing maps of Eretz Yisrael? With respect to the simple diagrams on Judges 21 and his introduction to tractate Gittin, the answer is rather straightforward. As a skilled exegete and teacher, Rashi embraced these tools in order to augment his readers' comprehension. This claim is supported by, *inter alia*, Rashi's aforementioned letter to one R. Samuel of Tzavyareh, in which the author mentions that he also sent a diagram for the purpose of improving his explanation.[35]

The diagrams for Numbers 33 and 34 necessitate a broader answer. Rashi's maps, especially those depicting the promised borders, are a product of his special relation to the Land of Israel, on the one hand, and the rise of the polemics between Christianity and Judaism, on the other.[36] In particular, Rashi underscored the commandments that are dependent on the Land—an emphasis that naturally requires a working knowledge of the expanse's boundaries. As noted, Rashi's entire generation contended with the on-going challenge of the Jewish-Christian ideological dispute. This polemic has indeed left a lasting mark on his oeuvre.[37] Touitou asserts that it was one of Rashi's prime motivations for writing his commentary on the Pentateuch. In addition, Grossman points to the polemical aspects of the exegete's work on Psalms and the Book of Proverbs, which were so severe that the Christian censure erased various passages from the early incunabula.[38] During Rashi's lifetime, a new front opened in the Jewish-Christian rivalry: the ideological competition between the three major faiths over the following question:[39] Did the Land of Israel belong to the Muslims who controlled it, to the Christians who aspired to take it over, or to the Jewish people who claimed that Eretz Yisrael was promised to them by God?

The Christian efforts to bring their aspirations to fruition would reach a peak during the Crusades. However, the opening shots in this campaign were fired during Günther of Bamberg's grand pilgrimage to the Holy Land between 1064 and 1065,[40] the speech of Pope Gregory VII in 1074, and above all Pope Urban II's sermon in 1096 (approximately ten years before Rashi's death). Urban's exhortation indeed constituted the immediate trigger behind the Crusades.[41]

Rashi's espousal of the Jewish people's divine right to the Land is quite evident in the way he chose to open his exegesis on the Hebrew Bible:

> Rabbi Yitzchak said: The Torah [which primarily teaches God's commandments] need only have begun with [the verse] "This month is for you [the foremost of the months]," which is the first

---

35 Penkower, "A Replica of a Diagram."
36 For an in-depth look on this relation, see Grossman, *Rashi*, pp. 170–180; idem, "Eretz Israel in the Thought of Rashi."
37 Rosenthal, "The Antichristian Dispute" [Hebrew]; Grossman, "The Wording of Rashi's Commentary on the Hebrew Bible and the Jewish-Christian Polemic" [Hebrew]; idem, *Rashi*, pp. 19–20.
38 Touitou, "What Motivated Rashi to Write a Commentary on the Pentateuch;" Grossman, "Revolutionism and Commitment—the Contours of Rashi's Image."
39 Grossman, *Rashi*, pp. 175–177; idem, *Rashi: Religious Beliefs*.
40 Hans H. Kortüm, "Der Pilgerzug von 1064/5 ins Heilige Land."
41 On the evolution of the Crusader ideology and the role of Gregory VII and Urban II in its genesis, see Prawer, *A History of the Latin Kingdom of Jerusalem*, pp. 75–87 [Hebrew].

commandment given to the Jewish people. For what reason, then, does it begin with Genesis? Because He wished to tell His people that [by virtue of] "the strength of His actions [i.e., the creation of the world] [God was entitled] to give them the inherited land of the nations" (Psalms 111:6). For if the nations of the world were to say [to the Jewish people], "You are bandits because you conquered the lands of the seven nations [of Canaan]," they could answer them, "The entire earth belongs to the Holy One, Blessed be He; He created it and gave it to whom He saw fit. When He wanted, He gave [this land] to them [the nations], and when He wanted, He took it away from them and gave it to us."

The entire purpose behind this *midrash* (homiletic story) is, of course, to demonstrate that the Land of Israel belongs to the Jewish people.⁴² In this particular instance, Rashi did not cite a pre-existing text, but combined two different homiletic stories: one from *Genesis Rabbah*, where R. Joshua of Sikhnin speaks in the name of R. Levi;⁴³ and the second from the old *Midrash tanḥuma* on *Beraishit* (the opening chapters of Genesis), which also cites R. Yitzchak.⁴⁴ This reference to the latter at the very outset of Rashi's commentary led some researchers to believe that the author sought to commemorate his father, whose name was indeed Yitzchak. That said, the weakness of this hypothesis is that Rashi's father was, in all probability, not a distinguished scholar.⁴⁵ On the other hand, the question over the rights to Eretz Yisrael figured prominently in Rashi's gospel. It thus stands to reason that the maps of the promised borders in his commentary on *Mas'ei* as well as the opening paragraph of his Biblical exegesis were intended to bolster the Jewish claim to the Land of Israel.

The second question that begs asking is thus: What is the relation between Rashi's maps and those of medieval Christian scholars? As an intellectual who was attuned to the outside world and the ebb and flow of daily life, Rashi was acquainted with artisans and familiar with their techniques. Therefore, it stands to reason that in crafting maps, he was influenced by thinkers from beyond the gates of the Jewish community. Grossman and Kedar suggest that he drew on earlier Christian and perhaps even Muslim maps.⁴⁶ Expanding on this hypothesis, Kedar highlights two pri-

---

42 Pearl, *Rashi*, pp. 47–48.
43 *Genesis Rabbah*, I.II: "R. Joshua of Sikhnin in the name of R. Levi commenced [this discourse by citing the following verse]: 'He has declared to His people the power of His works, in giving them the heritage of the nations' (Ps. 111:6). What is the reason that the Holy One, blessed be He, revealed to the Jewish people what was created on the first day and what on the second? It was on account of the nations of the world. It was so that they should not ridicule the Israelites, saying to them, 'Are you not a nation of robbers [having stolen the land from the Canaanites]?' It allows the Israelites to answer them, 'And as to you, is there no spoil in your hands? For surely: "The Caphtorim, who came forth out of Captor, destroyed them and dwelled in their place!" (Deut. 2:23).'"
44 *Midrash tanḥuma*, *Beraishit* (ed. Buber) 1. 11: "R. Yitzchak said: It was only necessary to write the Torah from [the words] in 'This month shall be for you.' Why did he write from 'In the beginning?' To make known his mighty power. Thus it is stated: 'He has declared to his people the power of his works in giving them the heritage of the nations.'"
45 Maimon, *The Book of Rashi*, p. viii; Pearl, *Rashi*, p. 104, note 26.
46 Grossman and Kedar, "Rashi's Maps."

mary common denominators between all these cartographic works: the use of rectangular compartments for denoting places; and the route of the Nile. In Rashi's map and some of the Christian and Muslim renderings, the river indeed proceeds from east to west, rather than south to north.[47]

The foundation of Kedar's thesis is that Rashi did not create his maps from naught and was well aware of other cartographic works. While this argument holds water, his claim as to a close resemblance between all the maps that testifies to the Christian and Muslim authors direct influence on the Jewish savant is less persuasive. To begin with, unlike his putative sources, Rashi's maps are exceedingly schematic. He also ignores central topics and motifs that one would have expected him to reflect on had he indeed borrowed from Christian works. In contrast to Rashi's cartographic repertoire, the vast majority of known tenth- and eleventh-century Christian maps—those that Rashi could have feasibly had access to—are representations of the entire world, not defined regions. Furthermore, his maps are rectangular, whereas the world maps are either round or elliptical (i.e., the latter fall under the heading of T-O, Beatus, and Salustius maps).[48] In the Christian works, Jerusalem is frequently denoted as the center of the Earth, the Red Sea is painted red, and the Jordan River (when rendered) always starts in Mount Lebanon and is fed by two sources: the Ior and the Dan (conflated in Latin to the *Iordanis*). However, all these elements are clearly absent from Rashi's maps. From its very inception, with the appearance of Eusebius of Caesarea's *The Onomasticon* (a fourth-century book of place names),[49] the Christian cartographic tradition was preoccupied with the tribal allotments.[50] Examples include some of the aforementioned maps of the world, the sixth-century Madaba Map, and a manuscript of a ninth-century commentary on the Book of Joshua.[51] While these Christian works depict various places in the Holy Land, they practically ignore its borders. Therefore, if Rashi had indeed emulated contemporaneous Christian mapmakers, then at least some of the above-noted characteristics would have surfaced in his own renderings, but this is rarely the case.

---

47 Kedar, "Rashi's Map of the Land of Canaan and its Cartographic Background; idem, "Rashi's Map of the Land of Canaan, ca. 1100, and its Cartographic Background."
48 Edson, *Mapping Time and Space*, pp. 62–63; Harvey, *Medieval Maps*, pp. 19–37; Williams, *The Illustrated Beatus—A Corpus of the Illustrations of the Commentary on the Apocalypse*, pic. 5, 19, 239.
49 Eusebius, *Das Onomastikon*. Also see the annotated Hebrew edition: *The Onomastikon of Eusebius*, vol. xix and xxi, tran. E. Z. Melamed. The most commonplace version in the West was Jerome's translation and adaptation thereof; Jerome, *Liber Locorum*. This work netted a host of print runs.
50 Rubin and Levy-Rubin, "The Early Cartographic Tradition of the Holy Land and the Origins of the Crusader Maps of Jerusalem." For a more comprehensive discussion, see Levy-Rubin, "From Eusebius to the Crusader Maps: The Origin of the Holy Land Maps."
51 O'Loughlin, "Map as Text: A Mid Ninth Century Map for the Book of Joshua;" Scafi, *Mapping Paradise: A History of Heaven on Earth*, e.g., p. 142. Also see the photographs of manuscript-based maps with partial depictions of the tribal allotments in Harvey, *Medieval Maps*, p. 26; and Edson, *Mapping Time and Space*, p. 8.

Even from a graphic-cum-stylistic standpoint, Rashi's maps are less adorned and more schematic than those of his European contemporaries. A case in point is the lavish world maps from the eleventh and twelfth centuries,[52] which feature fortified cities, a wide range of flora, and a variety of other ornamental elements. With respect to the Nile, Kedar has observed that the so-called Map of Jerome and map of Ibn Hawqal indeed depict most of the waterway running from east to west. However, the part of the Nile closest to the Mediterranean Sea, which is the most relevant to the discussion at hand, takes a sharp turn northwards before spilling into the Great Sea. The majority of Christian maps and virtually all the Muslim maps from Rashi's time also refrain from presenting the Land of Israel as a distinct entity; instead, it is depicted as part of much larger regions. What is more, the orientation in all the extant Muslim maps is southward, and they cover entirely different topics than those of Rashi.[53]

Sed-Rajna makes a strong case that Rashi's maps for the Book of Ezekiel were the object of emulation on the part of a similar work in the manuscript of an exegesis by Richard of Saint Victor (d. 1173), a distinguished theologian who lived after Rashi's time.[54] This manuscript contains two maps that interpret Ezekiel 45 and 48, respectively,[55] each of which is very close in style and content to those of Rashi. A comparison indicates that Richard (see Fig. 8) indeed copied from the distinguished rabbi. Delano-Smith, whose own comparison encompassed additional manuscripts of exegetical maps on the vision of Ezekiel, reaches a similar conclusion.[56]

This dovetails neatly with the general consensus whereby the commentaries of Rashi and other Jewish scholars had a substantial impact on Christian exegetes, particularly those of the St. Victor Abbey school of thought. The latter even dubbed the Jewish-authored works the collections of the "Hebrews" (*Hebraei*).[57] Rashi's direct influence on Christian scholars is evident, above all, in the rise of a new exegetical outlook that preferred a straightforward interpretation of the Hebrew Bible to an allegorical one.[58] The epitome of this approach is the writing of Nicholas of Lyra (1270–1340), who is considered the most important Christian exegete to predicate his work on a literal interpretation (*peshat*). In Nicholas' oeuvre, there are indeed a

---

[52] Harvey, *Mappa Mundi: The Hereford World Map*.
[53] For a disquisition on Muslim cartography and its history, see Sezgin, *The Contribution of the Arabic-Islamic Geographers to the Formation of World Map*; idem, *Mathematical Geography and Cartography in Islam and their Contribution in the Occident*.
[54] See Sed-Rajna, "Some Further Data," pp. 155–157.
[55] MS Paris, Bibliothèque Nationale de France, lat. 3483, fol. 81; idem, nouv. acq. lat. 1791, fol. 39r.
[56] Delano-Smith, "The Exegetical Jerusalem."
[57] Smalley, *The Study of the Bible in the Middle Ages*, pp. 83–111; Hailperin, *Rashi and the Christian Scholars*.
[58] Funkenstein, *Styles in Medieval Biblical Exegesis: An Introduction*, p. 27–28 [Hebrew]; Grossman; *Rashi*, pp. 15–17.

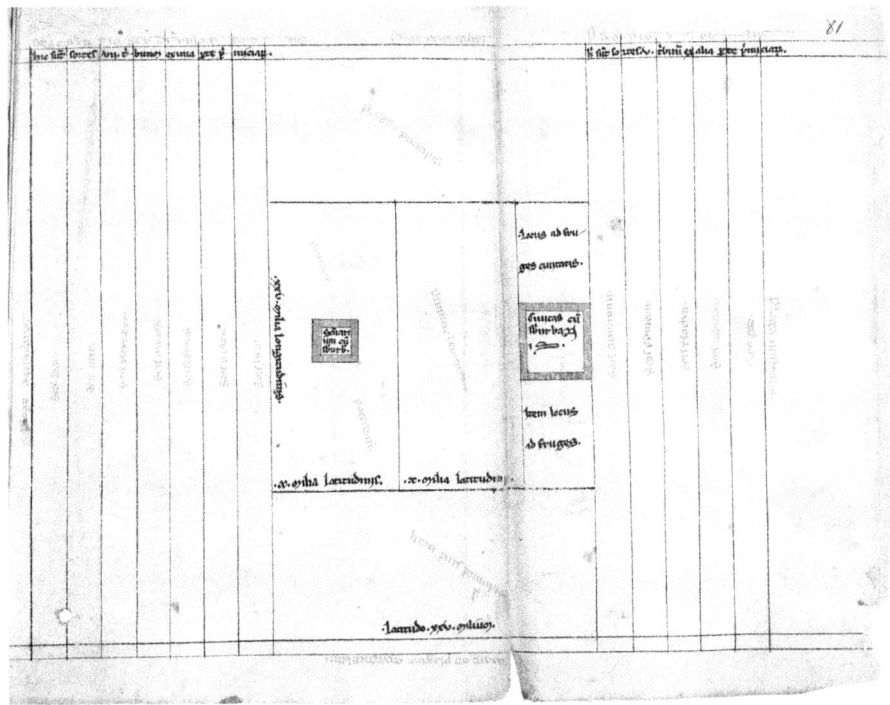

Fig. 8: A Latin map for Ezekiel 58 that draws on Rashi's map, the Paris manuscript, Bibliothèque Nationale de France (Latin 3438, fol. 81).

host of direct citations and close paraphrases, albeit in Latin translation, of one R. Salomonis—also known as Rabbi Shlomo Yitzhaki (Rashi).[59]

Conclusion

In all likelihood, Rashi authored at least six geographic diagrams that merit the distinction of being the earliest group of known Hebrew maps. Although two of these works are plain schematic diagrams, they were copied time and again and were later incorporated into print editions of Talmudic commentaries. The other four works—half of which pertain to the portion of *Mas'ei* and two on Ezekiel 45 and 48, respectively—left an indelible mark on the development of Jewish as well as Christian maps.

---

[59] Geiger, *The Commentary of Nicholas of Lyra on Leviticus, Numbers and Deuteronomy: A Study of the Jewish Sources of the Literal Commentary on Leviticus-Deuteronomy and in the Moral Commentary on Deuteronomy* [Hebrew]; Kaczynski, "Illustrations of Tabernacle and Temple Implements in the *Postilla in Testementum Vetus* of Nicolaus de Lyra;" Rosenau, "The Architecture of Nicolaus de Lyra's Temple Illustrations and the Jewish Tradition."

As discussed in the next chapter, the maps for *Mas'ei* became a pillar of Jewish cartography. They were copied in numerous manuscripts and subsequently underwent changes and additions. Upon the advent of the printing press, an adaptation of these two works would become the source of the most prevalent model of Jewish exegetical maps. The maps that had the greatest impact on the Christian commentaries were the tandem on the Book of Ezekiel. Owing to the central place of the prophet's vision on Christian thought, these two maps were destined to grace a wide array of Christian works on the Old Testament.[60] Towards the end of the 1700s, one Yehiel Hillel Altschuler copied a map pertaining to Ezekiel 48 from the works of Augustin Antoine Calmet, an eighteenth-century French Catholic scholar. In so doing, Altschuler paradoxically returned Rashi's archetype of this map to the realm of Hebrew cartography.

---

[60] Delano Smith, "Maps and Religion in Medieval and Early Modern Europe," pp. 186, 188.

2 Following in Rashi's Footsteps

Rashi's maps depicting the borders of Eretz Yisrael and the neighboring region as per the Torah portion of *Mas'ei* (Numbers 33 and 34) became the foundation stone of Hebrew cartography in the generations that followed. At first, these maps were copied in manuscripts of Rashi's commentary on the Hebrew Bible. Thereafter, they were included, with varying levels of change, into other works. Some of the versions are informed by a high level of draftsmanship, while others are less impressive. Diversity aside, all these maps share a strong affinity with those of Rashi. Like them they are rather schematic and plain in all that concerns design. Content-wise, though, the maps reflect an ongoing discourse concerning the promised borders. While grounded on the information in *Mas'ei*, this multi-generational discussion occasionally hints to other sources as well—both from other parts of the Bible and the literature of the Sages.

Manuscript Maps

The Ḥazzekuni

The first map in this survey is part of the Oxford manuscript of the Ḥazzekuni Bible commentary, which was written by Hezekiah ben Manoah of France in the mid-1200s.[1] In Neubauer's estimation, this manuscript was penned by the author himself.[2] The text is divided according to the Torah portions. An illustration of Jacob's ladder can be found in the portion of *Vayetse* (Genesis 28:10–32:3) and a sketch of the Tabernacle in *Trumah* (Exodus 25:1–27:19). Moreover, there is a map tracking the Israelites' peregrinations through the wilderness and depicting the borders of the Promised Land as delineated in *Mas'ei*. The map is rectangular and faces east. While leaning on the Rashi cartographic tradition, the author did take some liberties. Similar to the Rashi map, it places Mount Hor on the northwestern (bottom-left) corner. Along the northern border are rectangular compartments denoting the entrance of Hamath, Zedad, Ziphron, and Hazar Enan. The Great Sea constitutes the western border, along which is written "[For your northern boundary,] runneth a line from the Mediterranean Sea to Mount Hor" (Numbers 34:7). Ain, the Riblah, and Shepham form the northeastern border, from left to right. To their right is the Jordan River. The Sea of Galilee is drawn as a rectangle that descends from the Jordan to the west. The river's southern extremity takes a sharp turn east, before spilling into

---

[1] "Hezekiah ben Manoah," *Encyclopaedia Hebraica* [Hebrew]; Joseph Ofer, "The Ḥazzekuni Commentary on the Bible and its Different Versions," *Megadim* 8 (1989): 69–83 [Hebrew]. The document under review is MS Bodleian, Mich. 568. Malachi Beit Arie dates it to the second half of the 1200s; Beit Arie, *Catalogue of the Hebrew Manuscripts in the Bodleian Library*, p. 35.
[2] Neubauer, *Catalogue of the Hebrew Manuscripts in the Bodleian Library*, p. 46.

the Dead Sea. North of the Jordan, on the upper lefthand side, are "the Land of Bashan," "the Kingdom of Og" and "the land of the Ammonites." Due south (right) is "the Land of Sihon that he took from the Ammonites," "the Land of Sihon from the inheritance of his forefathers," and "the Land of Sihon that he took from the king of Moab." Further to the right are Arnon and the Land of Moab. Above these sites is a passage that is broken into four intervals: "And from the wilderness they went to Mattanah: And from Mattanah to Nahaliel: and from Nahaliel to Bamoth: And from Bamoth in the valley" (Numbers 21:18–20). Lining the top of the page is the Dead Sea, as though it were the border of Sihon and the Land of the Ammonites in eastern Transjordan. Consequently, the author was forced to abruptly divert the Jordan upwards for the purpose of connecting it to the Dead Sea.

The southeastern (right) border of Eretz Yisrael initially slants to the northwest from the southern tip of the Jordan. Upon reaching the Ascent of Akrabbim, the border veers back to the west in the direction of Zin, before proceeding to Kadesh Barnea, Hazar Addar, and Azmon. At this point, it hooks south and then west towards the Great Sea. Running along the length of the final intervals is a passage reading "From Sihor, which is before Egypt and its westerly results." The first part of this passage derives from the account of "the land that still remains unconquered" in Joshua 13:3, so that tying it to the border of Eretz Yisrael is mildly unconventional. Just south of this border is Mount Seir to the east and Egypt to the west. Compared to Rashi's map, the most interesting addition to the Ḥazzekuni version is on the south (righthand) side: the stations of the Israelite's journey through the wilderness, in accordance to Numbers 33. Lacking compartments or illustrations, the stops appear in ascending order from Raamses in the west to Oboth in the east. A verse that is cited in the first Rashi map—"Then they went along through the wilderness, and compassed the Land of Edom, and the Land of Moab, and came by the east side of the Land of Moab" (Judges 11:18)—traces the entire length of the stations. Another passage, comprised of two non-contiguous verses from the Book of Joshua, was interjected just above the Great Sea: "This is the land that yet remaineth: all the regions of the Philistines, etc. From Sihor, which is before Egypt, even unto the borders of Ekron northward, which is counted to the Canaanite" (13:2–3); and "[A]ll the land of the Hittites, and unto the Great Sea toward the setting of the sun, shall be your coast" (1:4). In citing these verses, Hezekiah ben Manoah was apparently emphasizing the fact that the coastal strip, Pleshet (the region of the Philistines), is the land that remains unconquered.

The Ḥazzekuni map is more elaborate than that of Rashi and contains quite a few new wrinkles, but the latter is most likely its primary source. This should come as no surprise, for the Ḥazzekuni is indeed a supercommentary on Rashi. Despite the beauty of the illustrations in the Oxford manuscript and the importance of this particular map, I did not find similar ornamentation in print editions of Heze-

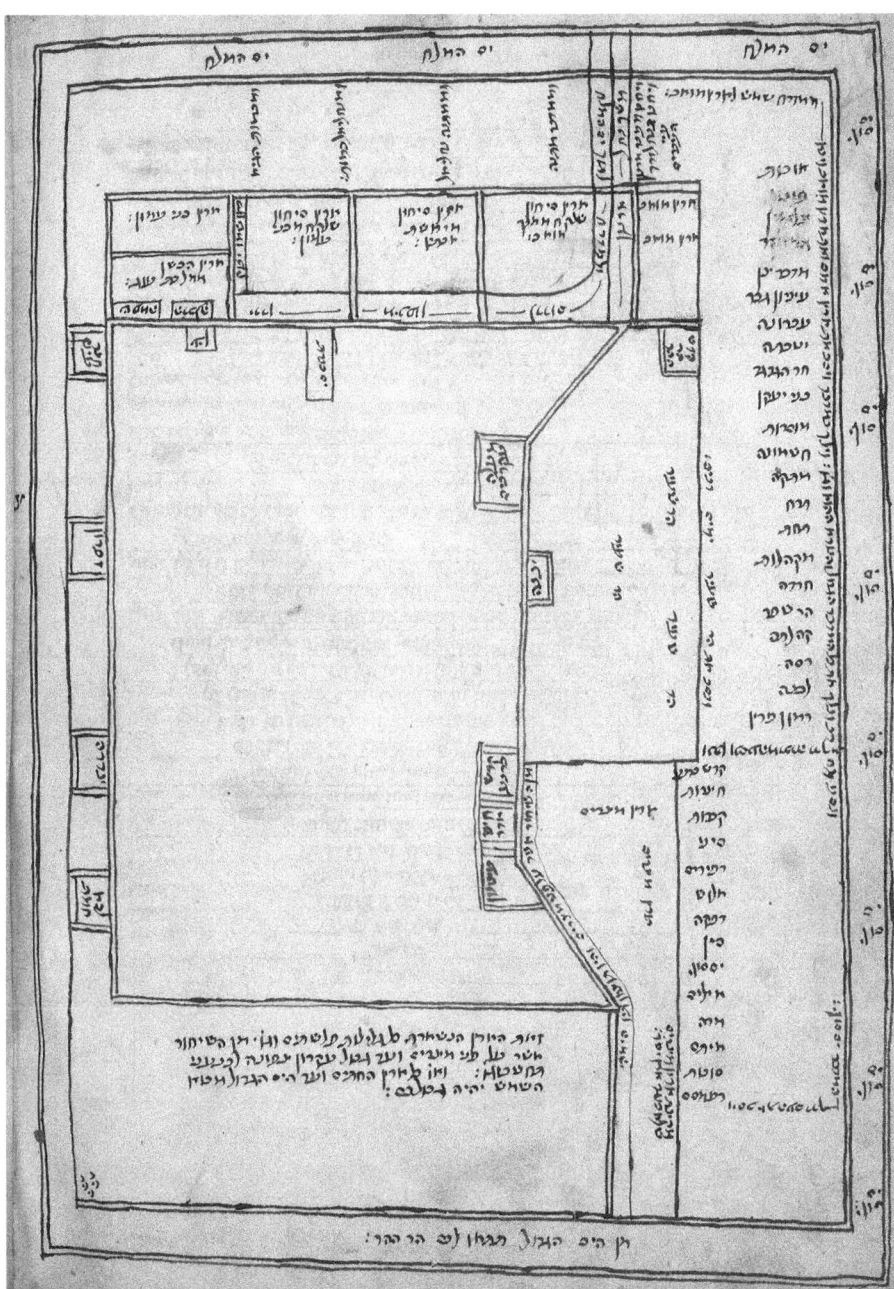

Fig. 9: A map in a manuscript of the Ḥazzekuni commentary on *Mas'ei*, the Oxford manuscript, The Bodleian Library, University of Oxford, ms. Mich. 568.

kiah ben Manoah's book.³ One possible reason for this discrepancy is that the print editions are adaptations of other manuscripts. Alternatively, the illustration may have been left out due to the steep cost of preparing the necessary plates.

Map from the Commentary of R. Joseph Kara

On the last page of a fourteenth-century manuscript of R. Joseph Kara's commentary on the Latter Prophets is the first map to integrate the tribal allotments into the traditional Rashi format.⁴ The page before the map ends with the words "This completes the commentary on Malachi and the entire commentary of the Prophets of our Rabbi Joseph Kara, a righteous man of blessed memory." A contemporary and personal acquaintance of Rashi, Kara was among the most important exegetes of the 1100s. In light of the above, it was only natural that he included a map that converses with that of Rashi in his work. At any rate, we cannot verify whether this map was originally included in Kara's commentary or was subsequently added by a copyist.

The map itself is oriented eastwards. Its captions are arranged in multiple directions, and the borders of Eretz Yisrael adhere to the Rashi map. Running along the borders are passages that are commensurate with the description of the Land in Numbers 34. However, several new details were incorporated into the basic Rashi template, the most prominent of which is the signification of all the tribal allotments within compartments.⁵ These areas do not form a geographical succession, as there are gaps between some of the allotments. Kara's layout of the tribes raises eyebrows, for it is incompatible with the accepted geographic distribution, not least the literal interpretation of the pertinent Biblical texts. For example, the area of Dan is rather capacious, extending over a substantial portion of the coast; Simeon is located to the north of Judah and to the east of Dan, rather than south of Judah and abutting the desert. In contrast to the description in the Hebrew Bible, Zebulun is located in the south (between the allotments of Simeon and Asher). What is more, it does not even approach the coast. The same can be said for Ephraim and Manasseh, which are situated in the east. Likewise, Asher is nowhere near the Mediterranean or Mount Hor to the northwest. Sed-Rajna has proposed that this map reflects the future allotments as per Ezekiel 48. However, given the differences with those maps based explicitly on this chapter, there are doubts as to the merit of this argument. What is more, as opposed to the Kara version, in the maps to Ezekiel 48 Gad, Reuben, and half of Manasseh are not located in Transjordan, but among the other tribes. There-

---

3 The editions that I checked were Cremona 1559, Venice 1567, Basilea 1606, Prague 1618, Wilhelmsdorf 1685, and Sulzbach 1741.
4 MS Paris, Bibliothèque Nationale de France, Hébreu, 162, fol. 174r; Sed-Rajna, "Some Further Data," pp. 350–351.
5 For an in-depth look at the standing of the tribal allotments in Jewish cartography, see chapters 4 and 5.

fore, Kara was most likely depicting the tribal allotments as per the Book of Joshua, but lacked sufficient geographical knowledge about Eretz Yisrael.

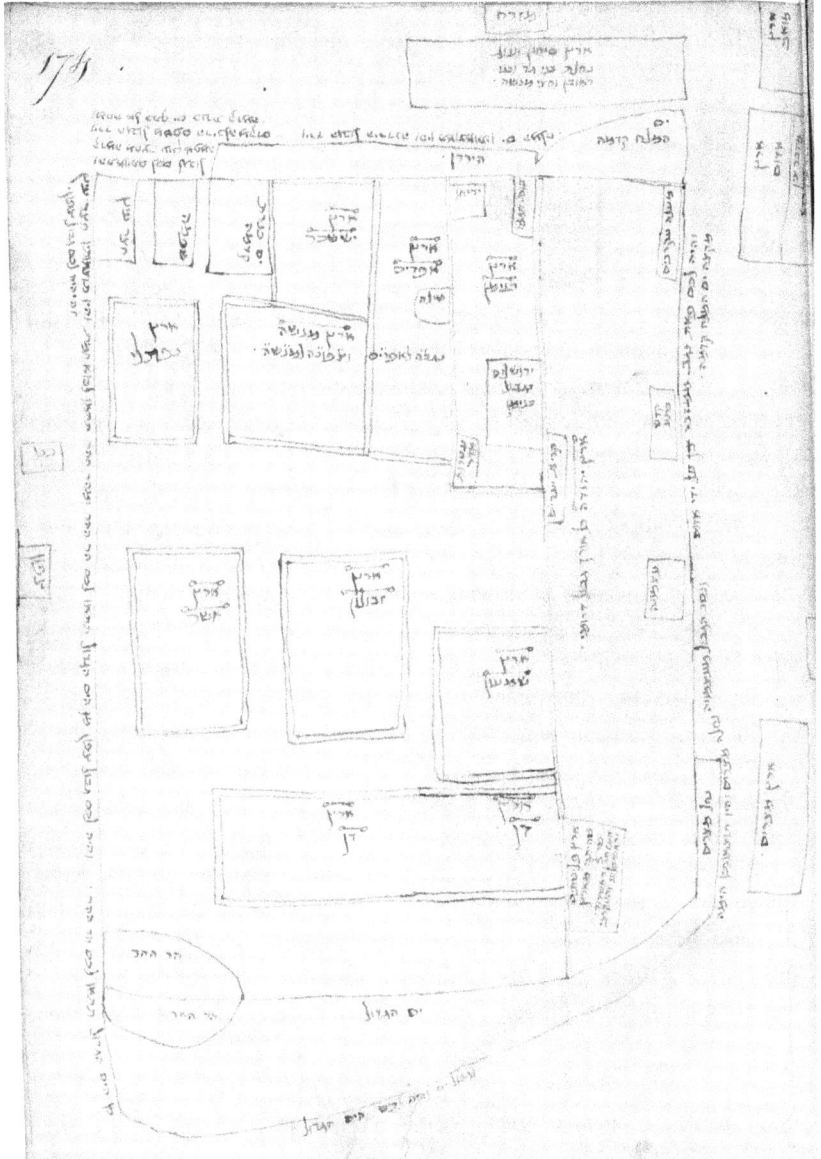

Fig. 10: The map in a manuscript of R. Joseph Kara's commentary on the Latter Prophets, Paris, Bibliothèque Nationale de France, Hébreu, 162, fol. 174r.

The map also contains a handful of settlements within the Land's borders: Shiloh in Ephraim; Ataroth Addar, Jericho, and Beth Hogla in Benjamin along with Kiryat Yearim in Judah.⁶ Jerusalem is depicted as a rectangle with the following text: "Jerusalem within the borders of Benjamin." On the top of the map is a long frame reading "The Land of Sihon and Og the allotment of the scion of Gad and the scion of Reuben and half of Manasseh." On the southeastern tip of the page are frames signifying the Land of Moab, the Land of Edom, and the Wilderness of Zin. Lastly, a box adjacent to the southwest corner of Eretz Yisrael circumscribes the following passage: "The Land of the Philistines that was the Gazan, the Ashdodian, and the Ashkelonian, and Ekronian Land of Israel."

Map from the British Library

In an anonymous sixteenth-century manuscript of Rashi's commentary at the British Library, there is an amalgam of illustrations that lack any clear correlation to the text. Among them are an illustration of the *menorah* (candelabrum), two sketches of the Tabernacle, and a somewhat unique map of Eretz Yisrael's borders.⁷ Although the map is nearly rectangular, its contours are less than straight. Consistent with the diagram's eastern orientation, the Great Sea appears toward the bottom and the Jordan darts across the top. The anonymous compiler added six boxes along the Mediterranean coast, five of which denote the Philistine cities and the other contains the word "Philistines." A little above the right edge of this strip is the caption "and Philistines from behind." On the upper righthand corner, between Moab and Edom, is the following text: "Aaron perished in Mount Hor." Just below are the captions Mount Hor and Kadesh. Pursuant to Rashi's commentary, the Jordan runs diagonally (a theme that recurs in Elijah Mizrachi's map below). For some reason, Jericho is situated to the north of the Sea of Galilee. Among the sites on the northeastern edge of the page is Aram-Naharaim. The Wajntraubs, who display this item in their above-cited book,⁸ ignored a piquant passage that extends along the northern border: "This is Aram the place of the Euphrates River that flows from Eden and circuits the Land of Israel from the north side." Yet another passage has been integrated on the bottom margin: "And these *nisin* in the sea are islands that from the west appear as though a string is stretched on them from Torai Amnon that is Mount Hor. . ."⁹ These texts are predicated on a discussion in the literature of the Sages regarding the precise contours of the western border.¹⁰ In this context, the ref-

---

6 As per Joshua 18:13–14, Ataroth Addar and Kiryat Yearim constituted the northwestern and southwestern edges of Benjamin's allotment, respectively.
7 *Rashi Commentary*, MS London, British Library, Board, BL Add. MS 27128, fol. 83. In addition, the map is exhibited and described in Wajntraub and Wajntraub, *Hebrew Maps*, pp. 27–29.
8 Ibid.
9 The term *nisin* or *nisin she'bayam* (islands in the sea) is quite commonplace in the post-Biblical Hebrew tradition. It is a distortion of *nesus* (νήσος), the Greek word for island, which penetrated the language of the Sages. The end of this passage is illegible.
10 See, for example, the Babylonian Talmud, Gittin 8a.

erence to the Euphrates River and to the *nisin* derives from a Talmudic interpretation of the Land's borders, rather than the description in *Mas'ei*.[11] As we shall see, these two topics also surface in printed Hebrew maps.

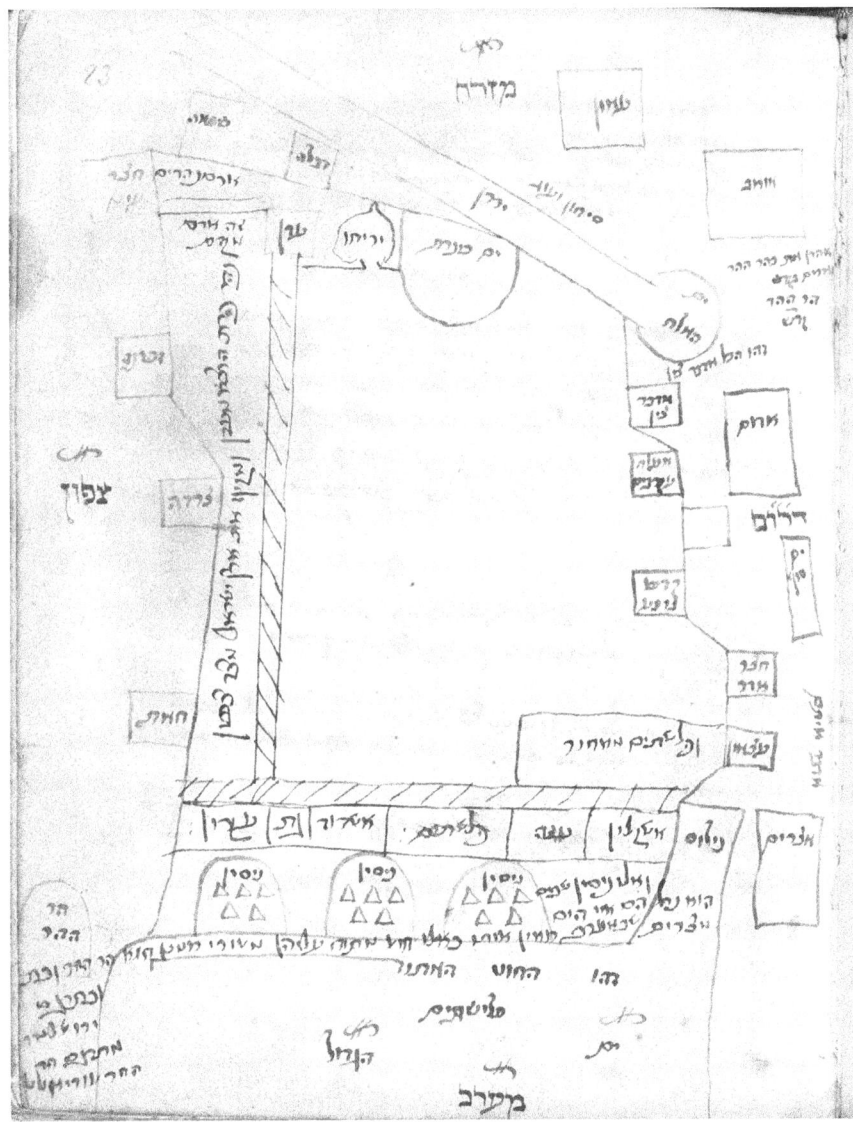

Fig. 11: Map of the Land of Israel in a manuscript of Rashi's commentary, the British Library, Add. MS 27128, fol. 83.

---

11 Demsky, "'From Kzib unto the River *near* Amanah' (Mish. Shebi't 6:1; Halla 4:8): A clarification of the Northern Border of the Returnees from Egypt" [Hebrew].

Map from Berlin

Another map was preserved in an anonymous thirteenth or fourteenth-century manuscript of Rashi's commentary on the Hebrew Bible, which is currently part of the Berlin State Library's collection. The workmanship and artistic quality of this map, which takes up the bottom two-thirds of the page, outshines that of the British Library. On the upper third of the page is a rendering of the showbread table, while the *menorah* of the Tabernacle graces the adjacent page.[12] The general format of this map ties it to the Rashi-inspired works, but it contains several unique elements: it faces north, not east; Ammon and Moab are located in the south, beside Edom, rather than in the east; and the text "Wilderness of the Peoples through which the Israelites journeyed for forty years" is lodged into the southeastern corner. The toponym "Wilderness of the Peoples," which Rashi identifies as the Sinai Desert, derives from Ezekiel 20:35. From a graphic standpoint, all the large repositories of water— the Red Sea, the Mediterranean Sea, the Sea of Galilee, and the Jordan—are adorned with wave-like lines. Similar to its predecessors, the Berlin map includes the following sites: the five Philistine cities; Jericho (between the Sea of Galilee and the Dead Sea); and the Euphrates. The latter runs parallel to the northern border of Eretz Yisrael and apparently connects the Sea of Galilee to the Mediterranean Sea.

---

12 Berlin State Library, Ms. or. fol. 121, 134a. The map is also displayed in Wajntraub and Wajntraub, *Hebrew Maps*, p. 36.

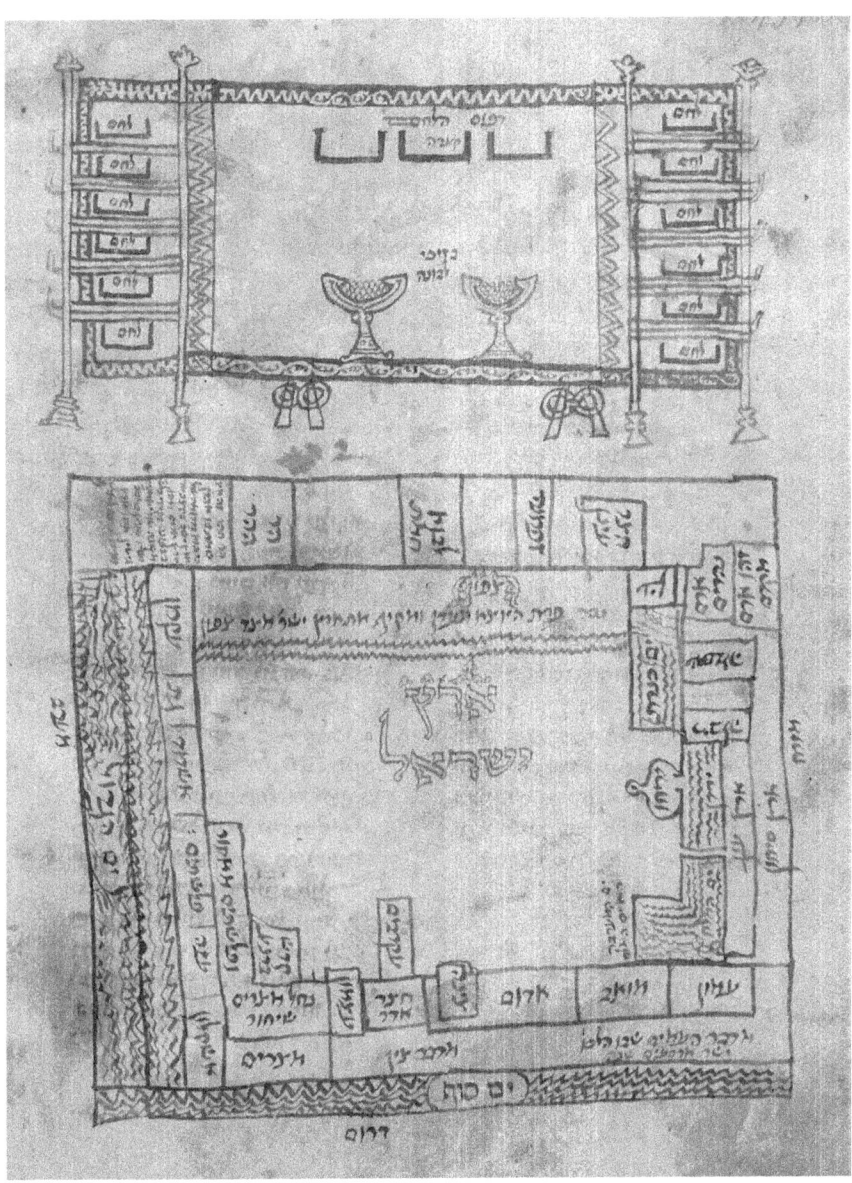

Fig. 12: Map in the Berlin manuscript, Berlin State Library, Ms. or. fol. 121, page 134b.

Map from the Vatican Library

Yet another Rashi-inspired map can be found, along with several other works, in a sixteenth-century manuscript at the Vatican Library.[13] The map takes up the last page of this manuscript, and there is no real connection between the former and the rest of the text's components. While the map preserves Rashi's basic template, there are also a few novelties. The map is quite disorienting, for some of its captions are arranged in unwieldy directions. At any rate the page has been bound into the manuscript with the eastern side on top, pointing to an eastern orientation. The borders of the Promised Land accord with the Rashi format and are highlighted with a thick, hatched ambit. Compartments bearing the names of the border sites are sprinkled along the Land's perimeter. In the northwest, there is an amorphous protrusion containing a partially illegible text beginning with the toponym "Mount Hor." The western portion of the map is mottled with sixteen triangles, each of which represents an "island." These islands in the sea—both here and in subsequent maps—allude to the Talmudic discourse over the question of Eretz Yisrael's western border.[14] Towards the bottom-left corner are the Land of Egypt and the River of Egypt. Below the river are four compartments bearing the captions Pithom, Raamses, the Land of Egypt, and the Land of Goshen. The Red Sea and Mount Sinai occupy a pair of triangles to the east, while the stubby, oval shape to the right denotes the other Mount Hor.

---

13 MS Vatican, Biblioteca Apostolica Vaticana, MS Borg. Hebr 4. This ms. was presented by Amy Phillips at the 16th World Congress of Jewish Studies in Jerusalem, 2013, within the framework of her lecture "The Cartography of Redemption." I would like to thank Ms. Phillips for sharing the pertinent information with me.

14 See, for example, the Babylonian Talmud, Gittin 8a; Jerusalem Talmud, Shevi'it, VI, halakha 1.

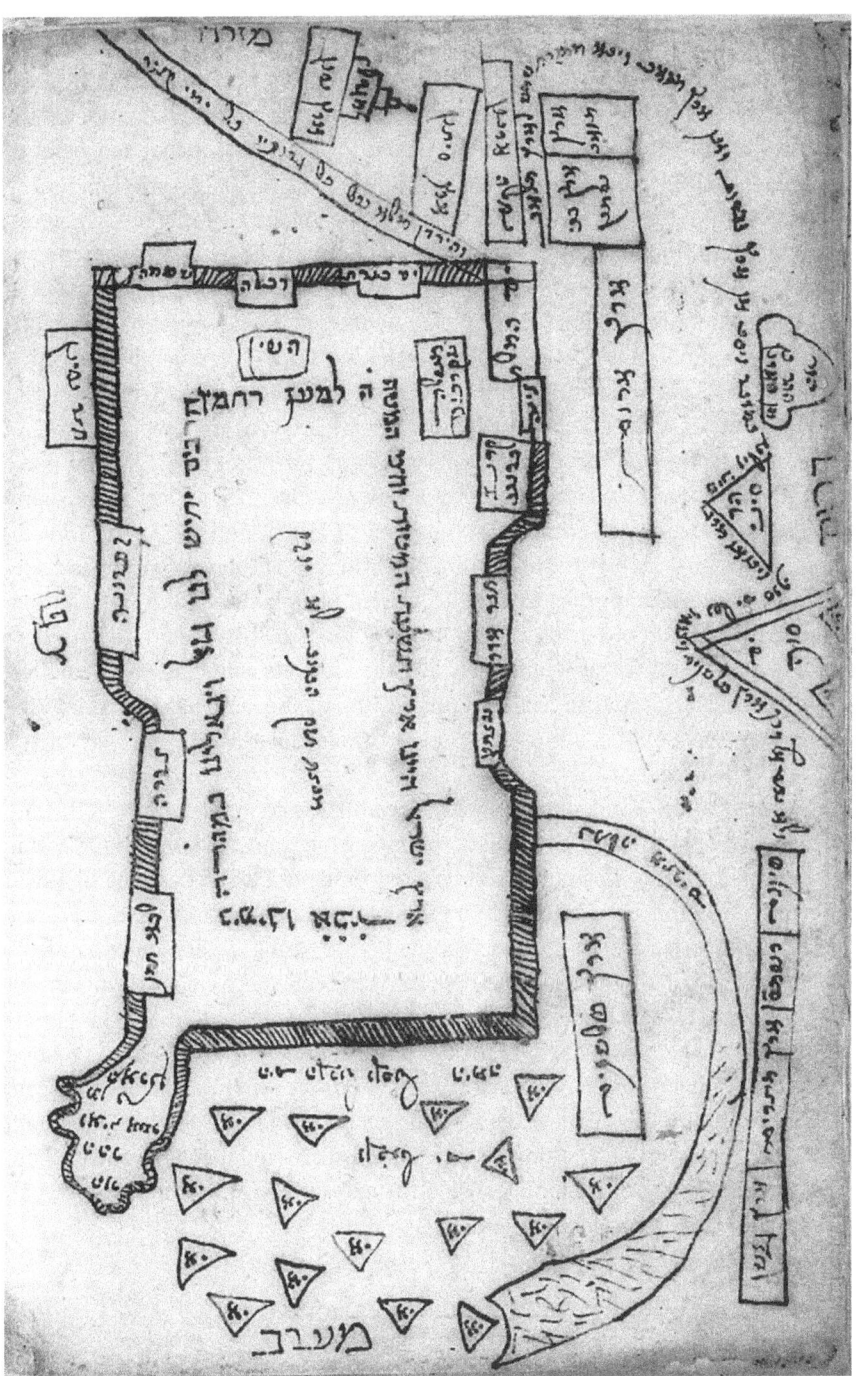

Fig. 13: Map in the Vatican manuscript, Biblioteca Apostolica Vaticana. MS Borg. Ebr. 4

Slithering along the lower and eastern sides of the map is the following succession of verses: "God led them not through the way of the Land of the Philistines [...] But God led the people about, by way of the wilderness by the Red Sea" (Exodus 13:17–18); "Then they went along through the wilderness, and compassed the Land of Edom and the Land of Moab and came by the east side of the Land of Moab" (Judges 11:18[15]); and "The Jordan overfloweth all his banks throughout the harvest season" (Joshua 3:15). This scriptural assemblage unfurls the Israelites' route through the wilderness, from the exodus until entering Eretz Yisrael via the Jordan River. Accordingly, the text hugs the coast of the Red Sea, continues along the slanted edges of Mount Sinai and Mount Hor, skirts the frames denoting the Lands of Edom, the Ammonites, and Moab, and threads past the Arnon River, before reaching the Land of Sihon and the Land of Og. Below and contiguous to the latter is a spired box with an indecipherable caption. We already encountered this sort of skewed writing in the Leipzig version of Rashi's map for Numbers 33, whose verse depicting the Israelites' peregrinations runs downwards: "Then they went along through the wilderness, and compassed the Land of Edom, and the Land of Moab, and came by the east side of the Land of Moab" (Judges 11:18). Conversely, the succession in the Vatican map repeatedly shifts course, thereby forcing the beholder to rotate the manuscript in order to continue reading the text. This element—both here and in other maps—is perhaps intended to make readers simulate the Israelites' route with their glance and movements. In so doing, the mapmaker underscores the opening verse of this continuum: "God led them not through the way of the Land of the Philistines."[16]

At the center of the Vatican map is a rectangle that is comprised of the following sentence: "Eretz Yisrael that is the land of the nine tribes and the half tribe by virtue of Your abundant mercy will summon for us a redeemer and will redeem us swiftly in our time amen may this be His desire." In the middle of this rectangle five words are written which are only partially legible, reading "... the strong ... will not be harmed". This seems to be similar to a formula known in Hebrew medieval manuscripts in colophons.[17] This text bears witness to the yearnings for redemption that pulsated in the anonymous mapmaker's heart—emotions that may well have spurred on the creation of this map. The inclusion of *"nisin she'bayam"* and the Jordan's diagonal descent into the Dead Sea tie this map to the one in the British Library. In all likelihood, this depiction of the Jordan is akin to Rashi's interpretation whereby the river "proceeds diagonally." This same template also turns up in Elijah Mizrachi's printed map and many of its subsequent adaptations.

---

15 As we have seen this verse is also cited in Rashi's map.
16 Daniel Connolly proposes a similar interpretation for the direction of the text in Matthew Paris' map, especially the pieces of parchment that were affixed to the margins of the main folio; Connolly, "Imagined Pilgrimage in the Itinerary Maps of Matthew Paris."
17 Avraham Berliner, *Selected Essays*, Jerusalem 1969, vol. 2 pp. 93–96 (Hebrew).

Map in Kaftor Va'ferech

A schematic, rectangular map of Eretz Yisrael surfaces on the last page of a manuscript of *Kaftor va'ferech* (Knob and Blossom), a treatise by Ishtori Haparchi on the laws dependent on the Land. This manuscript was copied in 1542 and is housed at Milan's Biblioteca Ambrosiana.[18] As opposed to the vast majority of Hebrew maps, this particular one is oriented southwards. Three of the four winds are noted on the margins. Conversely, the west side suffices with the Great Sea and two iterations of *nisin*. The map delineates the Land of Israel with four straight lines. In the upper lefthand corner is a modest niche, next to which is written "The eastern tip of the Dead Sea." A small rectangular protrusion denoting Mount Hor is lodged into the bottom-right (northwestern) corner. The word "Jordan" is written three times along the eastern boundary. Inside the Land's contours is a surfeit of toponyms. While most are from the Bible, there are also later names; a few might even refer to toponyms that were in use during the author's lifetime. Some of the elements therein share an affinity with Rashi's commentary and map, such as Mount Hor, Antioch, and Hamath to the north; "the eastern tip of the Dead Sea" in the southeast; and the islands in the sea to the west. That said, many of the names are connected to neither Rashi's map nor the account of the borders in *Mas'ei*.

---

18 Manuscript of *Kaftor va'ferech*, Veneranda Biblioteca Ambrosiana, Milano, De Agostini Picture Library, B 76, sup., fol. 452. I am grateful to Zur Shalev for pointing out this manuscript to me. It is described in the following catalog: Bernheimer, *Codices Hebraici Bybliothecae Ambrosianae*, pp. 150 – 151.

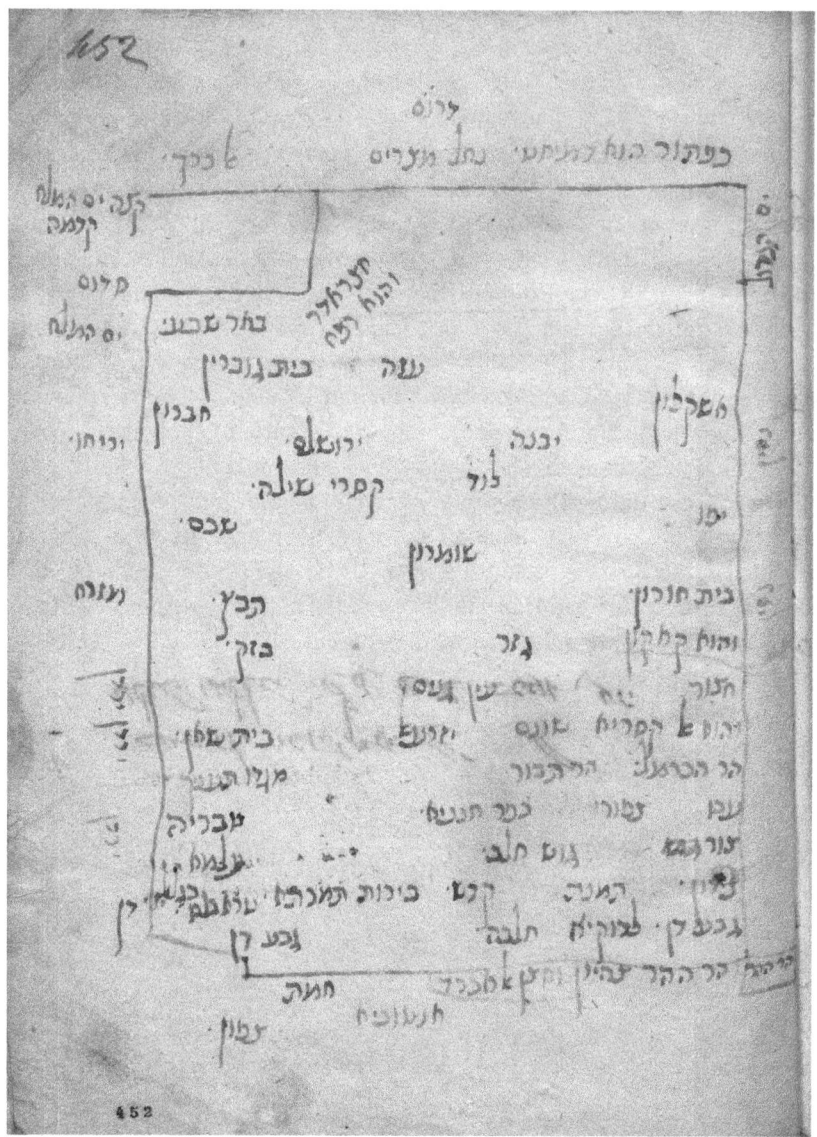

Fig. 14: Map from the manuscript of *Kaftor va'ferech*, Veneranda Biblioteca Ambrosiana, Milano, De Agostini Picture Library., B 76, sup., fol. 452.

The majority of these names come up in chapter 11 of *Kaftor va'ferech*, where the book indeed expounds on Eretz Yisrael and its settlements. Most salient are the Biblical toponyms that Haparchi enumerates within the framework of his description of the Land's borders, not least Megiddo, Ta'anakh, Jezreel, Shunem, and En Gannim in the center of the Land; Be'er Sheva, Beit Guvrin, Gaza, Yavneh, and Ashkelon in the south; and Tyre, Sidon, and Kadesh up north. What is more, the map includes toponyms from the exegete's own age. For instance, one caption reads "Hazar Addar that is Raphiaḥ [Rafah]," which he located in the southeast and should be in the southwest, of all places; on the northern border are the contemporaneous toponyms Beirut-Tamnata and Latakia. Another unique identification is "Bethoron that is Qaqun." Although this information is ostensibly anchored on knowledge of the Land, it entails a two-pronged error. Not only does Haparchi situate Bethoron in the plains, rather than the mountains, but the coastal location is inappropriate also for Qaqun, which is inland. Among the other elements of this map are the Arabic-derived captions "ṣ'hiyon ḥaṣan al Kard" and "al-Karakh" on the northwestern and southeastern edges, respectively. The names Gush Halav, Ṣippori (Sepphoris), Kfar Hananya, and Alma in the Galilee and "the tombs of Shiloh" possibly hint to sixteenth-century pilgrimage destinations. These toponyms also surface in the text of *Kaftor va'ferech*, so that the map is apparently grounded more on the latter than on a familiarity with Eretz Yisrael. Against this backdrop, the information was in all likelihood added by the manuscript's Europe-based copyist (or the copyist of an earlier manuscript that stood at his disposal). In any event, the map is devoid of any direct and unmediated knowledge of the Land of Israel.

Print Versions

With the emergence of the Hebrew printing press in the latter stages of the 1400s and, all the more so, in the early sixteenth century, many editions of commentaries on Rashi were authored and brought to press. It also bears noting that this literary genre has endured into the present era.[19] Over the centuries, numerous of square and schematic, Rashi-inspired maps of the Land of Israel have been published. Similar to the manuscript maps, the following elements were added to the original format in the print versions: toponyms, citations from the Bible and the exegetical literature, and even a few artistic illustrations. Over the next few pages, I will survey these maps as well as related illustrations that were published in similar contexts.

The Map of Elijah Mizrachi

The first exegete to write an unabridged, print supercommentary on Rashi's work that included a map of the Promised Land's borders was Elijah Mizrachi, who is also known by the acronym the ReEM (1450–1526). Born and educated in Istanbul, Mizrachi was interested in both the general sciences and Halakha. In time, he be-

---

19 Krieger, *Parshan-Data: Supercommentaries on Rashi's Commentary on the Pentateuch* [Hebrew].

came a central rabbinical authority in the Ottoman Empire, and his above-mentioned exegesis gained traction throughout the Jewish Diaspora.[20] At his son's initiative, the supercommentary was first printed a year after Mizrachi's death in Venice, at the press of Daniel Bomberg; and a second, more popular, edition came out nineteen years later.[21] Bomberg, a native of Antwerp, established his printing house in 1515. For three decades, he specialized in printing Hebrew works, especially the Bible and Talmud.[22] What is more, a template of the Talmud page designed in Bomberg's press remains the standard to this very day. The fact that the son of a prestigious rabbi who served in Istanbul—a city that then boasted its own Jewish-owned press, which even produced one of Mizrachi's books[23]—preferred to print the said commentary in Venice attests to the importance of Bomberg's establishment in the sixteenth century.

---

20 Hacker, "Mizrahi, Elijah."
21 Elijah ben Abraham Mizrachi, *Commentary on Rashi on the Torah*, Venice: Daniel of Bomberg, 1527, 1545, p. 282a.
22 Bloch, *Venetian Printers of Hebrew Books*; Penkower, "Bomberg's First Bible Edition and the Beginning of his Printing Press;" Rosenthal, "When did Daniel Bomberg Begin to Print?" [Hebrew]; Raz-Krakotzkin, *Censorship, Editing and the Text: Catholic Censorship and Hebrew Literature in the Sixteenth Century*, pp. 121–125 [Hebrew]; Baruchson, "On the Trade in Hebrew Books between Italy and the Ottoman Empire during the XVIth Century" [Hebrew].
23 Elijah ben Abraham, *Sefer (melekhet) ha'mispar*, Constantinople: Soncino Press, (1533) [Hebrew].

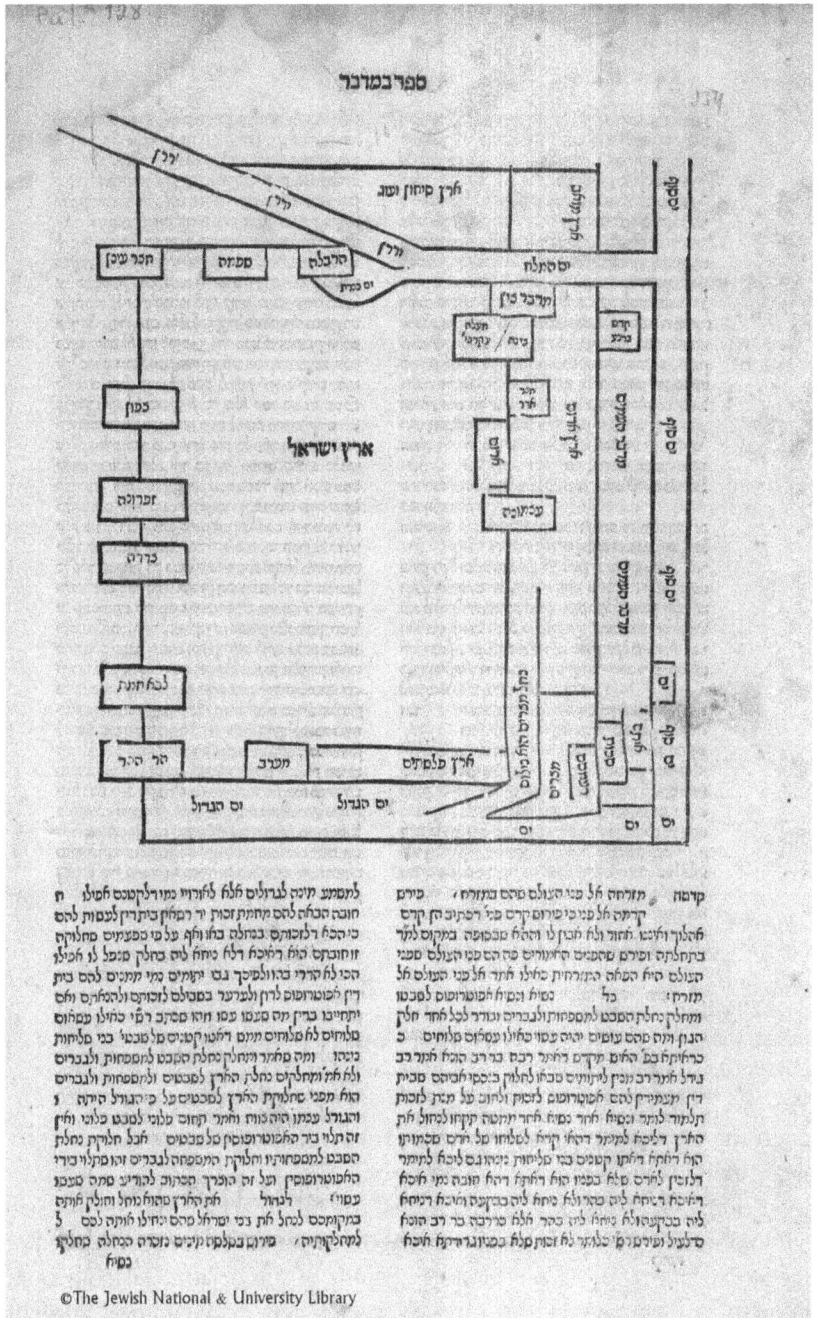

Fig. 15: Elijah Mizrachi, *Commentary on Rashi on the Torah* (Venice: Daniel Bomberg, 1545), p. 282a, 170 × 140 mm.

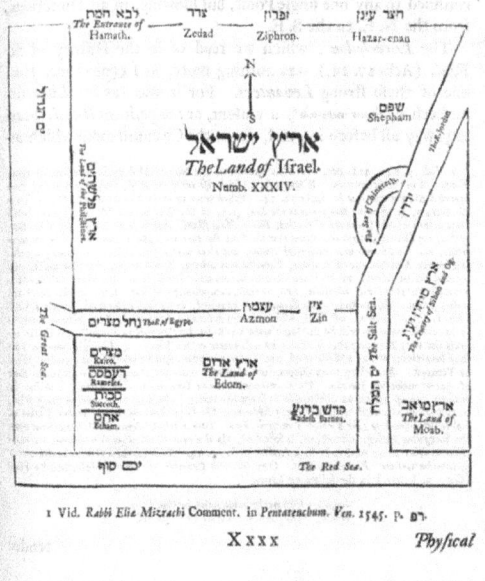

Fig. 16: Mizrachi's map as rendered in Thomas Shaw, *Travels, or Geographical, Physical and Miscellaneous Observations Relating to Several Parts of Barbary and the Levant*, 150 × 145 mm.

Although the maps in the first two editions of Elijah Mizrachi's commentary are identical, the 1527 version is on the lower half of the page and the 1545 is on the upper portion.[24] In essence, Mizrachi united the information from Rashi's two diagrams into one map. While omitting the lengthy explanation on the location of Mount Hor in Rashi's map, he preserved the spirit of the original. Facing east, Mizrachi's map is designed in the form of a rectangle with simple, straight lines. The words "Eretz Yisrael" grace the center in large, bold letters. The two

---

24 The 1525 edition of the map is part of the late Dr. Alfred Moldovan's collection. I would like to thank him for allowing me to examine this map.

major rectangular frames along the upper edge are the Land of Moab and the Land of Sihon and Og. The adjacent, eastern border of the Promised Land consists of the Jordan, the Sea of Galilee, and the Dead Sea, which form a single inter-connected unit. North of the Sea of Galilee are compartments denoting the Riblah and Shepham. Three conspicuous discrepancies vis-à-vis Rashi's map have become the trademarks of the Mizrachi version. To begin with, the Jordan runs diagonally from northeast (the upper-lefthand corner) to southwest. This element is apparently a graphical representation of Rashi's determination that the "Jordan comes at a diagonal" (Rashi for Numbers 34:11). Needless to say, we have already seen the first buds of this template of the river in the British Library and Vatican manuscripts. Assuming the outline of a spoon, the Sea of Galilee in the map at hand derives from the Jordan and dips towards the west. The Dead Sea heads south before spilling into the Red Sea. These three elements were destined to appear in numerous copies of the Mizrachi map as well as its more sophisticated adaptations.

The northern border (to the left) constitutes a vertical line, along which are hung horizontal boxes with toponyms from Numbers 34: Hazar Enan, Ziphron, Zedad, the entrance of Hamath, and Mount Hor (the other box reads "North"). The Great Sea forms the western (lower) border, above which are the Land of the Philistines and a compartment labeled "West." The right side of the map, which encompasses Egypt and southern Eretz Yisrael, is comprised of three parallel strips. The rightmost strip is made up entirely of references to the Red Sea. The middle one contains the Sinai Desert and two instances of the toponym "Wilderness of the Peoples." On the upper part of this column is a box denoting Kadesh Barnea; and towards the bottom are Etham and Succoth. The final and widest strip includes the Land of Edom on top and Raamses and Egypt below. Due left is "the River of Egypt that is the Nile;" and further up are five, unevenly aligned rectangular frames denoting "to Azmon," Edom, Hazar Addar, "to Zin," "the Ascent of Akrabbi[m]," and the Wilderness of Zin. These sites straddle the southern borders of Eretz Yisrael. For example, the Ascent of Akrabbim juts out to the north (left), just as Rashi described in his commentary.

The importance of the Mizrachi map to Jewish authors is evidenced, first and foremost, by its wide circulation in both new editions of his commentary and other works.[25] Isaac of Prostitz, the publisher of the Krakow edition, apprenticed at presses in Italy. It is not inconceivable that Mizrachi's exegesis first reached his hands while serving in this capacity.[26] No less significant is the map's replication and its inclusion in the exegetical works of other authors. A case in point is the fact that Thomas Shaw (1694–1751)—an English priest who organized an expansive

---

25 Mizrachi, *Elijah Mizrahi [on] the Commentary of Our Rabbi Shlomo Yitzhaki* [...], Isaac ben Aaron of Prostitz, Krakow 1595. The map is on p. 280, side b, at the end of the Book of Numbers.
26 Elbaum, *Openness and Insularity*, pp. 44, 87 [Hebrew]

journey to the "Levant" in 1722—included Mizrachi's map in his book and presented it in a bilingual, Hebrew-English format.[27] Shaw oriented this edition to the north and dubbed it a "rabbinical geography"—a term that was anathema to his Jewish contemporaries. The map's presence in Shaw's highly popular book indeed exposed it to many educated Europeans.[28]

An early, if not the first, replica of Elijah Mizrachi's map graced a Pentateuch that was printed in Riva di Trento, Italy in 1561. This work includes Rashi's exegesis, an abridged Mizrachi commentary, and the five scrolls (with a commentary by Isaac ben Moses Arama).[29] Above the map is the following heading: "The Land of Canaan which is the land of the nine tribes [...] To all that study this form [i.e., map]."

The Map of Simeon Halevi

Simeon ben Isaac Halevi of Aschaffenburg (a town in Bavaria) included three cartographic works in his book *Devek tov* (Good Glue). First printed in Venice in 1588, *Devek tov* subsequently netted many other editions.[30] In the section on Genesis, next to the portion of *Vayetse*, is a schematic diagram of Jacob's trek from Be'er Sheva to Bethel. On top of this succession of toponyms is a rendering of the ladder in the patriarch's dream. This simple drawing resembles the illustration of Jacob's ladder in the above-mentioned manuscript of the *Ḥazzekuni*. In the portion of *Bo* (Exodus 10:1–13:16), Aschaffenburg provides another schematic diagram that explains the location of the Red Sea relative to Eretz Yisrael; more specifically, the sea flanks the Land from the south and east. The final cartographic work in *Devek tov* takes up a full page in the section on *Mas'ei*. This map is highly reminiscent of the Mizrachi version, with one conspicuous exception: a dotted line signifying the Israelites' route through the wilderness. This rendering of the journey completely ignores the fact that the Israelites passed through eastern Transjordan and crossed at the eponymous river. Instead, Aschaffenburg's line traverses the southern border of the Land next to Hazar Addar and reaches the Sea of Galilee from the southwest. On the margins of the page before the map, there is a notation

---

27 Shaw, *Travels, or Geographical, Physical and Miscellaneous Observations Relating to Several Parts of Barbary and the Levant*, p. 357.
28 For a disquisition on Shaw and his book, see Rée, "Shaw, Thomas (1694–1751)."
29 Elijah Mizrachi, *The Five Books of the Torah with the Translation and Com[mentary] of Rashi and the Abridged Mizrachi [Supercommentary]and the Five Scrolls with the Com[mentary] of Rashi and the Supercommentary of Ba'al ha'akedah R. Isaac Arama [...]* Riva di Trento, 1521, page 228 [Hebrew].
30 Simeon Halevi Aschaffenburg, *Sefer devek tov*, Venice: Joan Di Gara Press, 1588 [Hebrew]. The illustration of Jacob's ladder appears on p. 28b; the diagram of the Red Sea on p. 53a; and the map of Eretz Yisrael on p. 113a. Also see Heller, *The Sixteenth Century Hebrew Book: An Abridged Thesaurus*, vol. I, pp. 758–759.

that begins with a critique of the printers who designed Mizrachi's map: "Make for yourself the drawing and not like it is drawn in the Mizrachi print because a mistake has been made therein on the part of the printers and I will give you short guidelines as [per] the words of Rashi of blessed memory..." In all likelihood, this comment refers to the Israelites' route. A similar remark also turns up in Isaac Cohen of Ostrowo's Biblical commentary, at the end of the Book of Numbers.[31] Under the influence of Aschaffenburg, the Israelites' journey was also incorporated into later editions of Elijah Mizrachi's exegesis.[32]

---

[31] Isaac Cohen of Ostrowo (Isaac Naftali Hertz Buking), *The Abridged Book of Mizrachi with All the Explanations that there are on the Rashi Commentary*, Prague: 1604. Unfortunately, the map was missing from all the editions that I checked.

[32] See the Krakow edition of Mizrachi, *Elijah Mizrahi [on] the Commentary of Our Rabbi Shlomo Yitzhaki*, esp. the map at the end of Numbers, p. 280 b. It is worth noting that the same press printed another edition of *Devek tov* in 1593.

2 Following in Rashi's Footsteps

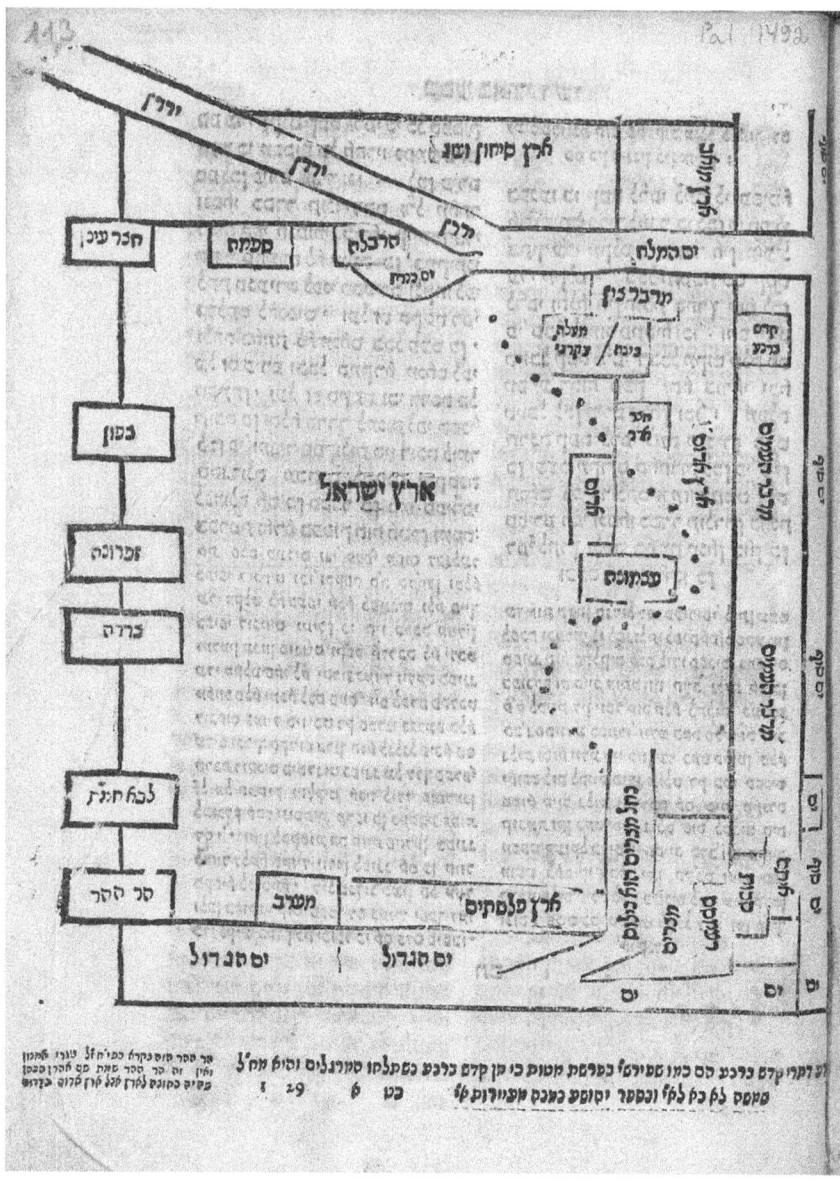

Fig. 17: Map of Eretz Yisrael's borders in Simeon Halevi Aschaffenburg, *Sefer devek tov* page 113a, 142 x 91 mm.

The immense popularity of the Mizrachi map and its replicas led to the printing of the Krakow edition in 1595. What is more, it was copied in the manuscript of a book that was written about two centuries after the map first came out.[33]

The Map of Natan ben Shimshon Shapira

Five years after the release of Aschaffenburg's book, an identical version of his map was printed in *Bai'urim al rashi* (Explanations on Rashi) by Natan ben Shimshon Shapira (d. 1577).[34] The illustrations in Shapira's book—the said map, a rendering of Jacob's ladder, and a diagram explaining how the spies carried the cluster of grapes—were clearly influenced by the Rashi tradition. In this work, the ladder initially slants upward before turning left and running parallel to the ground. Beneath the ladder is Jacob's escape route from Esau, as the diagram tracks his progress from Be'er Sheva to Bethel. The sketch of the cluster graphically depicts Rashi's interpretation of the spies' mission to Canaan. According to the savant, eight of the Israelites bore the cluster of grapes on their shoulders with the help of a sophisticated apparatus. Two of the other spies carried a pomegranate and fig, whereas Joshua and Caleb ben Jephunneh splintered off from the rest of the company. Of course, this interpretation contravenes the literal interpretation of the famous Biblical passage "and they bear it between two upon a staff" (Numbers 13:23). Similar illustrations turn up in other supercommentaries on Rashi.

---

[33] The Krakow edition is cited in the previous note. For the manuscript, see *Sefer toldot avraham* (the Book of Abraham's Annals) compiled by Abraham ben Joseph Kass of Pinczow, Poland, 1749. The map is located on the inner side of the back cover. The manuscript is housed at the National Library of Israel (MS. Laor 900).

[34] Natan (Neta) ben Shimshon Shapira, *Bai'urim al ha'eshel ha'gadol Rashi [Interpretations on Rashi, the Great Tamarisk] of Blessed Memory*] [...], Venice: the Matteo Zanetti and Comino Presigno Press, 1593, p. 150 [Hebrew]. Also see Margalioth (ed.), *Encyclopedia of Great Men in Israel*, vol IV, p. 1183 [Hebrew].

Fig. 18: Illustrations in *Sefer devek tov:* Jacob's ladder on page 28b; diagram of the Red Sea on page 53a.

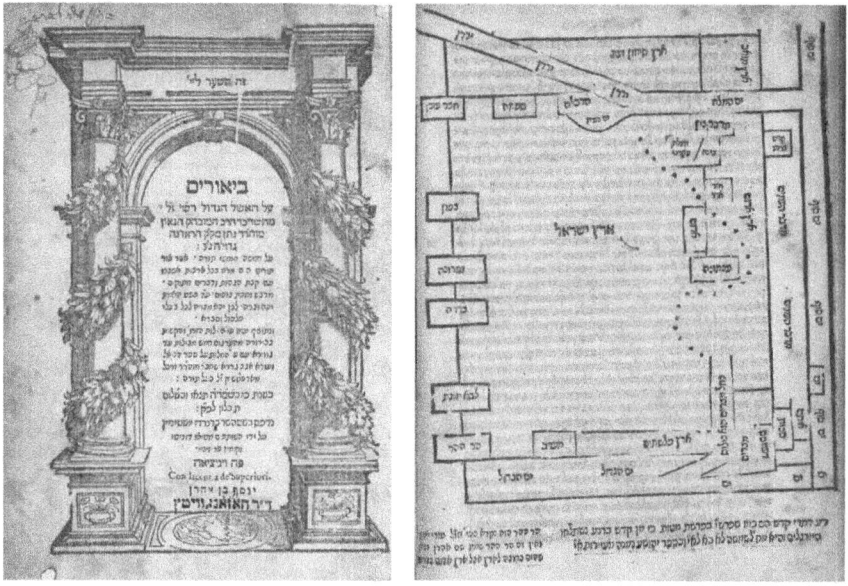

Fig. 19: Map of Natan ben Shimshon Shapira and the cover of his book *Bai'urim al ha'eshel ha'-gadol,* p. 150, 165 x 143 mm.

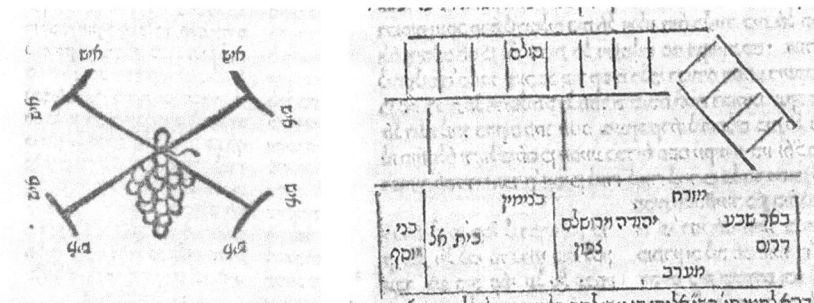

Fig. 20: Illustration of Jacob's ladder and the cluster of grapes, in Natan ben Shimshon Shapira, ibid. pp. 28b and 53, respectively.

All these versions of Elijah Mizrachi's map were conservative from the standpoint of both content and form, as the original design was practically untouched. Upon incorporating the dotted line tracking the Israelites' peregrinations through the wilderness, Aschaffenburg opined that this addition is warranted because the original printer had erred.

Mordecai Yoffe

Over the course of his lifetime, Mordecai Yoffe (1535–1612) shuttled between his hometown of Prague, Poland, and Venice. The rabbi's interest in Torah, philosophy, and astronomy all came to expression in his ten-book series—*Levush malkhut* (Raiment of Royalty). Each of the volumes was given the name of a certain attire, thereby meriting Yoffe the moniker *ba'al ha'levushim* (the owner of vestments).[35] The cover of one of the books, *Levush ha'orah* (Raiment of the Light),[36] is written in rhymes and offers the following information about the author and publisher: "An explanation on [the commentary of] Rashi on the Torah, authored by an awe-inspiring man possessing good deed[s] and the light of the Torah; look and spread the word, he [i.e., Yoffe] understood it [Rashi's exegesis] and also researched it; the sublime genius the rabbi my rabbi Mordecai Yoffe, a yeshiva head and court judge in the magnificent holy community of Poznań. Printed here in the holy community of Prague the capital under the government of our Lord Emperor Rudolf may his glory be exalted. In the lofty house of our honorable teacher and my rabbi Haim bar Yaacov Hacohen."

---

[35] "Yoffe, Mordecai," *Encyclopaedia Hebraica* [Hebrew]; Elbaum, *Openness and Insularity*, p. 20; Kupfer, "Jaffe, Mordecai ben Abraham," *Encyclopaedia Judaica*.
[36] Rabbi Mordecai Yoffe, *Levush ha'orah*, Prague: 1604, 82b [Hebrew].

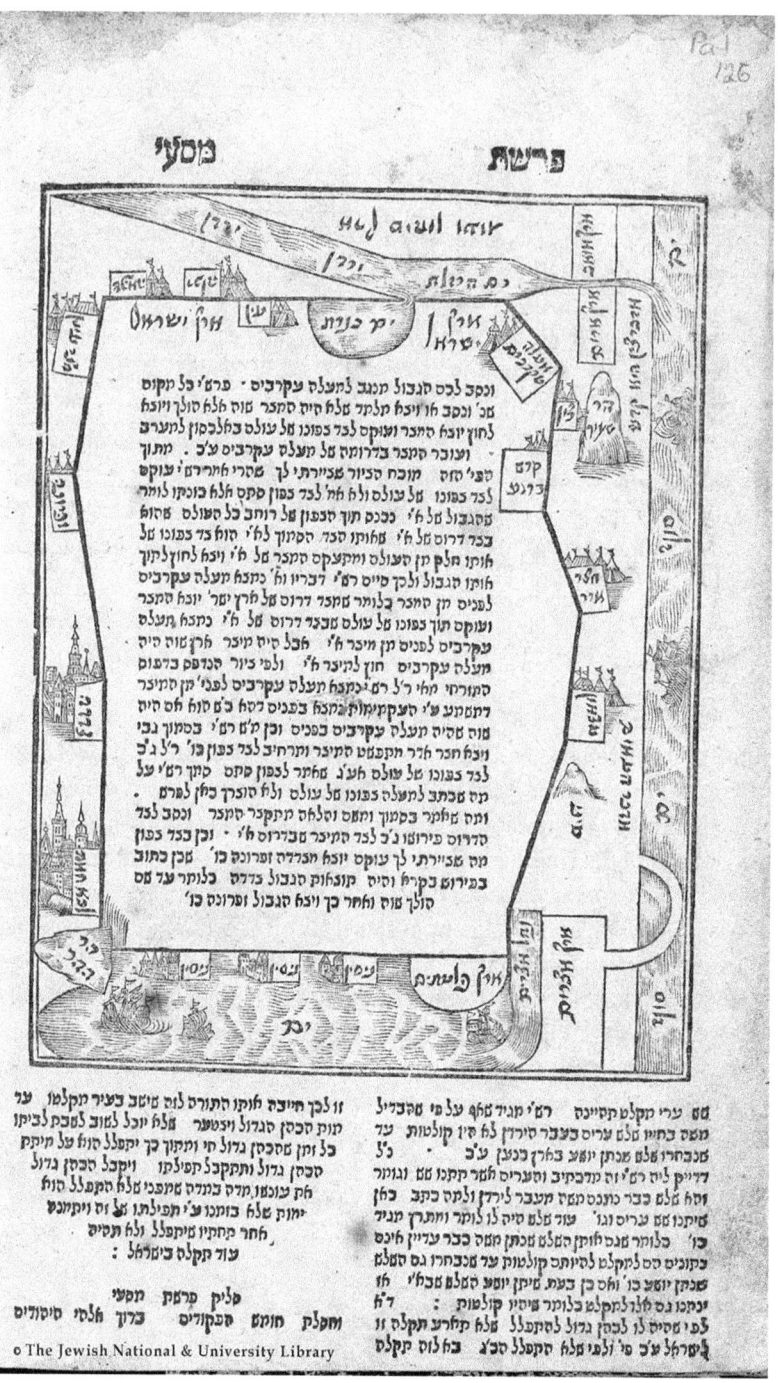

Fig. 21: The map of Mordecai Yoffe in *Levush ha'orah*, page 82b, 185 x 145 mm.

Yoffe adopted the basic model of the map that was familiar to his predecessors and hardly altered its content, including the standard inventory of toponyms, but he did add many new design features. In the first place, the scholar affixed a long Hebrew text interpreting the Biblical account of the Promised Land's borders to the center of the map. Moreover, he removed the captions Etham, Succoth, Raamses, and "Wilderness of the Peoples." Yoffe was undoubtedly familiar with *Sefer devek tov*, as Yoffe's father, Rabbi Avraham ben Avigdor, wrote the introduction to Aschaffenburg's book.[37] Unlike the map in *Devek tov*, Eretz Yisrael's northern and southern borders do not constitute straight lines. In this respect, Yoffe's version resembles the one in the British Library. Most importantly, he was the first to embellish a Hebrew map with illustrations that drew heavily on the era's cartographic norms. Yoffe's openness to ornamentation and the general spirit of cartography apparently stemmed from the fact that he was current on astronomy and philosophy. In addition, the liberal and erudite atmosphere in Prague at the time also came to bear. Two of his generation's most distinguished astronomers, Tycho Brahe and Johannes Kepler, indeed lived in Prague. What is more, the assistant of these two scholars was David Gans, a Jewish intellectual.[38] In turn, Gans was acquainted with Yoffe and even praises the latter in his works.[39]

In Yoffe's map, Mount Sinai and Mount Seir were indeed rendered as mountains. Most of the sites along Eretz Yisrael's borders are portrayed as tent camps, while Zedad and the entrance of Hamath on the northwestern border (to the left) were drawn as fortified European cities. The seas are peppered with ships featuring up to three masts—a design element that was quite prevalent in the era's cartography. Off the coast, he drew islands that are labeled with the Talmudic term *nisin*.[40] Each of the islands is portrayed as a built-up urban landscape. Artistic novelty aside, Yoffe viewed his map as an embodiment of the Rashi tradition:

> Here before you is a drawing of the boundaries of Eretz Yisrael as per Rashi's commentary as I received it, and it is true and correct in accordance with all the words of Rashi of blessed memory[.] As I will explain with God's help and as we have merited to draw it so too may we merit to see it [i.e., Eretz Yisrael] built and settled with our own eyes when God returns the exiles of Israel; and may the redeemer come to Zion and build Ariel amen, so may it be uttered by God the Lord.

---

37 "Avraham ben Avigdor," *Encyclopaedia Hebraica* [Hebrew].
38 For an in-depth look at the relations between the two Christian scientists and David Gans, whose primary field of endeavor was indeed astronomy, see Neher, "New Material concerning David Gans as Astronomer," *Tarbiz* [Hebrew]; idem, Neher, *David Gans (1541–1613) and His Times: Jewish Thought and the Scientific Revolution of the Sixteenth Century* [Hebrew].
39 See ibid, p. 54.
40 The term *nisin* or *nisin she'bayam* means islands in the sea.

Put differently, Yoffe believed that his map was a direct continuation of Rashi's and that he leaned on the illustrious exegete's authority. Furthermore, he hoped that rendering a map of the Holy Land would presage the return to Zion.

The Map of Haim ben Bezalel

A contemporary of Mordecai Yoffe, Haim ben Bezalel (1520–1588) was the eldest brother of the MaHaRaL (Judah Loew ben Bezalel, 1520–1609) and served as a rabbi in Worms and Friedberg. His supercommentary on Rashi, *Be'er mayim ḥayim* (the Well of Living Water),[41] has been preserved in several manuscripts from the sixteenth and seventeenth centuries. Moreover, a version of this book has netted a handful of editions since 1965.[42] With respect to the Rashi cartographic tradition, there are two complementary Oxford-based manuscripts of *Be'er mayim ḥayim*, both of which are dated to the 1500s or early 1600s. In the introduction to one of these manuscripts, the Hebrew acronym "may his memory be [destined] for the next world" is juxtaposed to the author's name, so that this version was evidentially copied after Haim ben Bezalel's passing. The section on the portion of *Mas'ei* contains the following passage: "The drawing [*my* i.e., the map] of all the borders of EY [Eretz Yisrael] as per what Rashi of blessed memory interpreted." However, the rest of the page was left blank, *sans* a map.[43] The second Oxford manuscript lacks the introduction, but does contain several illustrations on the following topics: the route of Jacob's escape and the ladder in his attendant dream; the *menorah* (candelabrum) and layout of the Tabernacle; the apparatus that the eight spies used to carry the cluster of grapes; and the star from the portion of *Balak* (Numbers 22–24). Last but not least, an fascinating map takes up a full page in *Mas'ei*.[44]

Content-wise, this map is quite similar to that of Rashi. To begin with, it faces east. On its southern (righthand) side are, from top to bottom, the lands of Edom, Moab, and Egypt as well as the River of Egypt. Just to the left are the settlements on Eretz Yisrael's southern border as per Numbers 34: Azmon, Hazar Addar, Kadesh Barnea, and Zina. Relative to the other two places, Azmon and Kadesh Barnea are situated slightly northwards (to the left). The Ascent of Akrabbim is drawn even further inland, adjacent to the Dead Sea. Lining the upper or eastern side are the Dead Sea, the diagonally-aligned Jordan River, the Sea of Galilee, the Ain, Riblah, Shep-

---

[41] Tobias and Derovan, "Ḥayyim ben Bezalel," *Encyclopaedia Judaica*.
[42] Haim ben Bezalel, *Be'er mayim ḥayim: Explanations on Rashi's C[ommentary] on the Torah*, part II, Leviticus [and] Numbers, with comments and references in the name of *Sheraga ha'mai'eer* [Sheraga the illustrator] by Sheraga Feyyish Shneebalg, London 1969. Among the other editions are London 1993 and Brooklyn 2011 [Hebrew]. These editions adhere to one of the Oxford manuscripts discussed herein.
[43] MS Oxford, the Bodleian Library, opp. 196, p. 254b.
[44] MS Oxford, the Bodleian Library, opp. 195, p. 234b.

ham, and Hazar Enan. Beyond the Jordan, to the east, sits the Land of Ammon. Along the northern (left) edge of the map are Mount Hor, Hamath, Zedad, Ziphron, and Hazar Enan. The Land of the Philistines is nestled into the southwestern corner of Eretz Yisrael. Lastly, three captions reading "Eretz Yisrael" form a triangle in the center. All these toponyms and their spatial alignment are indeed consistent with the cartographic tradition of Rashi and Elijah Mizrachi.

Graphically speaking, though, Haim ben Bezalel's map is completely different than the Rashi model, as it is quite ornate from an artistic standpoint. To begin with, all the settlements are portrayed as buildings. The places that the artist deemed to be of secondary importance are rendered as small stone buildings with domes or pointy roofs, whereas the more significant settlements take the form of large towers. Flags fly atop many of the spires. Towards the middle of the page is a monumental church-like building that consists of a triangular gable flanked by a pair of identical towers. Moreover, odd-looking male figures are stationed by the main entrance and bottom-level windows of this grand structure. Ships of various sizes ply the bodies of water in this map, while the Mediterranean Sea and the River of Egypt are also stocked with fish. Some of the boats even have human figures on board.

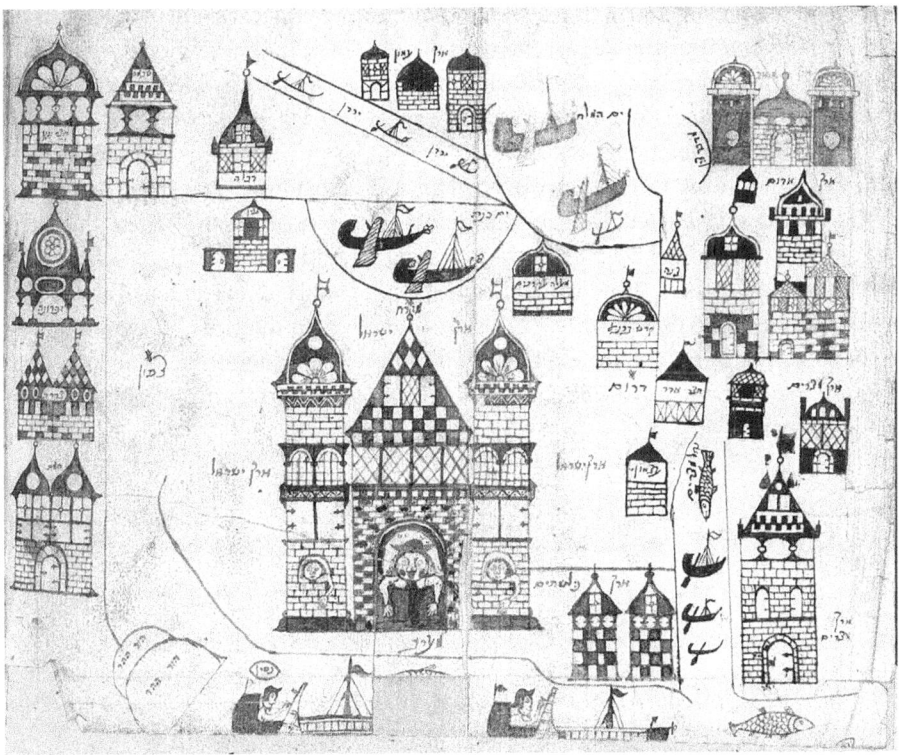

Fig. 22: A map of Eretz Yisrael in the Oxford manuscript of *Be'er mayim ḥayim*, opp. 195, 234b.

The aesthetic richness of this map clearly sets it apart from the other, Rashi-inspired maps in this chapter, as the ornamentation of its illustrations far surpasses that of Mordecai Yoffe. As evidenced by the church-like edifice signifying the Land of Israel, the bevy of European-style urban towers, and the figures sailing in the boats below, the sources of Haim ben Bezalel's graphic elements are to be found well beyond the ambit of the Jewish community.

The Map of Yosef ben Issachar Miklish

Yosef ben Issachar Miklish (1580–1654)[45] was a student of both Mordecai Yoffe and the MaHaRaL, a revered Jewish scholar. Miklish's book *Yosef da'at* (He who will Increase Knowledge) came out in Prague four years after Yoffe's *Levush ha'orah*. It includes several illustrations, among them a map that closely resembles that of Yoffe.[46] As hinted by the book's full title, Miklish endeavored to rectify various mistakes that were made by his predecessors. In addition, he evidently possessed an early manuscript of Rashi's commentary.[47] The map itself centers around the following text: "This drawing is an example of the drawing of my master my unequivocal teacher and rabbi the great genius our teacher and our rabbi Mordecai Yoffe may his candle illuminate[;] and it is true knowledge without any doubt and hesitation of the words of Rashi of blessed memory." Artistically speaking, Miklish's graphic elements, especially the cities, were a cut above those in *Levush ha'orah*.

Tucked into the bottom-right corner of the map is an arch spanning the Red Sea. This element already turns up in Yoffe's map and is, perhaps, connected to the rendering in the above-cited Vatican manuscript. However, Miklish's labels this arch "The parting of the Red Sea," thereby indicating that it is meant to depict the Israelites' route over the sea. Given the arched design, it stands to reason that Miklish was alluding to the question of whether the Israelites crossed the Red Sea from one side to the other, or merely made some headway before returning to the same side from whence they departed. We will delve into this question and its cartographic expressions in chapter 4.

---

[45] The information on Yosef ben Yissachar Miklish is quite sparse. However, according to the pertinent tombstone inscriptions, he belonged to a large family in Prague, served as the community's judge, and died in 1654. Hock, *Die Familien Prags. Nach den Epitaphien des Alten Jüdischen Friedhofs in Prag*, pp. 210–211.

[46] Miklish, *Yosef Da'at [...] to Fix the Distorted and Mistaken that Occurred [...] in the Commentary on Rashi of Blessed Memory on the Fives Books of the Torah [...]*, Prague: Gershom Katz Press [1609], 128a [Hebrew].

[47] Heller, *The Sixteenth Century Hebrew Book*, vol. II, pp. 254–255.

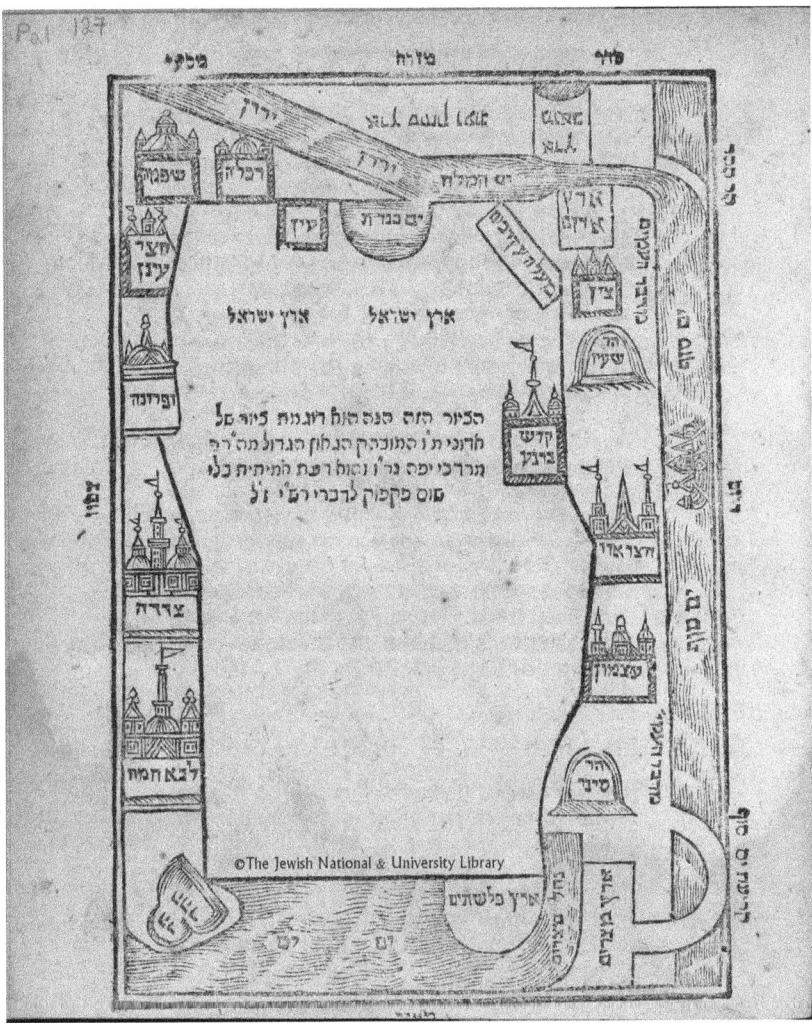

Fig. 23: Map of Yosef ben Yissachar Miklish, *Sefer yosef da'at*, p. 128a, 180 x 130 mm.

Over 250 years later, Mizrachi's supercommentary was incorporated into a widely-circulated compendium[48] that encompassed the following commentaries: *Gur Aryeh* (Lion's Cub) by the MaHaRaL, Yoffe's *Levush ha'orah*, and Shabbethai Bass' *Sif-*

---

48 One edition of this compendium was *Gur Aryeh, Levush ha'orah, Siftai ḥakhamim Four Commentaries on Rashi the Exile's Source of Light [...] All these have been Collected and Brought into a Single Booklet [...]*, part 4: Numbers, Warsaw: Isaac Rathauer Print, 1864 [Hebrew].

tai ḥakhamim (Sages' Lips).[49] In this work, the map appears in both Mizrachi's original format and in Yoffe's illustrated adaptation.

Maps of the Me'am Lo'ez

Me'am lo'ez was an immensely popular translation-cum-commentary on the Hebrew Bible.[50] Spiced with *aggadot* (rabbinic non-halakhic homilies), this work was a cultural pillar of Ladino-speaking communities.[51] The first stage of this expansive, multi-generational enterprise was the commentary by Yaakov Culi (1689–1732) on the Book of Genesis, and various other commentaries were integrated into this work over the years.[52] The section on the Book of Numbers, which pertains to our subject matter, was translated and annotated by Yitzhak Magriso.[53] All told, *Me'am lo'ez* was printed in several editions, the first of which came out in 1730. A few of the editions did not cover the entire Pentateuch, while others did. Some even included exegesis on the Prophets and *Ketuvim* (Hagiographa). Moshe David Gaon, who documented the history of the *Me'am lo'ez*, refers to the following editions: six on Genesis and eight on Exodus, a handful of which came out in two parts; six editions on Leviticus, five on Numbers; and six on Deuteronomy. A perusal through catalogs of academic libraries raises the possibility that, excluding the Book of Numbers, the tally is even higher.

In the section on *Mas'ei* (Numbers 33–34), there is a map of Eretz Yisrael that is based on the Rashi-Mizrachi versions. Most of the editions feature a uniquely designed map: instead of forming the borders and toponymical compartments with straight lines, they are comprised of a series of print characters. In the first edition, which was printed in Constantinople in 1764, the characters resemble fleurs-de-lis.[54] The map and its entire page remained exactly the same in the Salonica edition,

---

49 Shabbethai Bass (1641–1718) also spent part of his life in Prague. He is best known for *Siftai ḥakhamim*, a supercommentary on Rashi, and *Siftai yeshai'nim* (Lips of the Sleepers), a bibliographic work.
50 Gaon, *Heart Lockets: The Different Versions and Fate of the Book that was Read for Generations in the Ladino Tongue*.
51 Ladino, or Judeo-Spanish, is spoken by Sephardic communities throughout the world. A derivative of Old Spanish, Ladino was also influenced by Hebrew and Aramaic and, to a lesser degree, by other languages of the speakers' host communities.
52 For a disquisition on Yaakov ben Maikeer Culi, see Gaon, *Oriental Jews in Eretz-Israel (Past and Present)*, pp. 305–308 [Hebrew].
53 For more on Yitzhak Magriso, see ibid, p. 361.
54 *Sefer me'am lo'ez*, [on the] Pentateuch, part four that is the Book of Numbers [...], [by] our honorable teacher Rabbi Yitzhak Magriso: Reuven and Nissim Ashkenazi [Press], Constantinople, *ve'shaveha be'tzdakah* [gematria for:] (1764), 158a [Ladino]. This is the fourth volume of the first edition, which was first printed in 1730.

לבוש　　　　　　　　　　　　　　　　　האורה

צא לך הציור רחומי ארץ ישראל על פי פירוש רש"י באער קבלתי הוא אמרתי ונכון לפי כל דברי רש"י ז"ל כאשר אבאר בע"ה. וכאשר זכינו
לציר אותו כן נזכה לראותו בנוי ומיושב עין בעין במ"ב ה' נהדי ישראל ובא לציון גואל ובנה אריאל אמן כן יאמר יי' האל :

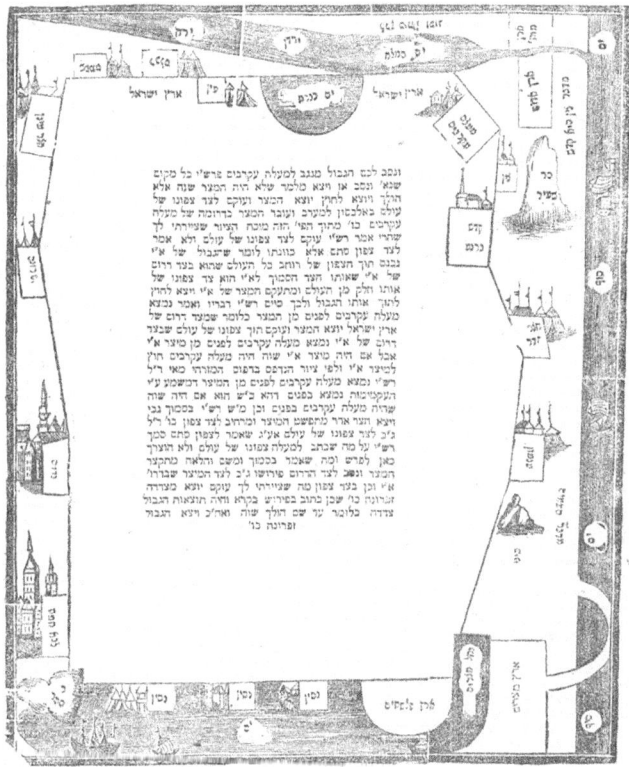

©The Jewish National & University Library

Fig. 24a: The Map as per Mizrachi (150 x 93 mm), p. 59b; and the map as per Yoffe (169 x 142 mm), *Four Commentaries on Rashi*, part 4: Numbers, p. 1a (Warsaw: Isaac Rathauer Print, 1864) [Hebrew].

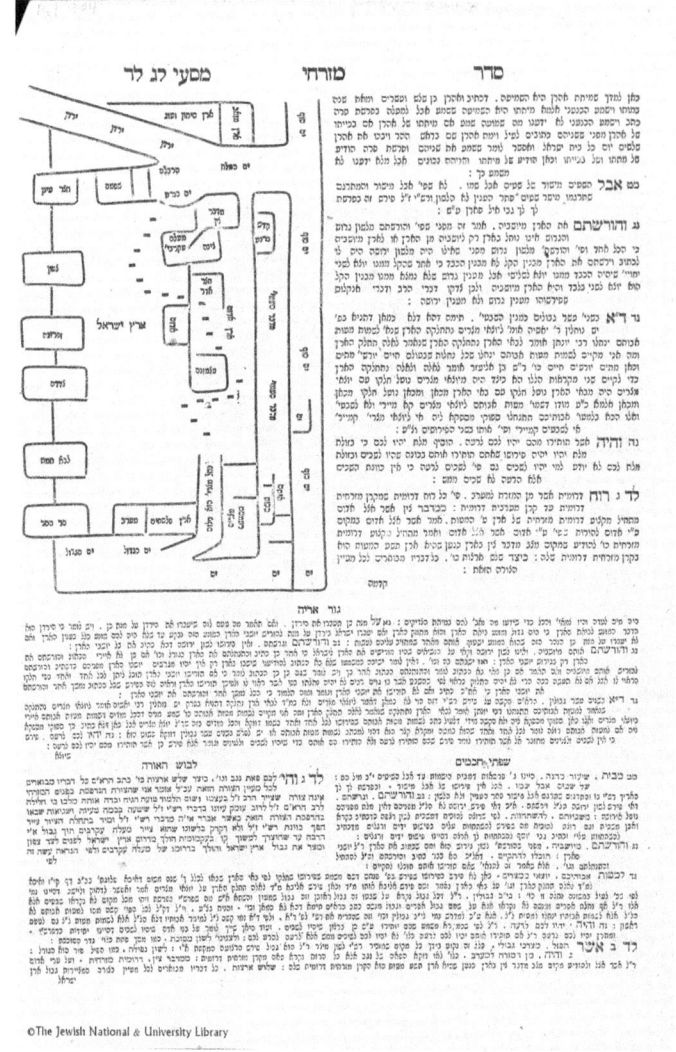

Fig. 24b

which came out approximately a hundred years later (1864).[55] In the mostly identical 1796 and 1815 Salonica editions,[56] the borderlines and frames are composed of several different print characters: the Greek letter Φ, horizontal number 8s, ornamental

---

55 *Sefer me'am lo'ez*, [...], fourth part Numbers, Salonica: Saadi Halevi Ashkenazi Press, 1863–1867 [Ladino].
56 *Sefer me'am lo'ez*, [...], fourth part, Salonica: M. Nachman and D. Israeligia, 1796–1803, p. 168 [Ladino]; *Sefer me'am lo'ez*, the fourth part that is the Book of Numbers, [...], Yitzhak Magriso [...], Salonica, [publishing house], 1815, page 168a [Ladino].

quasi-parentheses, and more.[57] The use of characters as typographic elements and substitutes for other marks also informs the cover page and sectional partitions of the *Me'am lo'ez's* early editions.

The exception among the versions of the *Me'am lo'ez* is the 1823 Livorno edition.[58] In all likelihood, this particular version was intended for Ladino-speaking Jews in Gibraltar and perhaps Spanish Morocco. Gaon cites the introductory words of its editor, Judah Koriat, who offered the following prayer: "the holy community HC Gibraltar may God protect it; all the community's members are brave, officers and dignitaries, righteous and devout...." This lavish praise attests to the fact that the Jews of Gibraltar funded the publication of this work.[59] As reflected by the spelling of various words in the sundry editions of the *Me'am lo'ez*, there were a few slightly different dialects of Ladino.[60] The Livorno edition differs from the others with respect to pagination and the map on *Mas'ei*. More specifically, the map fills the bulk of the page and is the only version to eschew the print characters for standard lines. The title of the map is in Ladino: "Here is the form [i.e., map] of the borders of Eretz Yisrael." Save for a dotted line that marks the route of the Israelites' exodus from Egypt, there are no significant discrepancies between the maps' content. All told, the Livorno map closely resembles its counterparts. In the identical Izmir editions from 1867 and 1872, the printer reverted back to characters. However, they are designed more intricately and aesthetically than in the previous works.[61]

In all the editions of the *Me'am lo'ez* at my disposal, the portion of *Shlakh lekha* (Numbers 13:1–15:41) contains a schematic diagram outlining the relation between the wilderness, the Land of Israel, and the borders thereof. This diagram is faulty from a geographic standpoint, as it confuses east with west and Hebron is located outside the southern border of Eretz Yisrael. On the same page, there is also an illustration of how the eight spies transported the cluster.[62] Based on Rashi's explanation, this rendering is a simplification of Shapira's above-mentioned diagram. Similar to *Me'am lo'ez's* map, these two illustrations were first designed with print characters; the marks were replaced by lines in the Livorno edition, only to have the characters reemerge in the Izmir edition (1867). Here too, the characters are arranged in a more elegant fashion than the earlier versions of these illustrations.

The Livorno edition contains another map that pertains to the Book of Numbers. It depicts the layout of the tribes' camp around the Tabernacle, "each under its ban-

---

[57] See Vilnay, "Maps of Palestine in Rabbinical Literature" [Hebrew]. It bears noting that Vilnay was less than precise with respect to the editions.
[58] *Sefer me'am lo'ez*, fourth part, the Book of Numbers, Livorno: Nachman Sa'adon Press, 1822–1823, p. 143b [Ladino].
[59] Gaon, *Heart Lockets*, p. 18.
[60] I would like to thank Dr. Dov Cohen for his insights on this matter.
[61] *Sefer me'am lo'ez*, the fourth part that is the Book of Numbers,[...], Yitzhak Magriso [...], Izmir: [..., 1872], page 158 [Ladino].
[62] *Sefer me'am lo'ez* (in both the 1764 and 1815 edition), p. 65a.

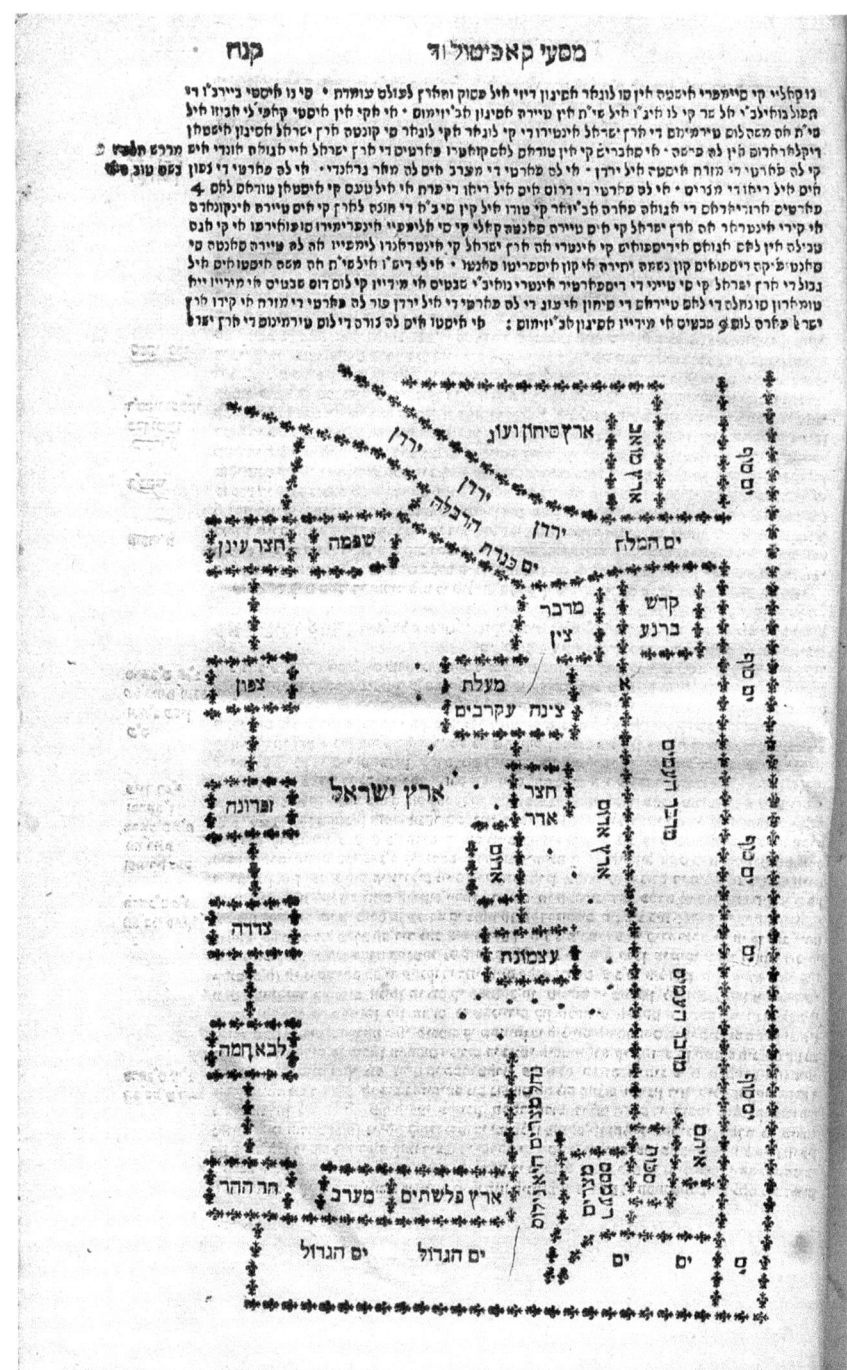

Fig. 25: Map comprised of print characters, *Me'am lo'ez*, Book of Numbers (Constantinople: 1764), p. 158, Yad Ben-Zvi Library, 185 x 300 mm.

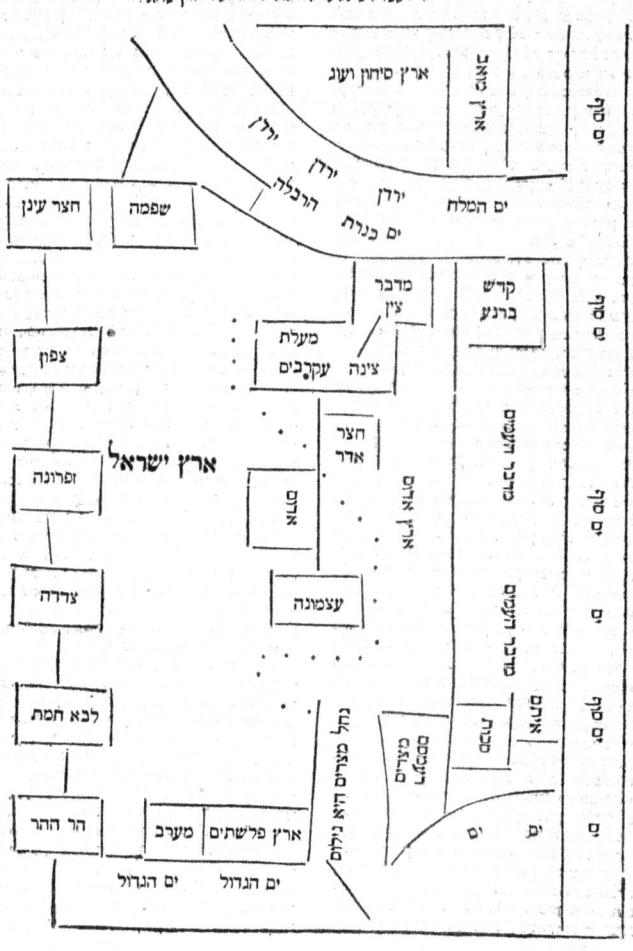

Fig. 26: The lined map of the Livorno edition of the *Me'am lo'ez* (1822–1823), Amnon Dor Collection, p. 143b, 155 × 230 mm.

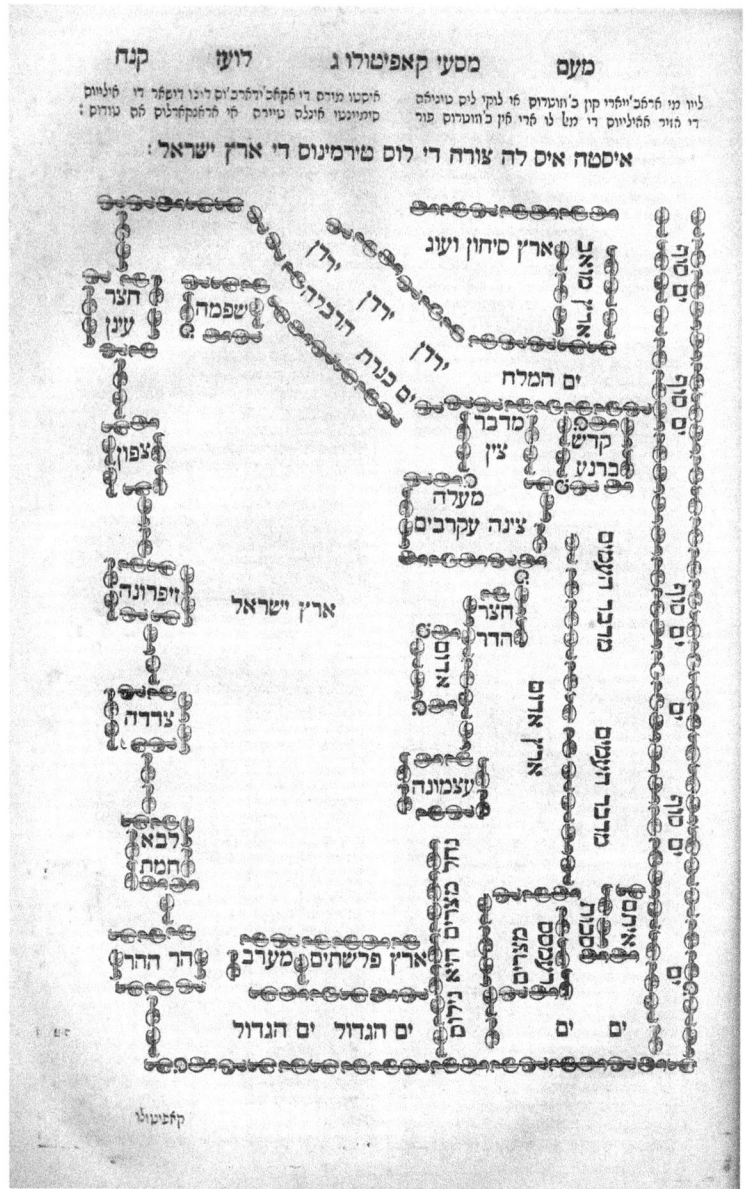

Fig. 27: Map in the Izmir edition of the *Me'am lo'ez* (1867), the Dov Cohen Collection p. 158a, 230 × 160.

ner."[63] The center of this map consists of three parts: the Tabernacle at the bottom; the ḥaṣer (plaza); and on top, "the well of Miriam." Moreover, the illustrator provides

---

63 *Me'am lo'ez*, Numbers, the Livorno edition, between pages 8 and 9. This map graces a copy of a

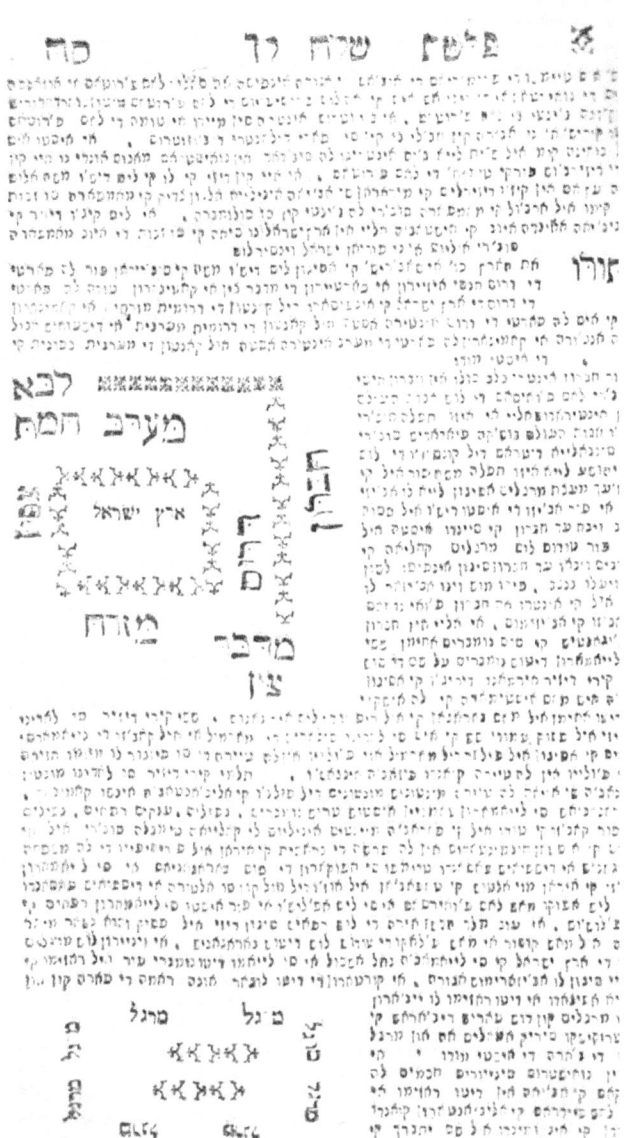

Fig. 28: Pair of illustrations from *Me'am lo'ez*, Book of Numbers (1715), p. 65a.

---

book that is housed at the Yad Ben-Zvi Library. I am indebted to Dov Cohen and the library staff for their help.

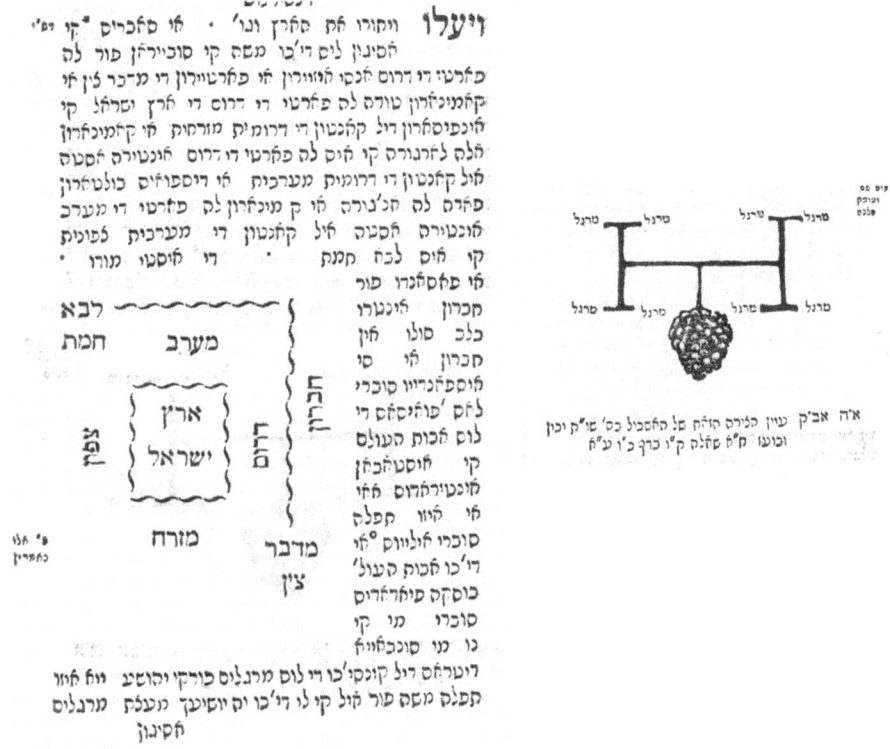

Fig. 29: Pair of illustrations from *Me'am lo'ez*, Book of Numbers (Livorno), p.59a.

the location of the three Levite clans: to the right and left of the well are the tents of Kohath and Merari, respectively; and behind the Tabernacle is the area of Gershon. Above the Tabernacle are the tents of Moses, Aaron, and Miriam. The middle of the camp is divided by four diagonal lines (the bottom two are longer than the upper two), which channel water from the well. Surrounding the inner zone are the camps of the twelve tribes, three per side. It is worth noting that this illustration appears exclusively in the edition under review. The absence of this map from the other versions of *Me'am lo'ez* and the fact that it was printed on a double sheet between pages 8 and 9, rather than meriting its own page number, betrays the fact that it was not an integral part of the book.

At the top of the illustration is a detailed text in Ladino describing the encampment and the water running from Miriam's well:

> This is the layout according to which the Israelites camped in the wilderness with the banners and how the well was situated at the entrance to the Tabernacle's court near the tent of Moses

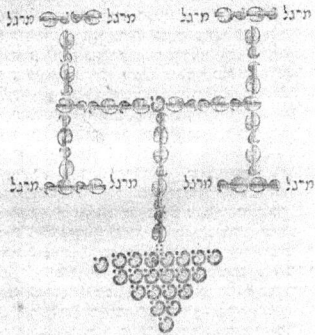

Fig. 30: Pair of illustrations from *Me'am lo'ez*, Book of Numbers (Izmir), p. 63a.

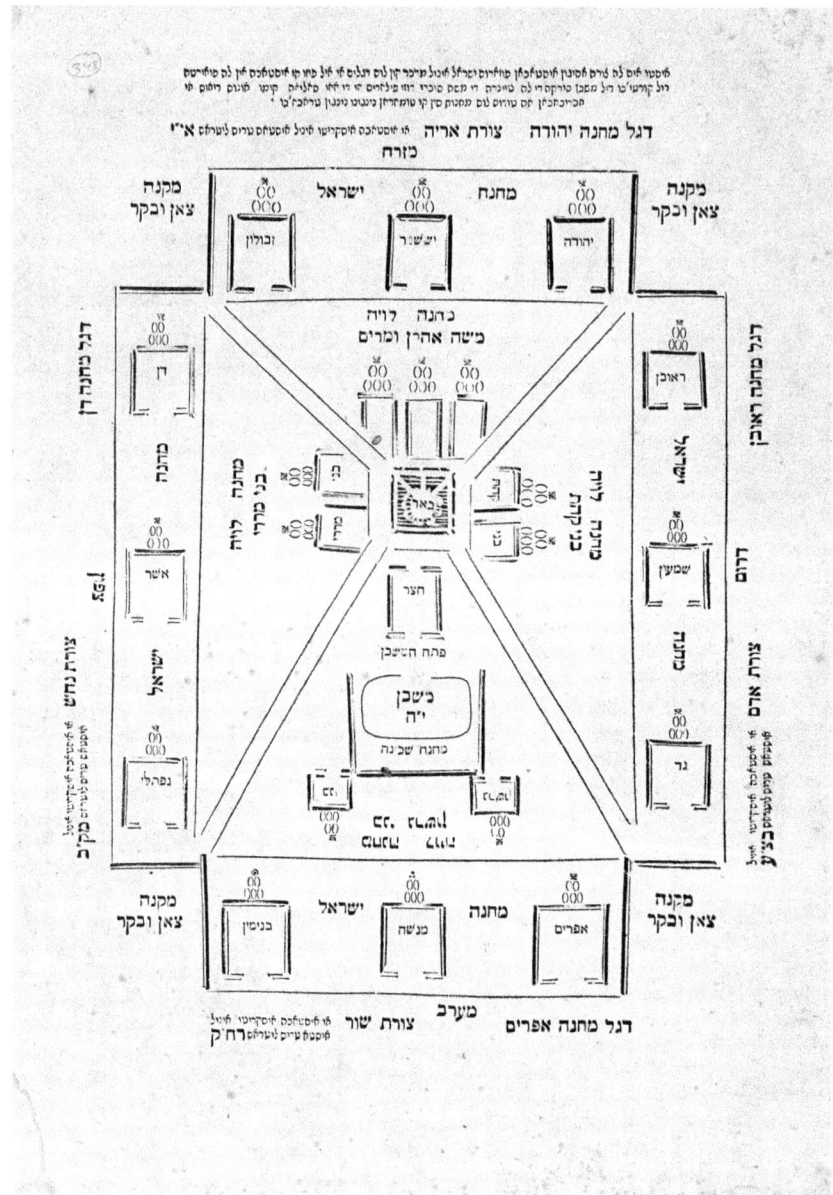

Fig. 31: The layout of the tribal camp, *Me'am lo'ez* (Livorno edition), Amnon Dor Collection, between pages 8 and 9, 230 x 315 mm.

on top of twelve poles. And from there, like rivers emanated as per their turn and nourished all the camps without [anyone] having to do any work.[64]

On each of the drawing's margins is information concerning the adjacent tribes. These texts are comprised of both Hebrew and Ladino. Above the area of Judah, Issachar, and Zebulun are the following captions: "east;" "the standard of the camp of Judah;" "the form of a lion;" and "and it was hereby written the three letters AYY [*aleph*, *yodh*, and *yodh*]. The text to the right, next to Reuben, Simeon, and Gad, reads thus: "south;" "standard of the camp of Reuben;" "the form of man;" and "and it was hereby written the three letters BṢA" [*beth*, *sadhe*, and *ayin*]. The captions elaborating on Dan, Asher, and Naphtali to the left are "north;" "the standard of the camp of Dan;" "the form of a snake;" and "and it was hereby written the three letters MKB [*mem*, *qoph*, and *beth*]." The text below Ephraim, Manasseh, and Benjamin reads "west;" "the standard of the camp of Ephraim;" "the form of an oxen;" and "and it was hereby written the three letters RḤK [*resh*, *ḥet*, and *qoph*]."

Miriam's well and the tribes' layout around the Tabernacle were prevalent subjects in Jewish art even before this map, and they come up in quite a few Christian works as well.[65] For instance, Zaddik's map, which was inspired by Adrichom, was printed in 1621 (see chapter 4).[66] The illustration at hand, though, stands apart from those renderings. Conversely, it may share an affinity with the map in Leon's book, which netted multiple editions in Amsterdam over the course of the seventh century.[67]

Map of Haim Isaiah Hacohen

Haim Isaiah Hacohen Halbersberg (1844–1910), the author of several halakhic and exegetical works, was born in Lublin. Following a long career in the Polish Rabbinate, he immigrated to the Land of Israel in 1907.[68] Among his books was *Sefer tshu'ot ḥayim*, an exegesis based on the commentaries of Rashi and his disciples. This work contains a map in the section on *Mas'ei* that is titled thus: "A drawing of the borders according to the opinion of the *Levush* [i.e., Mordecai Yoffe] of blessed memory, and

---

64 I owe a debt of gratitude to Dr. Yaron Ben-Naeh and Dr. Michal Held for helping me translate the map's Ladino text into Hebrew.
65 E.g., Adrichom, "Situs Terra Promissionis;"
66 Zaddiq, *Hebrew Map of Palestine*. This map is discussed at length in chapter 4. For more on this map, see Nebenzahl, "Zaddiq's Canaan."
67 Leon, *Retrato del Tabernaculo* (Amsterdam: Gillis Joosten, 1654), including the illustration between pp. 18 and 19. For more on Leon, his works, and the sundry editions thereof, see Offenberg, "Bibliography of the Works of Jacob Jehuda Leon (Templo);" idem, Offenberg, "Jacob Jehuda Leon (1602–1675) and his Model of the Temple."
68 Reisen, *Leksikon fun der Yidisher Literatur, Prese un Filologye*, vol. 3, pp. 12–13 [Yiddish].

just as we merited to draw it [the Land of Israel], so too may we merit to see it built up and settled, when God returns the exiles of Israel and may a redeemer come to Zion and swiftly build Ariel in our day amen.[69]

Content-wise, the map adheres to the cartographic tradition of Elijah Mizrachi. In other words, it faces east, the Jordan River assumes a diagonal course, and it enumerates the various border sites of Eretz Yisrael that are noted in *Mas'ei*. From a graphic standpoint, the map is quite simple and diverges from its predecessors in several respects. To begin with, the Jordan River and the Dead Sea are drawn as one long box, while the Sea of Galilee is mentioned only in a caption. Second, it lacks so much as a hint of the Red Sea. Third, neither the southern nor northern borders constitute a straight line. The former is comprised of two diagonal lines that form an angle pointing to the south, and the northern border loops to the left. Perhaps this rendering is a simplification of the crooked lines in Yoffe's map—the source of Halbersberg's illustration.

A Map in an Eighteenth-Century Pentateuch from Frankfurt

Our next subject, a Pentateuch that was printed in Frankfurt an der Oder in 1746, offers two juxtaposed maps on the same folio.[70] The first is a small sketch bearing the title "A drawing of the borders of Eretz Yisrael according to Rashbam," who was Rashi's grandson and student. The same name is also written on the lower part of the map, probably added in handwriting some time after the printing.

Oriented to the north, this diagram is shaped as a rectangle. Its upper lefthand side consists primarily of the Land of Canaan, while the entire western (left) edge is taken up by the Great Sea. To the south are boxes denoting the Nile, Egypt, Edom, and Moab. The bottom row contains a verse from the Book of Judges: "Then they went along through the wilderness and compassed the Land of Edom and the Land of Moab" (11:18). Due east of Canaan is a partitioned column signifying the Jordan and the Dead Sea. The rectangle to their right also contains two toponyms: the plains of Moab and Sihon. The space above Canaan suffices with the word "North" in large fonts. This diagram resembles Rashi's map for Numbers 33.

Just below this diagram is a map that was drafted along the lines of Rashi's cartographic representation of Numbers 34. This work is practically square, faces east, and is lined with toponymical boxes. Each of the four winds is denoted by a "boundary thread." The one to the south also contains the following caption: "Within these four threads is the Land of Canaan." In the middle of this illustration is a lengthy

---

[69] Haim Isaiah Hacohen, *Sefer tshu'ot ḥayim*, Lublin: Shneidermasser and Hirschenharen Press, 1894, p. 76 [Hebrew].
[70] *Five Books of the Torah: [...] Adorned and Arrayed in Ten Raiments of Light and Commentaries of Precious Glory: [...]*, in the press of the Lord, Hasid, Doctor, and Professor Garilla, volume 4, Book of Numbers, 1746, Frankfurt an der Oder [Hebrew]. The map is on page 152b.

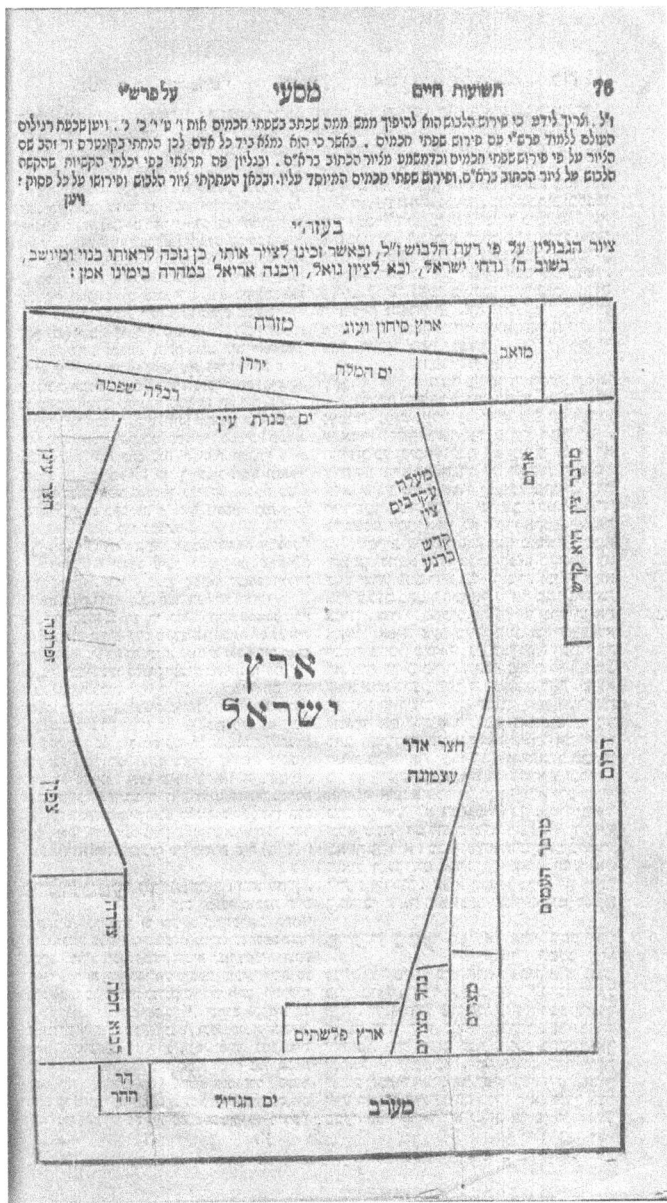

Fig. 32: Map of Eretz Yisrael's borders by Haim Isaiah Hacohen, *Sefer tshu'ot ḥayim*, p. 76, 156 × 112 mm.

passage that is similar to the one in Rashi's maps on the difference between the two Mount Hors. For some reason, this text is upside down. This map also possesses an unusual distortion of Numbers 34:3–4: "And a little of the Dead Sea and will encompass you from the Negev [i.e., the south] and beyond ... (which is the eastern border.)"

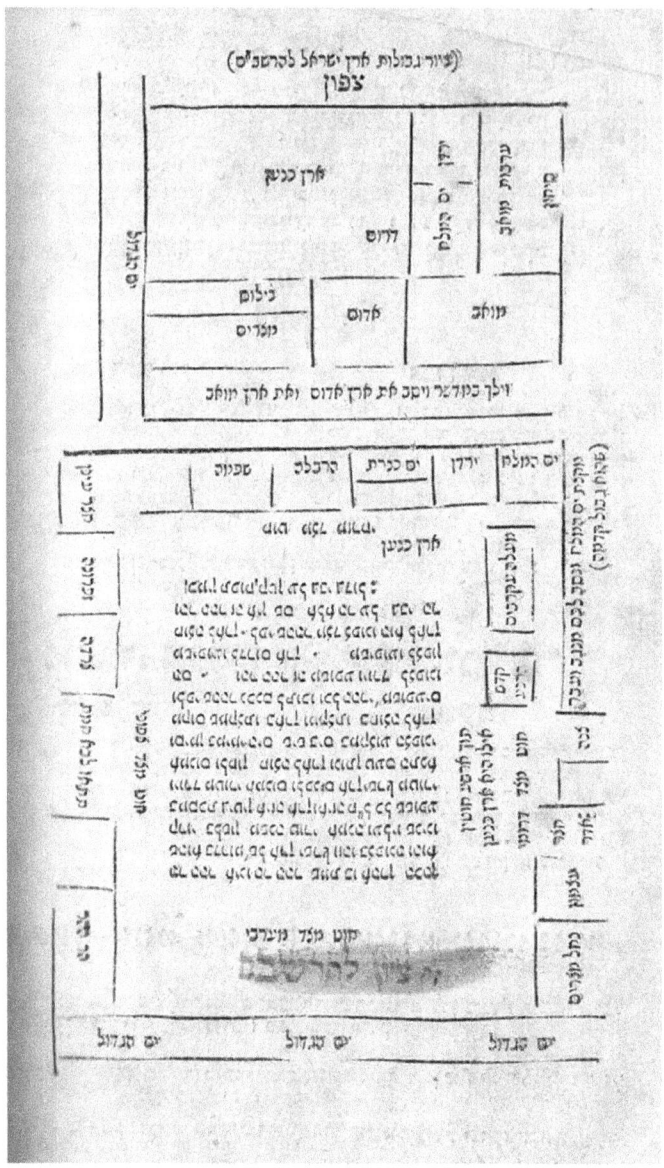

Fig. 33: Rashbam and Rashi's maps, *Five Books of the Torah [...]*, volume 4, Book of Numbers, page 152b, 165 x 95 mm and 205 x 125 mm, respectively.

Schematic maps as learning tools for the Mishna and Talmud

As discussed in the first chapter, Rashi included a schematic diagram of the route from Bethel to Shechem (Nablus) in his commentary on the Book of Judges and a sketch of the area between Acre and Kziv in his commentary on the first chapter of tractate Gittin. The latter was also printed in myriad editions of the Talmud and Talmudic exegeses.[71] The principal topic of Gittin's opening chapter is the conditions for receiving a divorce from a messenger who arrives in the Land of Israel from abroad—an issue that pertains to the question of whether Acre is part of Eretz Yisrael.[72] While on the topic, the Babylonian Talmud (Gittin 7b) cites the following verse: "Then they said, Behold, there is an annual feast of the Lord in Shiloh in a place which is on the north side of Bethel, on the east side of the highway that goeth up from Bethel to Shechem, and on the south of Lebonah" (Judges 21:19). The inclusion of this verse also ties the second diagram to this text. Over the next few pages, I will review several of the many versions of these two diagrams.

The MaHaRShA's Diagrams for the Talmud

Born in Krakow, Shmuel Eliezer ben Judah Eidels (1555–1631) served as a rabbi in Chełm, Poznań, Lublin, and Ostrog. Known by the Hebrew acronym MaHaRShA, his work *Ḥiddushei halakhot* is deemed to be an important commentary on the Babylonian Talmud and indeed graces many editions of this corpus.[73] Eidels included both the diagram of the Bethel-Shechem road and the sketch of the expanse between Acre and Kziv in this book. I located these simple maps in these, relatively early editions of *Ḥiddushei halakhot:* the Fürth editions from 1706 and 1721; the 1755 Berlin edition; and the 1792 Nowy Dwor edition. The latter was printed in the press of Johann Anton Krieger.[74] We will return to Krieger's establishment later on, for the German Christian printed numerous devotional and Halakhic books in Hebrew, some of which also included maps.

---

[71] It is worth noting that in print editions of the Talmud, there are also diagrams on other topics, some of which are attributed to Rashi as well. See, for example, the illustration of the "crooked alley" in the Babylonian Talmud, Eruvin 6a.

[72] This topic is also raised in tractate Shevi'it, within the framework of a discussion on the Land's borders. See the discussion on Acre's location in chapter 1, including the reference to Ne'eman, "Is Acre Part of Eretz Israel?" (note 33).

[73] "Samuel Eliezer ben Judah Ha-Levi," *Encyclopaedia Judaica*; Katz, *Rabbinate, Hasidism and Enlightenment*, pp. 70–75 [Hebrew]; Elbaum, *Openness and Insularity*, p. 21.

[74] As noted, this work came out in several editions, of which I had access to the following: MaHaRShA, *Sefer ḥiddushei halakhot* (Fürth: Reuven Fürst Press, 1706), p. 96b; ibid (Wilhelmsdorf: Hirsch ben R. Haim of Fürth, 1721), Gittin 123b; ibid (Berlin [...], 1755), p. 10b; ibid (Nowy Dwor: J. A. Krieger Press, 1792), p. 123b.

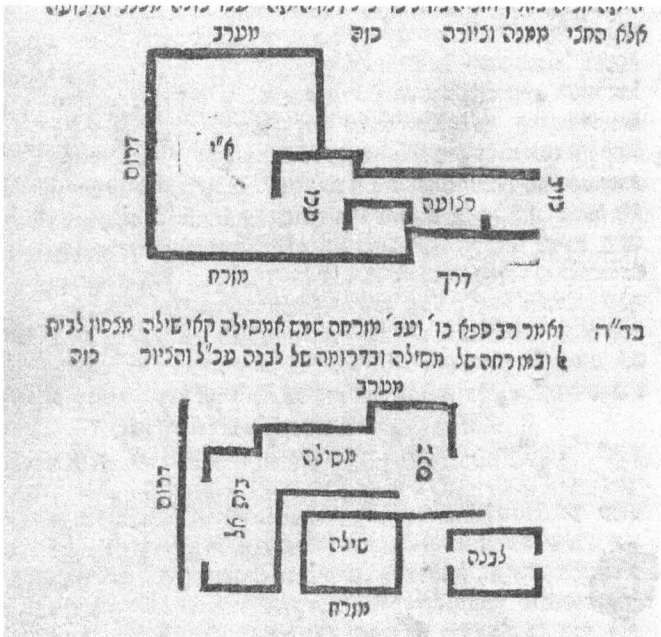

Fig. 34: Eidels' diagrams, *Sefer ḥiddushei halakhot*, chapter I, p. 123b, 36 x 60 mm; 34 x 52 mm.

The Diagrams of the MaHaRaM from Lublin

Meir ben Gedalia (1558–1616) of Lublin, known by the Hebrew acronym the MaHaRaM (our teacher Rabbi Meir), was a *rosh yeshiva* (seminary head) in Krakow, Lublin, and Lwów.[75] A contemporary of Eidels, his commentary *Meir einai ḥakhamim* (Illumines the Eyes of the Sages) was first printed by his son in Venice in 1619 and subsequently netted many more editions.[76] In the section on the first chapter of tractate Gittin, the MaHaRaM interjected a small adaptation of Rashi's diagram of Acre and its surroundings. Next to this map is the following explanation: "I was obliged to draw this, in order to reconcile what is written in the Tos[afot] at the top of the chapter in the statement beginning And Ashkelon etc. that part of Acre is EY [Eretz Yisrael] and part of it is abroad." The diagram graces the upper half of the page in

---

75 Eidelberg, "Lublin, Meir ben Gedaliah," *Encyclopaedia Judaica*; Katz, *Rabbinate, Hasidism and Enlightenment*, pp. 65–69; Elbaum, *Openness and Insularity*, p. 21.
76 Meir ben Gedalia of Lublin (MaHaRaM...), *Sefer meir einai ḥakhamim*, Venice: Bragadin Press, 1619 [Hebrew]. Among the many other editions are Frankfurt: N. Weimann Press, 1709; Fürth: R. Avraham Beng and Boneft Shene'ur Press, 1724; Wilhermers dorf: Hirsch ben Haim of Fürth Press, 1737. The illustration is on page 115b in some of the editions and on page 94a in others. Also see Heller, *The Sixteenth Century Hebrew Book*, vol. II, pp. 1090–1091, note 127.

some editions and the bottom part of others. However, there are no substantial discrepancies between the sundry versions of this illustration.

Fig. 35: Diagram of Acre and its environs in Meir ben Gedalia, *Sefer meir einai ḥakhamim* (1709), p. 115b, 70 × 40 mm.

The above-cited works of the MaHaRShA and the MaHaRaM were not only produced in the same generation; they were included in the same volume of exegetical literature that was compiled by the former. In fact, their respective maps for tractate Gittin appeared on the very same page, with the MaHaRShA's diagrams at the top and his colleague's below.[77]

Diagram of Yisrael ben Gedalia Lipschitz

A slightly different diagram of the Acre region surfaces in *Tiferet yisrael*, a Mishna commentary by Yisrael ben Gedalia Lipschitz (1782–1860)—a rabbi who served in Wronki, Dessau, Colmar, and Danzig.[78] The illustration is located in the section on tractate Shevi'it, which also ponders the status of Acre.[79] Compared to earlier ver-

---

77 Shmuel Eliezer Eidels, *The MaHaRShA's Book Ḥidushai halakhot [Novellae of Halakhas]: and with Ḥidushai aggadot [Novellae of Aggadah] [...] the Book of the MaHaRShaL [Solomon Luria] Ḥokhmat shlomo [...] and Sefer meir einai ḥakhamim, which was Authored by [...] R. Meir [...] from the Holy Community of Lublin.* Slavita, Moshe Shapira [Press], 1816, vol. 2, Gittin, page 3 [Hebrew].
78 David, "Lipschutz, Israel ben Gedalia," *Encyclopaedia Judaica*.
79 This diagram was printed, with minor revisions, time and again. For example, *Mishna [...] with the Commentary of Our Rabbi Obadiah of Bertinoro and the Primary Portion of the Tosafot yom tov [...] and a Commentary [...] has been added [...] by the name Tiferet Yisrael [...], Authored by Our Rabbi Yisrael Lipschitz Son of Our Rabbi Gedalia [...] with Numerous Additions [...]* Warsaw: Y. Goldman Press, 1873, 140b; *Mishna, the Order of Zeraim, with All the Exegetes and Many Additions as Explained in the Second Title Page*, Vilnius: the Widow Romm and Sons Press, the publishing house of the rabbi our rabbi Haim Noah Eisenstadt may his candle illuminate in Warsaw, year [..., 1878], 163a (325); *and Mishnas*

sions, Lipschitz adds a box denoting "the river" in the southwest. Put differently, he seems to consider the River of Egypt (*nakhal miṣrayim*) to be the southern border of Eretz Yisrael. Above the Great Sea, the phrase "and inside [Eretz Yisrael]" appears three times. Furthermore, the mapmaker continues the erroneous tradition of situating Acre and Kziv inland.

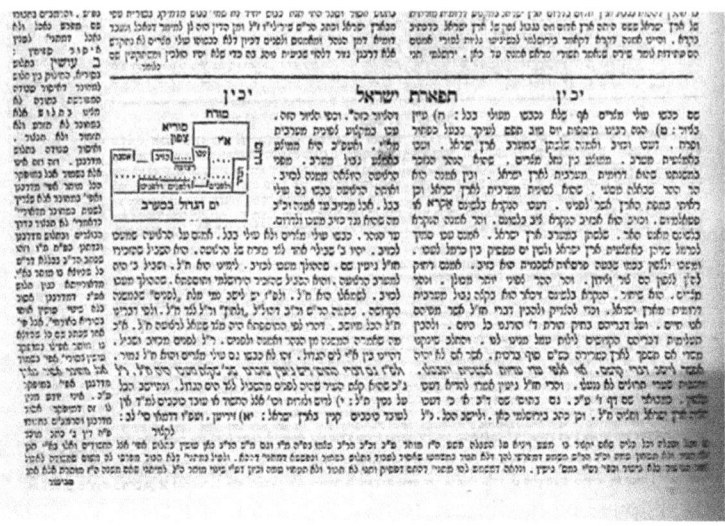

Fig. 36: Diagram of Acre and its environs in Yisrael Lipschitz, *Tiferet Yisrael*, p. 163a (i.e., 325), 40 X 30 mm.

Additional versions of these sketches appear side-by-side in numerous print editions of the Talmud. For example, they were incorporated into the Amsterdam edition, which came out between 1752 and 1765.[80] What is more, contemporary print editions,[81] such as the Steinsaltz Talmud, continue to run these diagrams.[82]

As noted, this paradigm situates Acre and Kziv inland, with the Mediterranean coast far off to the west. This depiction naturally contravenes Acre's status as a port town, even though the geographic reality was expressly known to Talmudic

---

[...] with the Commentary of Our Rabbi Obadiah of Bertinoro and with the Primary Portion of the Tosafot yom tov [...] and with the Commentary Tiferet yisrael, Authored by [...] Our Rabbi Yisrael Lipschitz son of Our Rabbi Gedalia [...], Vilnius: the Romm Press, 1913, volume 1 tractate Shevi'it, 143b. It also turns up in this modern edition: *Six Orders of the Mishna with the Commentaries of the Rishonim and Aḥaronim*, photographic print as per the Romm Publisher 1909, Jerusalem 1955–1958, Order Zeraim, tractate Shevi'it, p. 163.

80 *Babylonian Talmud with the Commentaries of Rashi, tosafot and Piskai Tosafot and Rabeinu Asher [ben Jehiel, namely the ROSH] and Piskei ha'rosh and the Commentary on the Mishna of Maimonides*, Amsterdam: Proops Press, 1752–1765, vol. VI, tractate Gittin [..., 1756], first chapter, 7a.
81 See the Oz V'Hadar Friedman edition of the Talmud, tractate Gittin, Jerusalem 2006, 7a, *inter alia*.
82 See, for example, the Steinsaltz edition, Gittin, p. 32.

commentators and stated on many occasions in the literature of the Sages. In all likelihood, the repeated use of this inaccurate representation is indicative of Rashi's venerated status as the "forefather" of these diagrams as well as the preeminent exegete of the Talmud. Consequently, the sketch has been preserved despite the knowledge that it constitutes a geographic fallacy.

Adeni's Map of Shalosh Arṣot le'Shevi'it

At the tender age of four, Shlomo Adeni (1567–1625) and his father moved from Sana'ah, Yemen to Safad, where the son eventually wrote *Melekhet shlomo* (Work of Shlomo)—a respected commentary on the Mishna to this very day.[83] In the section on the order of Zeraim, there are a couple of illustrations and a map was included in tractate Shevi'it. An early manuscript of this work, which may have been written by Adeni himself, is on display in the Braginsky Collection's website (the said map included).[84] Adeni's commentary and map were reprinted in various editions of the Mishna[85] and continue to adorn modern editions.[86]

Chapter 6 of tractate Shevi'it opens with the words "*shalosh arṣot le'shevi'it*" (literally three lands for the seventh, or zones for the sabbatical year). This discussion grapples with the division of Eretz Yisrael into areas where the commandments that are dependent on the Land are in force.[87] Adeni basically took the square frame of Rashi's map of the Promised Land's borders and adapted it to the needs of the topic at hand. Although his map differs from those surveyed above, I have included it in this chapter for two reasons: it possesses certain elements of the Rashi model; and it is related to the maps that were printed in annotated volumes of the Mishna and Talmud.

Adeni's map is oriented to the north and presents the Land of Israel and its borders in a square framework. The influence of the Rashi cartographic tradition on the Safadian scholar is quite evident. To begin with, Zin and the River of Egypt are located in the southeast and southwest, respectively. Dots signifying islands line the Great

---

[83] "Adeni, Solomon Bar Joshua," *Encyclopaedia Judaica*.
[84] Adeni, *Malekhet shlomo*, p. 144b, the Hebron MS, p. 149 [Hebrew]. See item 45 in the Braginsky collection: www.braginskycollection.com, last accessed January 2018.
[85] *Mishna, the Order of Zeraim, with All the Exegetes and Many Additions as Elucidated in the Second Title Page*, Vilnius, in the Widow Romm and Sons Press, the rabbi our rabbi Haim Noah Eisenstadt may his candle illuminate in Warsaw, year 1878, p. 162b (324), [Hebrew].
[86] *Six Orders of the Mishna with the Commentaries of the Rishonim and Aḥaronim*, photographic print as per the Romm Publishing House, 1909, Jerusalem 1955–1958, order Zeraim, tractate Shevi'it, VI, p. 324 [Hebrew].
[87] For an in-depth look at the roots of the Halakhic outlook regarding *shalosh arṣot le'shevi'it* in the literature of the Sages and the commentaries thereof, see Ben Eliyhau, *National Identity and Territory: The Borders of the Land of Israel in the Consciousness of the People of the Second Temple and the Roman-Byzantine Periods*, esp. pp. 139–180 [Hebrew].

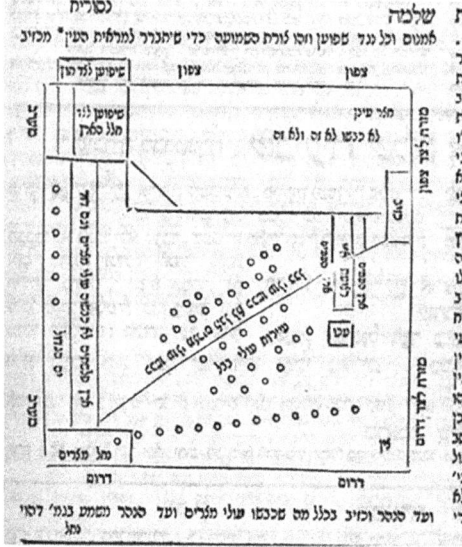

Fig. 37: Map of *shalosh artzot le'shevi'it* in Shlomo Adeni, *Malekhet shlomo*, p. 162b, 75 × 75 mm.

Sea. Abutting the Mediterranean is a rectangle with the following text: "Neither the returnees from Egypt nor David conquered the Land of the Philistines." The map distinguishes between "the north side of the east" and "the south side of the east." Hazar Enan constitutes the northernmost toponym. Below this site is a caption reading territory "that was conquered by neither this nor that." The brunt of Eretz Yisrael is taken up by a quasi-square that is split by a diagonal line running from southwest to northeast. Its lower half contains the towns of "the returnees from Babylon;" and its upper part signifies the territory that was "conquered by the returnees from Egypt but was not conquered by the returnees from Babylon." In each of these areas, small circles denote settlements. Adeni's diagram thus graphically expresses the concept of *shalosh arṣot le'shevi'it* by partitioning Eretz Yisrael into three zones.

Near the diagonal are remnants of Maimonides and Rashi's oversight, for Acre and Kziv are situated in the northeast of the Land, rather than on the northwestern coast. The fact that Adeni, who lived in Safad, included this error on his map even though he was certainly aware of these sites' actual location, bears witness to the staying power of these two venerated scholars.

In 1876, a rather similar version of this map was printed by Nahman Natan Koronil (or Koronel) in *Piskei ḥallah* (Rulings on Challah).[88] Koronil (1810–1890) was

---

88 *Piskei ḥallah* [...] *Our Rabbi the RaShBA* [R. Shlomo ben Aderet]. Printed one time [...] in Constantinople [in the] year 1518 [...] and at the end I affixed two rulings on customs and conventions regarding the laws of ḥallah from the genius Rabbi Yaacov [so]n of Tzahal among the sages of the generation before me and more from me [...], Nahman Natan Koronil, Jerusalem, Elijah and Moshe Chai Sasson [Press], 1876.

born in Amsterdam and immigrated to the Land of Israel in 1830. At first, he resided in Safad, but relocated to Jerusalem seven years later due to an earthquake. In Jerusalem, he befriended Joseph Schwarz, whose maps are examined in chapter 7. Koronil's literary enterprise consisted primarily of bringing old manuscripts that he discovered in the yeshivas of Jerusalem and Hebron to print.[89] However, he also insinuated his own works into these same texts. For instance, Koronil attached a paper that he called "Research the Matter, on the Overseas Custom of Allotting Ḥallah in Places Far from EY [Eretz Yisrael]" to *Piskei ḥallah*, a famous Halakhic work. He incorporated an exact replica of Adeni's map to his text, but also added a warning to the location of Acre:

> For if you peruse the map of EY that is made in accordance to the wisdom of geography [from the] y[ear] 1816 with the approv[al] of the genius R. Moshe Minz in Oven [i.e., Óbuda] and moreover according to a map from the y[ear] *teḥezenah* [gematria[90] for 1695] you will find there Acre to the NE [of] Jerusale[m] and the reason is that the Great Sea is not a straight line but it curves like a diagonal and as such it turns out that the words of the rabbi are correct. And in a new map that was printed in Berlin [in] 1864 and in a map from London, it appears that Acre is to the north side of Jerusalem and leans a bit to the west side and the drawing of the rabbi [who wrote] *Tiferet yisrael* is correct.[91]

Koronil felt the need to bridge the gap between his dedication to exegetes like Adeni and his first hand knowledge, as an inhabitant of Eretz Yisrael, that Acre and Kziv are located on the northwestern coast of the Great Sea. As Koronil put it, his knowledge of the geographic reality derives from concrete information on several maps. We can definitively identify two of these works, both of which will be explored in the fourth chapter: the map from "the y[ear] *teḥezenah*" is part of the Amsterdam Haggadah (1695); and the map from "1816" was crafted by Yaakov Oispitz.[92] On the other hand, I was unable to identify what he dubs the "new map that was printed in Berlin" or the "map from London." At any rate, it stands to reason that Koronil's abovementioned friend (Joseph Schwarz), who had an academic background in geography, exposed him to modern maps in European languages.

Diagrams of the MaHaRShA and the Choriner

Over the next few pages, our attention will turn to a completely different geographic model in which the Land of Israel is depicted as a square with Jerusalem at its center.

---

[89] Gelis, *Among the Greats of Jerusalem*, pp. 71–76 [Hebrew].
[90] The letters of the Hebrew alphabet possess numerical values. Practitioners and aficionados of gematria search for correlations between words and messages by dint of their sum numeric totals.
[91] Koronil, *Piskei ḥallah*.
[92] Koronil erred with respect to the date of Avraham bar Yaakov's map. More specifically, he thought that the entire word "*teḥezenah*" was tantamount to the year of publication, but in the first edition only the letters *taw, nun,* and *he* are highlighted. Therefore, the correct date is 1695.

These sort of diagrams face east and highlight the set of directions. To the south of Jerusalem is the Land of Judah; and to the city's north is Benjamin and the Galilee. The original version of this paradigm was printed in the 1756 edition of the MaHaR-ShA's *Ḥiddushei halakhot*, which also includes the commentary of the MaHaRaM.[93] At any rate, the sketch itself is exceedingly laconic. Therefore, its meaning must be inferred from the adjacent text, which revolves around the prohibition in tractate Berakhot of the Babylonian Talmud (51b) against people relieving themselves in the direction of the Temple and Jerusalem: "The relievee in Judah shall not face east and west, but north and south." Conversely, "the relievee in the Galilee shall exclusively face east and west." With this in mind, the apparent objective of this diagram is to enlighten those unfamiliar with the geography of Eretz Yisrael as to the relative location of Jerusalem, Judah, and the Galilee.

Fig. 38: Diagram in *Ḥiddushei halakhot*

A nearly identical diagram along with a more intricate version on the ensuing page were printed in a booklet by Áron Chorin titled *Sefer emek ha'shaveh* (the Book of the Vale of Shaveh).[94] The first pertains and indeed refers to the same Ha-

---

93 See Eidels, *Sefer ḥiddushei halakhot* (Berlin: 1756).
94 Áron ben Kalman Chorin (Choriner), *Sefer emek ha'shaveh: It is Agreeable to Every Soul, for it Includes Three Articles: The Article Rosh Amanah [Head of the Perennial stream], the Article Neshama ḥayah [Living Soul], and the Article Dirat aharon [Áron's Apartment]*, Prague: the Alzenwagner Press, 1803, 68b, 69a [Hebrew].

lakhic question as the MaHaRShA (Eidels). The second illustration (to the right here below) bolsters Chorin's assertion that the MaHaRShA "was incorrect in his drawing." To this end, Chorin replaces the latter with a more complex rendering in which the corridors denoting *isur medina*—the areas in which it is prohibited to relieve oneself in the direction of Jerusalem—are much narrower than those in the first diagram.

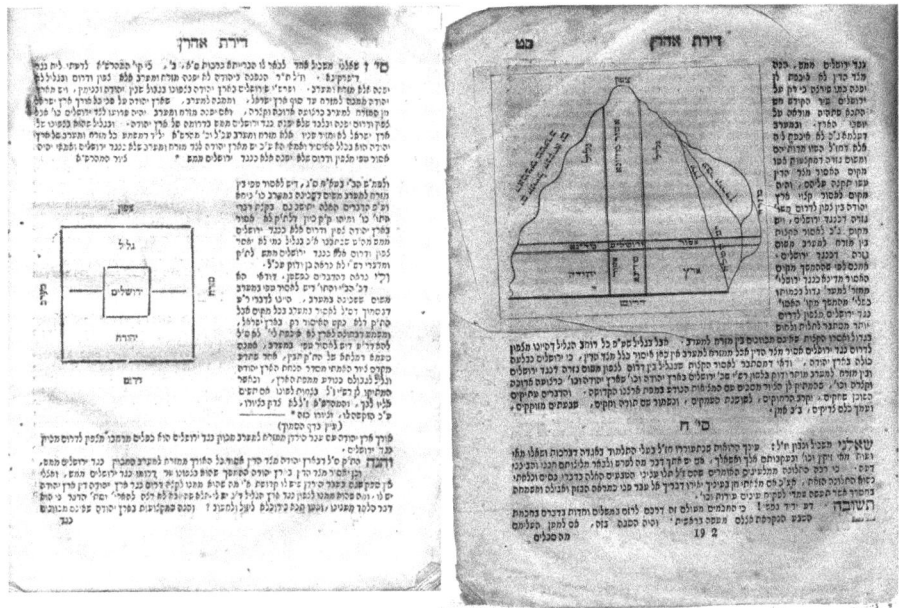

Fig. 39: Two diagrams in Áron Chorin, *"Dirat aharon"* (Áron's Apartment), *Emek ha'shaveh:* the first as per the MaHaRShA (p. 68b, circa 70 x 70 mm); and the second as per Chorin (p. 69a, 120 x 90 mm).

The substantial differences between these two cartographic tools stem from the general outlook of each author. Áron Chorin (1766–1844), also known as the "Choriner," was a rabbi in Transylvania, which at the time was part of the Austro-Hungarian Empire. A pioneer of Reform Judaism in Hungary, he was one of the first Jews to claim that the faith of Moses was in need of an overhaul—views that indeed drew the ire of Orthodox Jewish leaders.[95] In his book *Emek ha'shaveh*, Chorin contends with strictures and customs that, in his estimation, lack any real Halakhic basis.[96] The gap

---

95 Pelli, "Rabbi Áron Chorin's Ideological and Halakhic War on Behalf of Religious Reform in Judaism;" Meyer, *Response to Modernity: A History of the Reform Movement in Judaism*, pp. 184–185 [Hebrew].
96 Marton, Schveiger, and Ioanid, "Arad," *Encyclopaedia Judaica*; Tolkes, "Chorin, Aaron," *Encyclopaedia Judaica*.

between the sweeping restriction in Eidels' diagram and the exceedingly more modest *isur medina* in that of the Choriner exemplifies his lenient approach.

Conclusion

Drawn in manuscripts from the 13th century and later and printed between the sixteenth and nineteenth centuries, the maps surveyed in this chapter were integral parts of the rabbinical-exegetical literature on the Bible, Mishna, and/or Talmud, which were intended to help readers comprehend the texts. The content of these maps points to a deep-seated conservative bent, as they barely changed over the years and their development was quite meager. Additionally, the topics of discussion—the borders of Eretz Yisrael according to *Mas'ei* and several related cases in the Talmudic discourse—cleave to the tradition of Rashi. With the exception of an accompanying remark on the part of Koronil, the compilers were unattuned to the geographical reality of the Land. From a graphic standpoint, most of the works perpetuate an extremely simple, schematic format and eschew auxiliary illustrations. According to scholars of cartographic history, the development of maps generally reflects the way in which geographic knowledge is produced, accumulated, updated, and transmitted. However, the maps exhibited in this chapter were practically immune to modification or progress, for they embody the conservative nature of the rabbinical exegesis on the promised borders. These maps were designed to reinforce the abstract outlook on this topic that was espoused by Jews who mostly resided in the Diaspora. This perspective also undergirded an erudite, abstract discourse on the commandments that are dependent on the Land—one that had no actual bearing on the lives of this same audience.

For the sake of comprehending these maps, it is incumbent upon us to glance over at the European cartography of the period in question. Artistically speaking, the contemporaneous European-Christian corpus was brimming with illustrations, both within the map and on its margins. For instance, they were laden with spires and fortresses, ships and monsters at sea, towering mountain ranges, and numerous miniatures depicting people and fauna. In comparison, the vast majority of the Hebrew maps under review, with the exception of Yoffe, Haim ben Bezalel and Miklish's offerings, are completely devoid of ornamentation. Content-wise, the European maps of the Renaissance and the early modern era stand in absolute contrast to their Hebrew counterparts. The authors of the former strived for innovation, objectivity, and functionality. To this end, they availed themselves of mathematics, astronomy, and the laws of perspective. Likewise, they exhibited the most up-to-date geographic information, which steadily accumulated thanks to the findings of geographic expeditions and the increased sophistication of cartographic methods and tools of measurement.⁹⁷ The Christian maps of the Holy Land that were drawn up and printed in Europe from the sixteenth century onwards were mostly predicated on the Scriptures, rather than firsthand geographic knowledge. Nevertheless, the authors employed modern cartographic tools and aspired to present the Land for all its mountains and valleys, brooks and rivers, cities and villages.

In sum, this chapter's assemblage of Hebrew maps indeed reflects, both in content and form, a complete detachment from the parallel Christian tradition. The Jewish authors turned a blind eye to the arts and sciences that generally informed the field of cartography in Europe and were not swayed in the least by this rich world. That said, these maps constitute but a single genre of Hebrew-Jewish cartography. In the chapters to come, I will discuss those Jewish mapmakers who were influenced by cartographic innovations in Europe and contended with these advances by dint of emulation, replication, and competition.

---

97 Harvey, *The Condition of Postmodernity*, esp. pp. 240–250.

3 Between Jewish and Christian Maps: The Sixteenth-Century Map of the Israelites' Peregrinations from Mantua

A special place in the annals of Hebrew Cartography is reserved for a map that was printed in Mantua, Italy during the 1560s. This map stands apart from the traditional maps that followed in Rashi's footsteps (the topic of the previous chapter). Moreover, to the best of my knowledge, it does not fall under the rubric of the artistic maps that Jewish mapmakers copied from their Christian counterparts (the topic of the next chapter). As we shall see, the Mantua Map blends the world of content undergirding Rashi and Elijah Mizrachi's maps on the borders of Eretz Yisrael with graphical elements from the maps that appeared in mid sixteenth-century Christian Bibles. Against this backdrop, it has the distinction of being the earliest original artistic, drawn, and printed Hebrew cartographic work. While scholars have touched upon this map,[1] their short discussions do not do justice to its rich content, unique design, and singular text. Consequently, the objective of this chapter is to take stock of this map, track down its sources, and ponder the circumstances behind its creation.[2]

Survey of the Map

The Mantua Map was printed from a woodcut onto a rectangular paper folio measuring 377 x 509 mm. Although it is a printed work, there is only one extant copy (at the Central Library of Zurich).[3] Above all, the map depicts the Pentateuchal expanse according to the traditional Jewish sources and from a religio-historical perspective. However, it also contains a few hints of the reality in the Holy Land during the 1500s. The frame on the right side of the page consists of a long text and a drawing of the *menorah* (candelabrum). The text's opening sentence—"These are the journeys of the Israelites, which went forth out of the Land of Egypt" (Numbers 33:1)— was rendered in large letters and also serves as the map's title. Oriented to the east, the map features a sizable caption indicating the direction on each side. With respect to the major areas, Egypt and Sinai are on the right of the page; Transjordan takes up the upper portion; Eretz Yisrael and its borders (as per Numbers 34) are located in the middle; the northern border of the Land is on the left side; and the Med-

---

[1] Hans Jaacob Haag, "Hebräische Karte des Heiligen Landes um 1560," pp. 66–71, 173–174; idem, "Die vermutlich älteste bekannte hebräische Holzschnittkarte des Heiligen Landes (um 1560)"; idem, "'Elle mas'e vene Yisra'el asher yatz'u me-eretz Mitzrayim:' eine hebräische Karte des Heiligen Landes aus dem 16. Jahrhundert." Also see Tishby (ed.), *Holy Land in Maps*.

[2] I previously addressed this topic in Rubin, "A Sixteenth-Century Hebrew Map from Mantua."

[3] I would like to thank the Zentralbibliothek Zürich and its management for their permission to publish this map.

iterranean Sea lines the bottom portion. The map is replete with Biblical sites (the majority of which merit captions) and illustrations of major scriptural and Midrashic events. Distortions mar the relative location of and distances between places and/or events, to the point where the map is utterly devoid of geographical precision.

Fig. 40: The Mantua Map, the Central Library of Zurich. 377 x 509 mm

The Mediterranean Sea is situated on the lower part of the map, with the Egyptian coastline bulging into the right side. Next to a small caption reading "the Red Sea" is a patch of reeds (Hebrew Yam-Suf means Sea of Reeds). The waters to the left of Egypt are identified as "the Great Sea from the Land of Israel[...]" This section of the map is adorned with a few boats and an island labeled "*nisin she'bayam.*" Venice is tucked into the bottom-left corner.

Egypt, the Red Sea, and the Sinai Desert take up the righthand portion of the map. A large city, which is identified as Egypt, is located on the lower part of the map. To its right is a much smaller settlement that bears the caption "The Land of Raamses the best of the land." Just outside of Raamses is an illustration of Pharaoh's army chasing the Israelites. Between the former and latter is a clump of amorphous balls, which evidently depicts the pillar of cloud trailing the Israelite camp. Southeast of the large city is an inhabited expanse bearing the caption "the Land of Goshen." Flanking Goshen are the Land of Egypt to the left and Raamses to the right. Further north runs the estuary of the Nile, which is identified as both "the Nile" and "the river of Egypt."

Above and to the right of this section are sites that pertain to the exodus from Egypt: Succoth, Etham, Pi Hahiroth, and Baal Zephon. The first is rendered as a tent encampment, Etham and Pi Hahiroth as cities, and Baal Zephon as a mounted sculpture of a man brandishing a spear. Opposite Baal Zephon is an illustration of Moses waiving his staff at the sea, with the Israelites behind him. The citation under this scene reads "To turn back and encamp" (Exodus 14:2). Looming over the Israelites is Migdol. Atop the tower roar the tempestuous waves of the parted Red Sea, which is denoted by the following caption: "Here [the Israelites] went through the sea on dry ground."

As opposed to the contemporaneous Christian maps that listed all the stations of the Israelites' journey (Numbers 33), the one at hand suffices with a mere handful of the sites. Moreover, it does not delineate the route to Canaan. The map does include several names for parts of the wilderness, such as Zin, the Peoples (Ezekiel 20:35), the Red Sea, and Etham. Among the wayfaring stations that turns up on this map are the tree-lined strip of Elim (Numbers 33:9); Marah beside a depiction of Moses near the water; Mount Sinai with the lawgiver on its peak and the Israelites by the foot of the mountain; Mount Seir and the attendant verse "and circled Mount Seir for many days" (Deuteronomy 2:1); Mount Hor; and a rendering of the battle at Rephidim (Exodus 17:8–16). To the right and left of Kadesh are two more settlements, whose names I failed to decipher. There is also a rendering of the serpent of brass surrounded by Israelites, some of whom are being attacked by snakes (Numbers 21:9).

Along the upper portion of the map is a depiction of Transjordan and the final stages of the Israelites' journey. To the left of the brass serpent is the following text: "As per the commentary of Rashi, they will have turned their faces to the north." In Numbers 34:3, Rashi indeed wrote "and from there they turned their faces to the north until they circuited the width of its entire eastern boundary" (statement beginning "From the Wilderness of Zin). Put differently, upon reaching southern Transjordan, the Israelites stopped heading east and turned north. This section of the map contains the Wilderness of Moab, the City of Moab, the "crevices in the boulders," and Zered Brook.[4] Moreover, the Israelites are shown approaching this tributary and swimming across it. The fording of Zered Brook marks the end of the peregrinations through the wild, as the tribes enter the inhabited regions of east Transjordan.[5] In the Second Temple period, the Zered was considered Transjordan's southern bor-

---

[4] The words "crevices in the boulders" appear in Isaiah 7:19 and Rashi's commentary on Job 30:6. Given the context of this particular map, it is unclear why it was included herein.
[5] Ben-Gad Hacohen, "The Southern Boundary of the Land of Israel in Tannaitic Literature and the Bible" [Hebrew].

der; and during the Roman and Byzantine eras, it served as the border between Provincia Arabia and Provincia Palaestina (and subsequently Palaestina Tertia).[6]

Beneath Zered Brook is a brick-lined well and the citation "O well, sing you unto it" (Numbers 21:17)—a site and verse that Rashi indeed expounds upon. Spread below is the Land of Moab. Its territory features a medium-sized city and four smaller towns; none of the five settlements are labeled. Between the adjacent Arnon River and the Jabbok River to the left are two more cities. The caption next to the northernmost town reads "Heshbon" and "Sihon;" the other city is identified as "Og," "Bashan," and "Gilezer[7] who called the land of the Rephaites."[8] On the southwestern border of Moab is an illustration of Moses gazing out at the Promised Land from atop Mount Nebo. The adjacent caption reads "the plain of Jericho," which is actually located to the west of the Jordan (not the east). Left of the Jabbok is the Land of the Ammonites, which consists of a large, eponymous city and a bevy of smaller habitations.

The following text winds through the famed river dividing Ammon from the Promised Land: "The Jordan that pulls and comes at a diagonal from the north side." This description is based on Rashi's commentary and Mizrachi's map. At the confluence with the Jabbok, the Jordan spills into the relatively diminutive Sea of the Galilee and continues toward the Dead Sea. As in the Mediterranean, boats ply both of these seas. The northern and northeastern borders of Eretz Yisrael adhere to Numbers 34. More specifically, the Ain, the Riblah, Shepham, and Hazar Enan comprise the eastern side, while Ziphron, Zedad, Hamath, and Mount Hor dot the northern border. All these sites are portrayed as fortified cities—a standard graphical template of sixteenth-century European cartographers. These places are tethered to one another by a quasi-fence of trees or bushes. Next to the border towns is the Biblical passage "[A]nd from Mount Hor to the entrance of Hamath. Then the boundary will go to Zedad. And the border shall go on to Ziphron, and the goings out of it shall be at Hazar Enan" (Numbers 34:8–9). Along the eastern border is another citation: "The boundary will go down [from Shepham] to Riblah" (ibid: 11). In addition, there are another two captions along the northern and northeastern border: "Hamath that is Antioch"—an identification that comes up in Rashi as well; and "To this point, the allotment of the scion of Gad and the scion of Reuben." While these allotments are located to the west of the Jordan on this map, according to the Biblical narrative they should be to the river's east. North of the Sea of Galilee is a simple box that is labeled "the place where they erected the altar." This clearly refers to the twelve-stone altar

---

[6] Tsafrir, "The Provinces of Eretz Yisrael—Names, Borders and Administrative Jurisdictions," pp. 361–362 [Hebrew].
[7] "Gilezer" should perhaps be read Gilead.
[8] These words apparently refer to Deuteronomy 2, 20, where the expression "land of the Rephaites" turns up in the context of this region, which is subsequently called Havoth Jair.

that Joshua built upon crossing the Jordan River (Joshua 4). However, the Bible places this event near Gilgal—a site in the vicinity of Jericho.[9]

The depiction of the Promised Land's southern border is similar from a design standpoint to that of the northern border. Here too, the settlements—[the Ascent of] Akrabbim, Zina (to Zin),[10] Kadesh Barnea, Hazar Addar, and Azmon—are depicted as fortified cities linked by a floral fence. Running along the southern border is the following passage: "And your border shall turn from the Negev to the Ascent of Akrabbim, and pass on to Zin: and the going forth thereof shall be from the Negev to Kadesh Barnea, and shall go on to Hazar Addar, and pass on to Azmon. And the border shall turn from Azmon unto the River of Egypt" (Numbers 34:4–5). At Kadesh Barnea, another sylvan row branches off from the southern border and partitions Edom from Egypt. Nestled between these lines is the realm of the Edomites.

Only a few settlements are listed within the boundaries of the Promised Land. The southeast corner of the Land is dominated by Jericho. The ancient city takes the form of a round maze comprised of seven concentric walls. This classical representation of Jericho was indeed quite prevalent in Jewish art,[11] as beautiful examples of this template grace the following manuscripts: a Hebrew Bible from 1294, which is part of the New York Public Library's Spencer Collection;[12] and the so-called Farhi Bible, which was crafted in Spain or Provence during the mid-fourteenth century by Elisha ben Abraham Crescas.[13] This template appears to be a synthesis between the traditional European labyrinth figure and the desire to portray Jericho as a city that was circuited seven times (Joshua 6:1–20). It also bears noting that this depiction of Jericho surfaces in Christian manuscripts from as far back as the late twelfth century. Therefore, it stands to reason that this element penetrated the Jewish art world from Christian sources.[14]

---

9 For a disquisition on the traditions connecting the Israelites' entrance into the Promised Land to the Sea of Galilee, of all places, see Reiner, "From Joshua to Jesus—the Transformation of a Biblical Story to a Local Myth (a Chapter in the Religious Life of the Galilean Jew)" [Hebrew].
10 While in Hebrew *Zina* means "to Zin," the Mantua Map seemingly refers to this word as a settlement in its own right.
11 Ya'ari, "Miscellaneous Bibliographical Notes: 37: The Drawing of the Seven Walls of Jericho in Hebrew Manuscripts" [Hebrew]; Sarfati, "The Illustrations of Yiḥus ha-Avot: Folk Art from the Holy Land." Also see Stein Kokin, "Entering the Labyrinth: On the Hebraic and Kabbalistic Universe of Egidio da Viterbo."
12 The Spencer Collection, New York Public Library, Hebrew MS 1, pt. II fol. 1.
13 The Farhi Bible was indeed in the possession of one Haim Farhi until his death in 1818. Thereafter, the book changed hands a number of times before being purchased in the early twentieth century by the Sassoon family. For this reason, it is "officially" known as MS Sassoon 368.
14 An example of this sort of Christian-authored depiction of Jericho graces the following manuscript: Honorii (Augustodunensis) de imagine mundi libri III [u.a.], Bavarian State Library, Clm 14731, fol. 83r. I am indebted to Dr. Penina Arad for bringing this important source to my attention.

To the left of Jericho are three cities. One is identified as the Land of Benjamin, even though it is erroneously located south of Jerusalem; and the second lacks a caption. Although part of the name came out poorly from the press, the third city is in all likelihood "the Allotment of Judah." At the center of Eretz Yisrael stands Jerusalem. It is portrayed as a walled city with a hexagonal or octagonal building at its heart. This graphical element, which of course signifies the Dome of the Rock, was commonplace in the period's Italian Jewish art. For instance, it turns up on the logos of Jewish printers, in Haggadahs, and on *ketubot* (marriage contracts).[15] The source of this template is apparently contemporaneous Italian-Christian art works.[16]

Fig. 41: Jerusalem and the Temple in the Mantua Map, and the Venice Haggadah.[a)a)]*Seder haggadah shel pesakh [Passover Seder Haggadah] [...]*, Venice: in the House of Yo'ani Ḳalioni the Printer, 1629.

Mount Zion, which is rendered as a towering mountain crowned by a spired compound, abuts the upper-left corner of Jerusalem. The above-noted distinction between Jerusalem and Mount Zion also warrants our attention, for it entails both a physical and a conceptual separation. Once an integral part of Jerusalem, Mount Zion has been outside the city's truncated wall from as early as the Crusader period. Furthermore, the Mantua Map gives voice to a conceptual distinction that was popular among Jews. A case in point is Ishtori Haparchi's fourteenth-century observation that "Jerusalem stands alone and Zion stands alone." This passage hints to the fact

---

15 Ya'ari, "Miscellaneous Bibliographical Notes 37: The Drawing of the Seven Walls of Jericho in Hebrew Manuscripts."
16 Sabar, "Messianic Aspirations and Renaissance Urban Ideals: The Image of Jerusalem in the Venice Haggadah, 1609"; Pamela Berger, *The Crescent on the Temple,* pp. 214–223.

that at the time, Mont Zion and Jerusalem proper were considered distinct entities.[17] Against this background, we can infer that this conceptualization was a by-product of the physical separation between the two sites.[18]

On the other side of the city are three burial sites: the Tomb of Zechariah, the Tomb of Samuel, and the Tomb of Eli. The first two are indeed established pilgrim destinations. In reality, Zechariah's Tomb is to be found in the Kidron Valley, while the traditional site identified with Samuel's final resting place is roughly five miles northwest of the Old City. Unlike the first two sites, the grave of Eli does not come up in any other cartographic or historical source in the vicinity of Jerusalem. According to various pilgrim accounts, this tomb was located in Shiloh, approximately twenty miles north of Jerusalem.[19]

Dominating the coastal plain below Jerusalem is a fairly large city bearing the caption "the Land of the Philistines." Between this expanse and the sacred city is an illustration of the twelve spies. Eight are transporting the cluster of grapes on a sophisticated apparatus and two others wield satchels containing, as per the Bible, a fig and pomegranate. I already noted that this description of the spies' mission is commensurate with Rashi's interpretation and that the apparatus is based on supercommentaries of the revered savant's work,[20] but the artistic rendering herein is more developed than the other maps. To the left of the ten spies, their two colleagues —Joshua son of Nun and Caleb son of Jephunneh—are walking towards a town and a juxtaposed cave. These sites are identified as Hebron and "Caleb [sic] cave," respectively. This scene alludes to the midrash, which Rashi indeed cites, that Caleb visited the Cave of the Patriarchs in order to pray for the resolve to avoid slandering the Land of Israel.[21]

In deciphering this map, it is incumbent upon us to point to the striking absence of two topics. To begin with, the makers chose to omit the nine and a half tribal allotments (west of the Jordan), even though they constitute the main topic of the Christian Holy Land maps, beginning with the six-century Madaba Map, on through the handful of known maps from the early Middle Ages,[22] and ending with the map of Marino Sanudo.[23] The latter was the primary source for such maps in the early

---

17 Ishtori Haparchi, *Kaftor va'ferech* [Knob and Blossom], Luntz Edition [1976], Jerusalem pp. 559–560. These words also serve as the title of the following article: Reiner, "'Since Jerusalem and Zion Stand Separately:' The Jewish Quarter of Jerusalem in the Post-Crusade Period" [Hebrew].
18 This distinction is noteworthy because "Zion" is often synonymous with Jerusalem.
19 Ilan, *Tombs of the Righteous in the Land of Israel*, p. 364 [Hebrew]; Ish-Shalom, *Holy Tombs: A Study of Traditions concerning Jewish Holy Tombs in Palestine*, p. 100 [Hebrew].
20 See the discussion in chapter 2.
21 Rashi's Commentary to Numbers 13:22; Ginzberg, *The Legends of the Jews*, vol. III, pp. 270–271.
22 O'Loughlin, "Map as Text: A Mid Ninth Century Map for the Book of Joshua;" Edson, *Mapping Time and Space: How Medieval Mapmakers Viewed Their World*; Levy-Rubin, "From Eusebius to the Crusader Maps: The Origin of the Holy Land Maps, Bianca Kühnel."
  Harvey, *Medieval Maps of the Holy Land*, pp. 94–127.
23 Sanudo, *Liber Secretorum Fidelium Crucis*.

 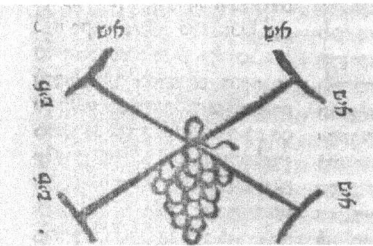

Fig. 42: Illustration of the spies in the Mantua Map and in Natan Shapira, Venice, 1593.

days of the print revolution, like those of Bendecitus Arias Montanus and Christian van Adrichom.²⁴ Secondly, the Mantua Map's authors completely ignored the geographic reality of the Land during their own time. For example, there is no reference whatsoever to the important coastal towns of Tyre, Acre, and Jaffe, which had penetrated the Italian consciousness following the Crusader Period. Likewise, there is no mention of Safad or Tiberias—towns within the ken of Italian Jewry during the 1500s. In light of the above, it is evident that the absence of these topics was not for lack of knowledge. The map's producers merely sought to depict the Israelites' peregrinations from the exodus until their arrival in Eretz Yisrael as laid out in the Pentateuch, especially *Mas'ei* (Numbers 33–34). For this reason, the allotments and contemporary toponyms were outside the map's purview.

There are ostensibly three conspicuous geographical errors in the Mantua Map: the location of Joshua's altar next to the Sea of Galilee, rather than in the vicinity of Jericho; the Land of Benjamin is situated south of Judah, in contravention of the Bible's literal interpretation; and the map leaves out the allotment of half the tribe of Manasseh beyond the Jordan, despite signifying its two Israelite neighbors. While I lack an explanation for the other two, the first "mistake" perhaps stems from a tradition that connects the end of the Israelites' journey with the Sea of Galilee and its environs. Most notably, a couple of sources in the Midrash and Talmud discuss the Midrashic tradition of "the well of Miriam." According to these sources, the well was miraculously hewn for the purpose of supplying the tribes with water in the wilderness and accompanied them from station to station over the course of their peregrinations. When the Israelites entered the Land, the well was deposited into the heart of the Sea of Galilee.²⁵ According to this tradition, then, the wanderings indeed come to a close in the northern part of Eretz Yisrael.

---

24 See, for example, the map "Situs Terrae Promissionis" in van Adrichom, *Theatrum Terrae Sanctae*; Bendecitus Arias Montanus, "Tabula Terrae Canaan Abrahmae tempore et ante adventum filior. Israel cum vicinis et finitimis regionib," *Biblia sacra Hebraice, Chaldaice, Graece, & Latine*, 8: *Liber Chanaan*. For a discussion on this book, see Shalev, "Sacred Geography, Antiquarianism and Visual Erudition: Benito Arias Montano and the Maps of the Antwerp Polyglot Bible."
25 *Numbers Rabbah* 1:2.

88 — 3 Between Jewish and Christian Maps

The Accompanying Text of the Mantua Map

As noted, there is a lengthy Hebrew text on the righthand side of the map. To follow is a full translation thereof:[26]

1. Here are the stages in the journey of the Israelites when they went forth out of the Land
2. of Egypt.[27]
3. When Israel went forth from Egypt,[28] which is in the southwest between the Red Sea
4. and the Nile, coming from the Great Sea along the western border and extending to the south side.
5. So God led the people roundabout by way of the wilderness,[29] hugging the Red Sea on its shore;
6. and they continued onwards: and they journeyed to Ra[a]mses; and to Succoth and Etham, whereupon they returned to the side
7. of Egypt, and they camped between Migdol and the sea; and behold [the army of] Egypt is going
8. after them,[30] and they went into the sea heading northwards.
9. Not to us O Lord, not to us[31] but in the name of the helmsmen HTRR [his honor our teacher the rabbi, Rabbi] Jacob Krup[?].
10. Give respect and gratitude, for he undertook to expound [by means of] this map[32] expounding on,
11. in his wisdom, the breadth [of the] Land[, and he] was the first to overcome all the hardships
12. that befell us along the way.[33]
13. The Lord has been mindful of us and will bless[34] him and his home and all that he possesses;[35] and
14. the cases of each of us, the undersigned, shall be brought before this man of God,[36] as
15. he is the first and foremost [candidate to handle] any sacred matter. This is of the essence,
16. while we are ancillary and praise the essential and dismiss

---

[26] I am indebted to Prof. Aharon Maman for his help deciphering this text.
[27] Numbers 33:1.
[28] Psalms 114:1.
[29] Exodus 13:18.
[30] Paraphrase of Exodus 14:1–22.
[31] Psalms 115:1.
[32] Paraphrase of Deuteronomy 1:5.
[33] Paraphrase of Numbers 20:14.
[34] Psalms 115:12.
[35] Genesis 12:20; Joshua 7:15; 2 Samuel 6:12; 1 Chronicles 13:14.
[36] Paraphrase of Exodus 22:8.

17. the inconsequential.[37]
18. I love the Lord, for he hath heard my voice[38] and he presided over the matter of all the stations of the journeys as depicted in
19. the drawing of our lord, the helmsman, for he turned his ear to deliberate over [the route and concluded]
20. that they [i.e., the Israelites] crossed the sea once and did not backtrack[39] as indicated in this drawing,
21. which leaves neither trace nor echo of the Mizrachi [map], whereupon they reached Edom and Moa[b],
22. for they did not repeat themselves twice and thrice; and his [i.e., Mizrachi's] reputation is beyond reproach because he did not come
23. to draw the peregrinations, only the borders.
24. In distress he [i.e., R. Krup] called for help[40] and an offering [i.e., donation] was contributed by one Master Aharon ben R.
25. Haim ha-Levi Shalit to bring them in the tradition of the covenant of the press.
26. And as for myself, I am his slave—a fox beyond his prime;[41] he called for the help
27. of the explanation and print setting, which are the last in a succession of gleaners.[42]
28. This is one of those places where the meager bolstered

---

37 This is a reference to "blessings before enjoyment." These sort of blessings are recited over the primary item, and there is no halakhic need to bless over ancillary items. For example, if one were to partake of stuffed chicken, it is only obligatory to recite a blessing over the chicken—not the stuffing. See *Shulkhan arukh*, Orekh Hayim, p. 212 [Hebrew].
38 Psalms 116:1.
39 In the Babylonian Talmud, Arakhin 15b, there is an *aggadah* on the Israelites' vacillations over whether to go into the Red Sea. According to one interpretation of this *aggadah*, the Israelites did not traverse the sea in one fell swoop; they entered the sea, backtracked, and got out, before circumnavigating the entire body of water on land. This *aggadah* stands at the heart of the debate over whether the Israelites crossed the sea once or twice. For more on this tradition and the cartographic expressions thereof, see chapter 4.
40 Paraphrase of Psalms 118:5.
41 This is an adaptation of an Aramaic expression, *ta'alah be'idanyah sagid laih*, in the Babylonian Talmud, Megillah 16b, which may be translated thus: "a fox in his prime—kowtow to him." In other words, it is advisable to pay deference to a lower-ranking or disreputable figure when he is in the catbird's seat. In the Talmud, this aphorism is used to explain why Jacob bowed down to his son Joseph (Genesis 47:31); and Rashi cites the expression in this same context. The saying was adapted in the text at hand to read *ta'alah be'loh idanyah*, namely a fox beyond his prime and thus, by all counts, unworthy of respect. By means of this inversion, the Mantua Map's artist and printer efface themselves before their esteemed collaborators—the exegete and philanthropist.
42 This expression comes up in the halakhic discussion on gleaning from a reaped field in the Babylonian Talmud, Bava Meṣia 21b. According to Jewish law, any stalks that are left in the field after the harvest are for the poor to glean. In referring to themselves as "the last in a succession of gleaners," the two artisans are, once again, expressing humility.

29. the abundant. Let the wise man listen and edify himself.⁴³
30. Blessed be he who comes and acquires in the name of God. He will open his bountiful
31. storehouse⁴⁴ and give money for expenditure[s] to those who merit a blessing⁴⁵
32. from God. The Lord is God, and he has cast his light upon us⁴⁶ until by the city gate will be known⁴⁷
33. all our deeds. In His mercy, He will reiterate that⁴⁸
34. all of Israel are friends,⁴⁹ soon in our days. Amen. May this be His will.
35. [???] A fox beyond his prime [is] Joseph son of the honorable Master Jacob
36. of blessed memory from Padua.
37. Furthermore, this drawing was made by Isaac son of R. Samuel
38. Bassan may his Rock [i.e., God] protect him and preserve the sexton of the synagogue of [...] the honorable Master
39. Port Katz, may his rock protect and preserve him.⁵⁰

My analysis of this text has yielded several insights. To begin with, its literary structure is based on the *Hallel* (Laudation) prayer.⁵¹ Every stanza begins with the opening line of a psalm from the *Hallel*, thereby drawing a correlation between this general prayer and our text's praise of the Mantua Map's creators. This link is hardly random, for the first line of the *Hallel* also constitutes the second sentence in our text: "When Israel went forth from Egypt." Needless to say, this is a reference to the exodus from Egypt and the Israelites' peregrinations, which are indeed the topics of the map.

The text lists four people who were involved in the production of the map: an exegete by the name of R. Jacob Krup[?], who I did not manage to identify; a philanthropist—Aharon ben R. Haim Halevi Shalit, who funded the entire project; an artist—Isaac ben Samuel Bassan; and a professional printer—Joseph ben Jacob of Padua.

---

43 Proverbs 1:5.
44 This sentence is an amalgam between Psalms 118:26 and Deuteronomy 28:12.
45 Paraphrase of Deuteronomy 28:8.
46 Paraphrase of Psalms 118:27.
47 Paraphrase of Proverbs 31:23. The "city gate" is where all the local elders and providers assembled.
48 The end of this line derives from the *Keter kedusha* (Sanctification of [the Divine] Crown) prayer, which is part of the Musaf service.
49 This phrase is taken from *birkat ha'hodesh* (the blessing of the new month).
50 The cover page of the Mantua Haggadah discloses Bassan's full title: "Isaac ben R. Samuel Bassan may his rock protect and preserve him, the sexton of the synagogue of the officer and donor, our honorable teacher, the rabbi Isaac Port Katz, may his rock protect and preserve him." This passage sheds light on the difficult lines toward the end of the text. See *Seder haggadah shel pesakh [Arrangement of the Passover Haggadah] with Illustrated Miracles and Wonders* [...] by Isaac ben R. Samuel Bassan, Mantua, Rufinalli [Press], 1560.
51 An assemblage of select Psalms, the *Hallel* is recited on Jewish holidays.

Fig. 43: Text of the Mantua Map.

Fig. 44: Cover page of the Mantua Haggadah and the biographical details of Bassan, *Seder haggadah shel pesaskh*.

The latter two are well documented in the literature on Italian Jewry. Bassan's drawings indeed grace other works that were printed in Mantua. Most noteworthy are his illustrations for a Haggadah that came out in 1560.[52] Joseph ben Jacob is also known by the pseudonyms *Shalit* and *Ashkenazi*. In addition, he may have been a relative of the above-mentioned patron. A known figure in the history of the Italian Hebrew press, Joseph ben Jacob was involved in printing and publishing throughout his lifetime, first in Mantua, then Venice, Sabbioneta, Riva di Trento, and back in Mantua.[53] For example, he printed *Igeret ba'alei ha'hayim* (Proverbs of Living Creatures) in 1557 and *Mishlei shu'alim* (Proverbs of Foxes) the following year. In 1562, he printed a spe-

---

[52] For the Mantua Haggadah see note 50 above. See also Simonshon, *History of the Jews in the Duchy of Mantua*, p. 655, n. 263.
[53] Amram, *The Makers of the Hebrew Books in Italy*, pp. 290, 323–324; Bloch, *Hebrew Printing in Riva di Trento*, p. 8; Busi, *Libri Ebraici a Mantova*, pp. 38–39, 54, 117, 132.

cial edition of the Book of Psalms, which included a Yiddish commentary.[54] Joseph also brought a Passover Haggadah to print six years later.[55] The two artisans' self-deprecating inversion of the Talmudic expression "a fox in his prime—kowtow to him" (see note 42) attests to their relationship with the donor and exegete. The identification of Isaac Bassan and Joseph ben Jacob allows us to date the map to the mid-sixteenth century, around 1560, the time when they worked in Mantua.

To a certain extent, this text enables us to reconstruct the circumstances behind the map's production. The exegete, R. Jacob, probably saw Elijah Mizrachi's map and reached the conclusion that it entails difficulties which must be resolved. While declaring that Mizrachi's "reputation is beyond reproach," the text criticizes him for only drawing the "borders," to the exclusion of the Israelites' journey. All the more so, Mizrachi is reproved for asserting that the Israelites "crossed the sea once and did not backtrack." In its makers' estimation, their work offers a more accurate description of the forty-year journey through the wilderness and the crossing of the Red Sea than their distinguished predecessor. The need to address these issues was indeed the primary catalyst behind this cartographic enterprise. However, as we shall soon see, this was not the only possible reason.

The Menorah

Under the lengthy text is an image that was quite common in the era's Jewish works of art: a model of the *menorah* (candelabrum) that graced the Tabernacle and Jewish Temple.[56] Atop the seven-arm *menorah* is a verse from Psalms: "For with Thee is the fountain of life; in Thy light shall we see light" (36:10).[57] Above three of the words are dots, which clearly indicate that this verse constitutes a code. Apparently, the marked letters encoded the name of the donor, Aharon ben R. Haim ha-Levi Shalit. According to this explanation the name Aharon is encoded in the word "shall we see" (נראה), the letters *beth* and *resh* should be read obviously as "son of" and the word "life" (חיים) stands for his father's name.[58]

---

54 *Psalms [... in] the Ashkenazic Tongue [...]* as per an expert and with an abridged commentary [by] a great, well-versed and pure rabbi, Mantua, [printed by] the partners Meir Sofer and Joseph ben Jacob of Padua, the year 1562 [Hebrew].
55 Heller, *The Sixteenth Century Hebrew Book: An Abridged Thesaurus*, vol. 1, pp. 442–443, 450–451; vol. 2, pp. 596–597.
56 For a disquisition on the candelabrum as a Jewish symbol, see Yisraeli, *To the Light of the Menorah* [Hebrew].
57 This verse was also featured atop the cover page of a book that was printed in Mantua at around the same time: Samuel Tzartza, *Sefer mekor ḥayim: An Explanation on the Torah*, Mantua 1559.
58 In the Hebrew edition of this book I interpreted these letters as a chronogram (see ch. Four, pp. ...), but this interpretation was problematic. I am indebted to Rabbi Menachem Silber for his suggestion to read this as a name code rather than a chronogram.

Adorned with knobs, blossoms, and a goblet, the candelabrum's central shaft is mounted atop a pedestal. Further up the shaft, three branches loop out to either side. The upper portion is lavishly festooned, and each arm is crowned by a flame. A caption at the base of the illustration reads "The height of the candelabrum is 18 handbreadths" (Tefah, a unit of measure). Above the pedestal is the following information: "The feet and the blossom [are] 3 handbreadths [long]." The next three captions read "And [this] part [is] two handbreadths;" "Goblet, knob, and blossom a handbreadth;" and an iteration of "And [this] part two handbreadths." Adjacent to both of the outer branches are the two parts of another descriptive passage: "And [the letters] *gimel teth* of the knobs and the six branches come out from it." Running parallel to the frame's left margin is the measurement "And 3 handbreadths until the top." Between the lower division points are the letters *teth* and *ḥet*, which in all likelihood stand for *tepakh* (handbreadth) and *ḥelek* (part). Alternatively, the upper bifurcation point reads "*beth teth*" and "*ḥet*." All told, the vertical measurements add up to "18 handbreadths."

These captions clearly deviate from the Biblical account of the *menorah*, for the relevant passages of Numbers 25 do not even contain the word "handbreadth" (Tefah). On the other hand, the captions are anchored to Rashi's commentary on the candelabrum, some via accurate quotations and others with light paraphrasing. A similar illustration, which is accompanied by similar captions and references to Rashi, can be found in Natan ben Shimshon Shapira's book[59] (see chapter 2).

Textual Sources

Like other works on Biblical topics, the Mantua Map refers to various strata of the Jewish exegetical literature. The primary source is undoubtedly the Pentateuch, especially the Books of Exodus and Numbers. That said, it is impossible to separate the Biblical text from Rashi's commentary, which was an exceedingly popular companion for Jewish Torah study. Accordingly, there are numerous references to Rashi in the map. For instance, the savant is expressly mentioned before the quotation "They will have turned their faces to the north." The cited verse "O well, sing you unto it" is merely a passing thought in Numbers 21:17, but merits a lengthy discussion in Rashi. The same can be said for the captions pertaining to the *menorah* and other topics that are beyond the limited purview of this book.

Another explicit source of this map is Elijah Mizrachi's commentary. In the first place, the authors of the Mantua Map take issue with him in the accompanying text. Two other allusions to Mizrachi are the captions "the Jordan that comes at a diagonal" and "The Wilderness of the Peoples"—a pair of topics that come up in his map. It also bears noting that they surface in the Rashi commentary, albeit in textual form.

---

59 Shapira, *Bai'urim al ha'eshel ha'gadol*, p. 150.

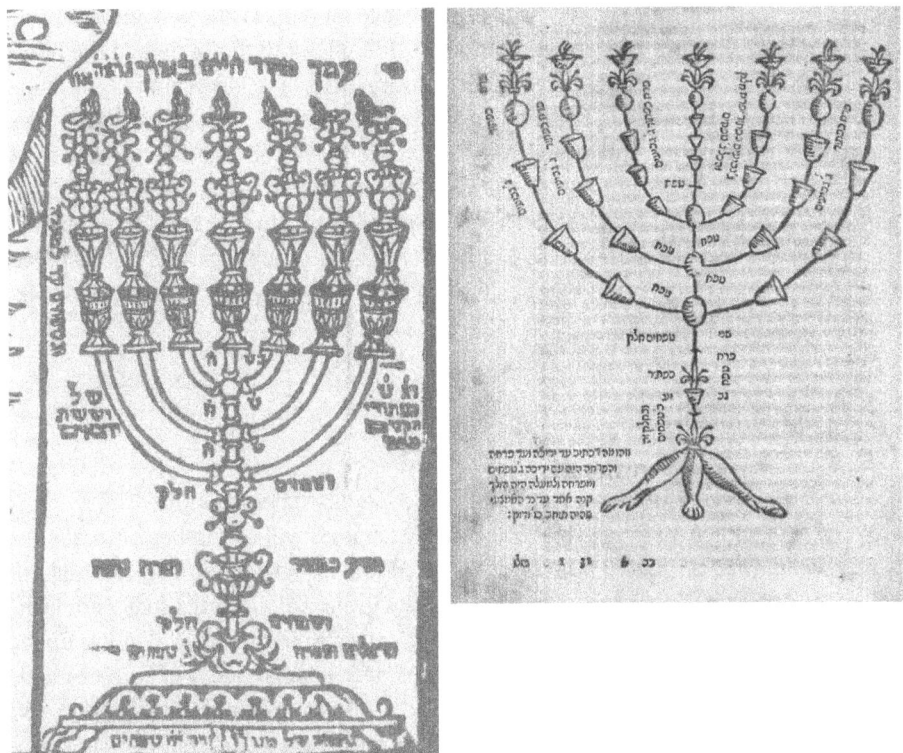

Fig. 45: A comparison between the candelabrums in the Mantua Map and Shapira's supercommentary.

In several places, the map refers to the Midrashic-cum-Talmudic tradition. Examples include the inversion of the aphorism "a fox in his prime—kowtow to him;" the reference to the aggadah (legend) of Caleb praying at the Cave of the Patriarchs; and the allusion to tractate Berakhot in the phrase "dismiss the inconsequential." Although scholars would be hard-pressed to determine whether the map's compilers were referring to the Midrashic sources themselves or merely quoting from Rashi, we may assume that they were familiar with these traditions (at least in their Talmudic raiment).

The Mantua Map also drops a couple of graphical hints as to the era in which it was created. A case in point is the aforementioned rendering of Venice in the bottom-left corner. The inclusion of the Serene Republic bears witness to the mapmakers' Italian identity as well as the city's status at the time as the port of departure for pilgrims to the Land of Israel; the capital of Hebrew printing; and perhaps the unrivaled hub of cartography and map production. Other contemporaneous topics are the tombs of Samuel, Zechariah, and Eli. The first site is obviously synonymous with *nebi samu'il*, a popular sixteenth-century pilgrimage destination approximately five miles northwest of Jerusalem. As in our day, the Tomb of Zechariah was a well-known site in the Kidron Valley. More surprising, though, is the reference to Eli's grave and its placement near the holy city. According to both the Bible and the influential pilgrim accounts from the late Middle Ages, this site is located amid the ruins of Shiloh—not in the vicinity of Jerusalem.[60] The very depiction of these venues ties the map to late medieval pilgrimage customs, which revolved around "the tombs of the righteous." However, unlike the period's accounts of contemporaneous pilgrimage routes, the Mantua Map does not expand on this topic; for instance, the tombs in the Galilee are completely ignored. Lastly, it stands to reason that the above-noted distinction between Zion and Jerusalem reflects the accepted worldview of sixteenth-century Italian Jewry.

Visual Sources

The Mantua Map is saturated with graphic descriptions that were influenced by multiple layers of source material. Over the next few pages, I will expand on several strata that I identified with certainty. However, a couple of layers and many details are bound to have escaped my eye. The first includes graphical elements that were well-established in sixteenth-century European art, cartography included. Falling under this category are the following templates: ships, tent camps, and walled cities with spired towers.[61] Given the prevalence of these themes in the period's European art, Bassan (the map's draftsman) undoubtedly borrowed from works that he was familiar with.

The second layer consists of sites that had long-standing design templates. Drawing Jerusalem in the shape of an ellipse and, all the more so, incorporating an octagonal building in the middle of the city were widespread in Italian art during the period following the Crusades. Needless to say, the building in question is the Dome of the Rock, which was meant to symbolize the Jewish Temple. This template of Jerusalem was a mainstay of Jewish art, especially its Italian wing.[62] For example, this pattern graced the headings of Italian *ketubot* (Jewish wedding contracts), where it gave graphic expression to the verse "If I do not place Jerusalem at the top of my joyous occasion" (Psalm 137:6).[63] Images of Jerusalem and of the Dome of the Rock were also commonplace on the banners of Jewish printers in Italy even before the time of the

---

[60] Ilan, *Tombs of the Righteous*, pp. 243–244, 361–362, 364; Ish-Shalom, *Holy Tombs*, pp. 63–65, 100, 113–116.
[61] Elliot, *The City in Maps—Urban Mapping to 1900*.
[62] See Sabar, "Messianic Aspirations."
[63] There are numerous examples of this theme in the *ketubbot* collection on the National Library of Israel's website, last accessed September 2012: http://jnul.huji.ac.il/dl/ketubbot/.

Mantua Map.⁶⁴ As above-noted, the depiction of Jericho as a labyrinth turns up in earlier Christian and Jewish manuscripts as well as the first Jewish prints.

Themes like troops wielding long spears in, say, the Israelites' flight from Pharaoh or the battle against Amalek at Rephidim were frequently displayed in maps that were incorporated into the earliest print editions of the Holy Scriptures. Among the many examples are the maps of Froschauer and Liesvelt,⁶⁵ the map in the 1535 edition of the Myles Coverdale Bible,⁶⁶ and Della Gatta's map.⁶⁷ At times, the affiliation between a map and its sources is only discernible from the selection of topics. On other occasions, it is also apparent from the graphical format that is used to present a certain theme.

In my estimation, the Mantua Map's illustrations of Moses on Mount Sinai and the battle against Amalek share much in common with those of Nicolas Barbier and Thomas Courteau in the 1559 edition of the Geneva Bible, which was printed a few years before the map in question. Their map subsequently came out in many others editions and in different languages, though some of the versions were modified.⁶⁸ At any rate, the 1559 Geneva Bible ultimately became the most widespread and influential Protestant version of the Holy Scriptures. Therefore, it comes as little surprise that the producers of the Mantua Map deemed Barbier and Courteau's map of the exodus from Egypt to be a challenge and an object of competition. Needless to say, the Jewish cartographers left out the scene of the golden calf and the letters of the Tetragrammaton from their representation of Mount Sinai. In any event, it appears that the Mantua Map was predicated on several artistic sources. Whereas some of the Christian-Italian artistic conventions that Bassan emulated had already penetrated Jewish art, others had yet to be adopted by his community. Against this backdrop, the Christian maps that began to appear in Bibles most likely inspired and were the motivation behind the design and printing of the Mantua Map, for its makers sought to provide a fitting Jewish response to their era's Christian cartographic depictions of the exodus and the borders of the Promised Land.

---

64 Ya'ari, "Miscellaneous Bibliographical Notes 25: Illustrations of Jerusalem and the Temple as Decoration for Hebrew Books," p. 25 [Hebrew].
65 For a disquisition on the maps of Christophe Froschauer (Zurich 1525) and Jacob van Liesvelt (Antwerp 1526), see Delano-Smith, "Maps in Bibles in the Sixteenth Century," esp. p. 4.
66 The map was printed *sub rosa* for a translation of the Bible that was smuggled into England in 1535. For more on this topic, see Ingram, "A Map of the Holy Land in the Coverdale Bible: A Map of Holbein?"
67 Della Gatta, *Palaestinæ sive Terre Sancte descriptio*.
68 Delano-Smith and Ingram, *Maps in Bibles, 1500–1600: An Illustrated Catalogue*; Berry (ed.), *The Geneva Bible, A Facsimile of the 1560 Edition Bible*.

Fig. 46: The map of *The Geneva Bible*, Richard Harrison edition (London: 1562), 154 x 210 mm.

Fig. 47: Details from *The Geneva Bible*.

The Milieu of this Enterprise

The formulation and printing of this intricate map can be attributed to the convergence of numerous factors in the cultural and public sphere of sixteenth-century northern Italy. Only a complex encounter between these elements (many of which will be enumerated below, but others that have certainly eluded my grasp) could have yielded such a compelling result. While some of these factors are endemic to Italian Jewry and its way of life, others are more general in character and scope.

To begin with, the cities of sixteenth-century Italy, particularly in the north, hosted important centers of Jewish education and scholarship. The development of Jewish communities where long-standing Italian families lived side-by-side with immigrants from Germany and Spain prompted a cultural flourishing. The Jewish printing press evolved hand-in-hand with these communities. On occasion, the venue for this phenomenon was printing houses under Jewish ownership, but in many cases the establishments belonged to Christians who focused on Hebrew books. To this end, these presses employed learned Jews, such as the aforementioned Joseph ben Jacob, to proofread and print texts.[69]

Secondly, the Jews of Italy enjoyed relative freedom during the period under review, and many came into contact with Christian scholars of the Renaissance. Besides their interest in Jewish scriptures, erudite Italian Jews occupied themselves with, among other pursuits, the sciences, medicine, and music. Figures like Abraham Farissol,[70] Azariah de' Rossi,[71] David Provinciali and his son Abraham (both of whom strived to establish a Hebrew university in Mantua[72]) are but a few of the Italian Jews who embraced the sciences during the Renaissance Period. These Jewish intellectuals had no compunctions against borrowing ideas and material from the general culture and integrating them into their own works.[73] Some of them were specifically involved in the study of geography and were thus influenced by the geographical expeditions and discoveries of their time. Abraham Farissol, who was among the first Jews to take an interest in this field, wrote about different foreign lands in his book *Igeret orkhot olam* (Epistle on Lifestyles of the World). De' Rossi explored the writings of Ptolemaios and Mercator. Joseph ben Joshua Hacohen's *Maṣiv gvulot olam* (Who Setteth the Borders of the World) is essentially a translation of a Latin work by Johann Boemus.[74] At the end of *Maṣiv*, the Jewish scholar added a chapter on "Other Islands in the Mediterranean Sea." In *Sefer Indiya ve'sefer meksico* (The

---

69 Amram, *The Makers of the Hebrew Books in Italy*, pp. 290, 323–324; Bloch, *Venetian Printers of Hebrew Books*.
70 Ruderman, *The World of a Renaissance Jew*.
71 Weinberg, *Azariah de' Rossi's Observations on the Syriac New Testament*.
72 Shulvass, *Jewish Life in Renaissance Italy*, pp. 239–240 [Hebrew].
73 Bonfil, *As By a Mirror*, pp. 119–136 [Hebrew].
74 Weinberg, "Yoseph ben Yehoshua Cohen and His Book Maziv Gevulot Amim" [Hebrew].

Book of India and the Book of Mexico), he edited an abridged translation of a work by López de Gómara on the New World.[75]

The third factor was the feverish messianic agitation that convulsed Italian Jewry in the first half of the 1500s, due to the arrival of David Reuveni. This development triggered, *inter alia*, a general interest in expeditions to exotic places in search of the ten lost tribes of Israel and strengthened the bond between Italy's Jews and Eretz Yisrael. For instance, there is evidence of Italian Jews sending contributions to, visiting, and even settling down in the Jewish community of Ottoman Palestine.[76] It is also worth noting that one of the Hebrew books on the Land of Israel's tombs of the righteous was first printed in Mantua during this same period.[77] In light of the above, Italian Jewry was characterized by an amalgam of internal factors that tied its members to general education, geography, cartography, printing, and the Holy Land—connections that served as critical infrastructure for the production of the Mantua Map.

Alongside these internal developments were external factors that also had an appreciable impact on the map's makers. Foremost among them was, perhaps, the special position of Venice as a prosperous city with a global reach. What is more, the Serene Republic boasted comprehensive systems for geographic education, garnering cartographic knowledge, and printing maps. Cosgrove has examined the factors that turned Venice into the world center of cartography.[78] Likewise, Woodward expounds on the city's role as the leading producer of and market for maps.[79] Over the course of the 1500s, Woodward asserts, more maps were printed in Venice than any other place in Europe. Only the plague of 1575 would end its run as the heart and soul of international cartography. By their very nature, printing presses served as prime venues for the exchange of knowledge between intellectuals from various backgrounds and regions. Consequently, we can infer that Joseph ben Jacob of Padua, who worked for some time in a printing press in Venice itself, was well aware of the magnitude of the city's cartographic enterprise. In addition, he was exposed to general cartographic knowledge and, most importantly, to the fundamental idea that the manufacture of a printed map combines geographic information with artistic design.

All these developments laid the groundwork for and encouraged the creation of a full-fledged artistic Hebrew map. That said, the main impetus behind converting this potential into action was, in all likelihood the debut of maps in Christian Bibles. The resemblance between some of the topics and elements in these cartographic

---

75 Shulvass, *Jewish Life*, pp. 295–300; Mintz-Manor, *The Discourse on the New World in Early Modern Jewish Culture*, pp. 110–160 [Hebrew].
76 Shulvass, *Rome and Jerusalem*, pp. 54–88 [Hebrew].
77 *Sefer yiḥus ha'ṣadikim ha'nikbarim be'eretz yisrael u'be'yerushalayim [The Book of the Pedigree of the Righteous Buried in Eretz Yisrael and in Jerusalem] [...]*, brought to print by Gershom ben (Moses) Asher, Mantua: Press of Jacob ben Naphtali Hacohen of Gazolo, 1561 [Hebrew].
78 Cosgrove, "Mapping New Worlds: Culture and Cartography in Sixteenth Century Venice."
79 Woodward, *Maps as Prints in the Italian Renaissance: Makers, Distributors and Consumers.*

works to those in the Mantua Map, especially with respect to artistic motifs, raises the possibility that the map's makers also endeavored to contend with the graphic dimension of Christian cartography. To this end, they created a Hebrew map with unequivocal Jewish content whose beauty and quality were on par with the maps that they encountered in the Christian Bibles. Be that as it may, the Mantua Map was neither a replica nor an adaptation of an existing Christian map. As opposed to later Jewish cartographers, the producers of the map in question forged an original work featuring a cornucopia of patently Jewish content.

Conclusion

The Mantua Map of the Israelites' peregrinations through the wilderness survived as a lone copy devoid of known affiliations with any specific work, so that its original purpose cannot be substantiated. In Yudlov's estimation, it was slated for the 1560 Mantua Haggadah, but was ultimately left out of this publication.[80] However, the map's dimensions suggest that it was not an integral part of any book. From its size and beauty, we may infer that it was meant to be hung on the wall of a synagogue, a *succah* (tabernacle), or an upper-class home. The content of the map's text and its *Hallel* structure bolster the synagogue or the Haggadah hypotheses.

This map was the first Hebrew cartographic work that made considerable use of the visual arts, while deviating from the Jewish tradition of square, schematic maps. As such, it is neither a refinement of a previous map nor does it belong to any established cartographic tradition. On the one hand, its makers were probably well-versed in Jewish Biblical exegesis. On the other hand, they kept close tabs on the world of Italian cartography, which was then at its apex, and pitted their creation against contemporaneous maps in Christian Bibles. Content-wise, the Mantua Map perpetuated the Rashi-Mizrachi tradition of concentrating on the portion of *Mas'ei*. At one and the same time, though, it expands the ambit of Jewish cartography southwards to the exodus from Egypt and the wanderings through the Sinai Desert—topics that stood at the heart of the Christian maps that were included in sixteenth-century Bibles.

In summation, the Mantua Map is neither a replica of an earlier map nor does it toe the line of a Christian or Jewish cartographic tradition. On the contrary, it is a novel undertaking from the standpoint of both its content and design. For this reason, it constitutes a unique and riveting chapter in the history of the Hebrew map. My comprehensive research notwithstanding, the following mysteries remain at large: Why is there only a single known copy of this impressive work? And why did this rich and intriguing map fail to merit replicas, adaptations, or even an echo in the Hebrew cartographic tradition?

---

80 Yudlov (ed.), *The Haggadah Thesaurus*, p. 3 [Hebrew].

4 Following in the Footsteps of Christian Cartographers

Unlike the schematic and unadorned maps that were surveyed in chapter 2, artistic Hebrew maps were drawn up and printed against the backdrop of extensive inter-cultural contacts between European Jews and Christians. The copying of maps, as well as the repeated use of copper-plates that were passed from one printer to the next, was quite prevalent among European cartographers. Given the abundance of non-Jewish traditions interwoven into the Christian-authored maps, their replication on the part of Jews was a completely different kettle of fish than the intra-faith variety, for content had to be translated into Hebrew and revised in accordance to the tastes and strictures of the Jewish community. For instance, distinctly Christian sites and traditions had to be erased from the Hebrew adaptations. Some of the Jewish copyists took great pride in the diligence with which they implemented these revisions, and a few even expressly noted these steps in the preamble to their map. In the pages to come, I will paint a portrait of these artistic Hebrew maps by analyzing a few representative samples and discussing the circumstances behind their creation.

Two of the earliest maps that exemplify this replication process were published in the beginning and end of the 1600s. The fact that both maps were produced in Amsterdam is tied to the city's character and the lifestyle of its Jewish residents during this period. To begin with, seventeenth-century Amsterdam was the center of a thriving Jewish community. Many of its members were prosperous and educated individuals who lived in relative freedom and maintained close economic and cultural ties with their Christian neighbors.[1] Secondly, Amsterdam was among the top centers, and arguably the worldwide capital, of the cartography industry, as maps and quite a few of the period's large atlases were designed and printed therein.[2] What is more, Jews were involved in the local printing trade, particularly the design of maps and atlases.[3]

Both of the works under review are predicated—be it directly or otherwise—on Adrichom's highly regarded map of the Holy Land, which served as the basis for numerous copies and adaptations by Christian cartographers throughout Western Europe for centuries to come. A native of Delft (a town in South Holland), Christian van Adrichom (1533–1585) was a Catholic priest based in Cologne. He wrote a thick volume on the Holy Land, *Theatrum Terrae Sanctae*,[4] which primarily consists of an annotated list of the toponyms that are mentioned in the Bible.

---

[1] Van den Berg and van der Wall, *Jewish-Christian Relations in the Seventeenth Century*; Kaplan, "*Gente Politica:* The Portuguese Jews of Amsterdam *vis-à-vis* Dutch Society."
[2] Harley and Woodward, *The History of Cartography, Cartography in the European Renaissance*, vol. III, part 2.
[3] De Boer, "Amsterdam as 'Locus' of Iberian Printing in the Seventeenth and Eighteenth Centuries."
[4] Adrichom, *Theatrum Terrae Sanctae*.

The sites are arranged according to the tribal allotments, from Asher to Zebulun (A to Z). Each entry is accompanied by a short description that leans on the Bible and other sources (from ancient to late sixteenth-century texts). Moreover, the book includes maps of the tribal allotments as well as a capacious (1010 x 354 mm) and extremely detailed map of the greater Land of Israel.[5] Adrichom also penned a similar work on the annals of Jerusalem, which was indeed accompanied by a comprehensive map of the city during "the days of Jesus."[6] It also bears noting that the works on Jerusalem were incorporated into several editions of *Theatrum Terrae Sanctae*.

As in the rest of Adrichom's Holy Land book, the breakdown into tribal allotments is a central feature of the map's spatial and cartographic presentation. Hundreds of toponyms from the Bible and other historical sources are interspersed throughout the map. Alongside many of these names is what amounts to the cynosure of this map—miniature drawings of related events. Next to each of these illustrations is a number that refers to a detailed entry in the book. In all likelihood, these cartographic elements stem from the map of Marino Sanudo and Pietro Vesconti. The latter, a Venetian cartographer, drafted the source map for Sanudo's *The Book of the Secrets of the Faithful of the Cross*—a book urging Christians to embark on a new crusade.[7] This map, which draws on even earlier works, also divides the Land into the tribal allotments. Other notable features of the Sanudo-Vesconti map include the following: the source of the Jordan (*Iordanis*) is the Ior and Dan Rivers; the three sizeable bodies of water that the Jordan River taps into—the Hula, the Sea of Galilee, and the Dead Sea; the Jordan juts sharply to the east between the Hula and the Sea of Galilee; the Dead Sea's elongated shape and pointy beak; and the fact that the Kishon River links the Sea of Galilee to the Bay of Haifa. These traits, *inter alia*, also surface in Adrichom's map, so that the connection between these two cartographic representations of the Holy Land is beyond doubt.

The Sanudo-Vesconti map was familiar to generations of readers. By virtue of several existing manuscripts of this map, it is evident that some of its characteristics inform other sixteenth-century maps, like those of Bendecitus Arias Montanus. In the centuries that followed, many European cartographers indeed drew heavily on the maps of Sanudo, Adrichom, Arias Montanus, and their disciples-cum-copyists.[8]

The Map of Yaacov ben Avraham Zaddik

A Jewish merchant and banker who belonged to Amsterdam's Portuguese-Jewish community, Yaacov ben Avraham Zaddik (also known as Justo) spearheaded the cre-

---

5 Adrichom, "Situs Terra Promissionis."
6 Adrichom, *Iervsalem, et suburbia eius, sicut tempore Christi floruit [...] descripta per Christianum Adrichom Delphum.*
7 Sanudo, *The Book of the Secrets of the Faithful of the Cross*.
8 Arias Montanus, *Biblia sacra Hebraice, Chaldaice, Graece, & Latine*, vol. 8.

Fig. 48: Christian van Adrichom, "Situs Terrae Promissionis," Theatrum Terrae Sanctae, 1010 × 354 mm.

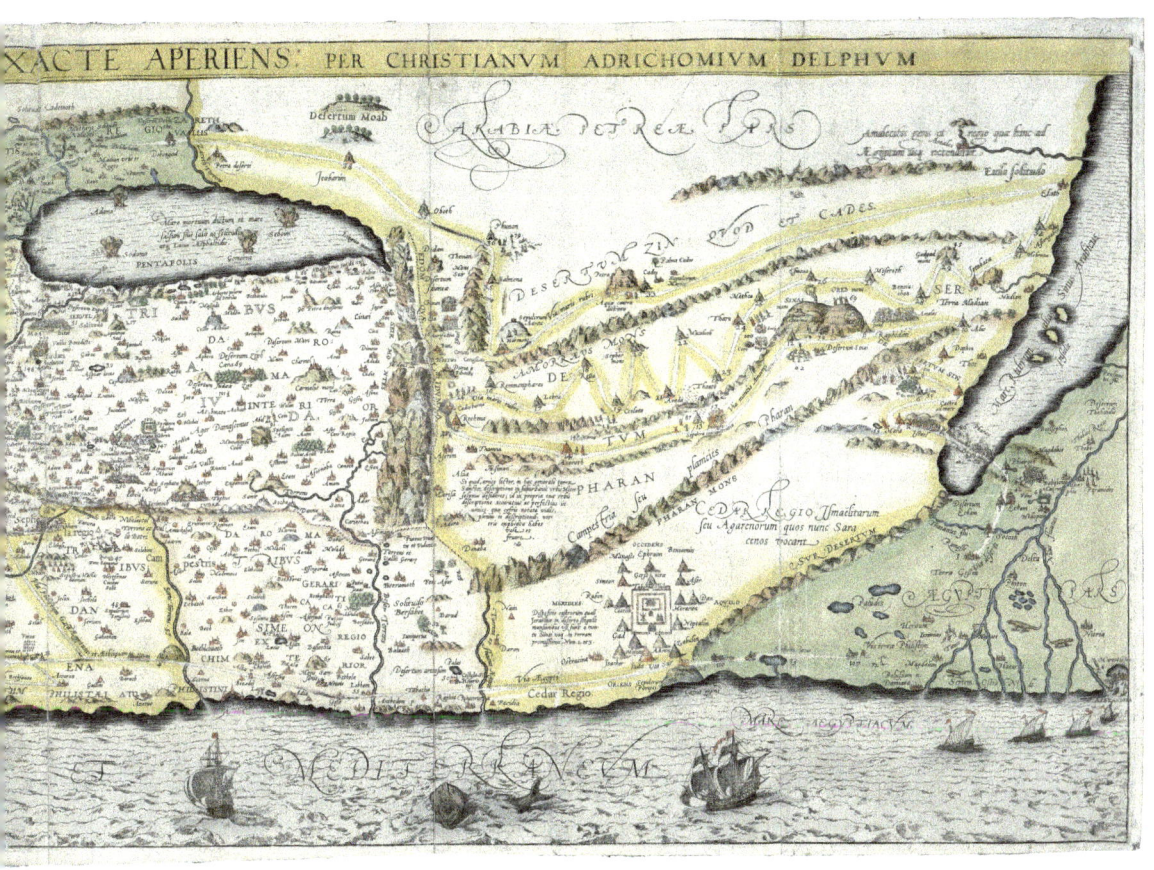

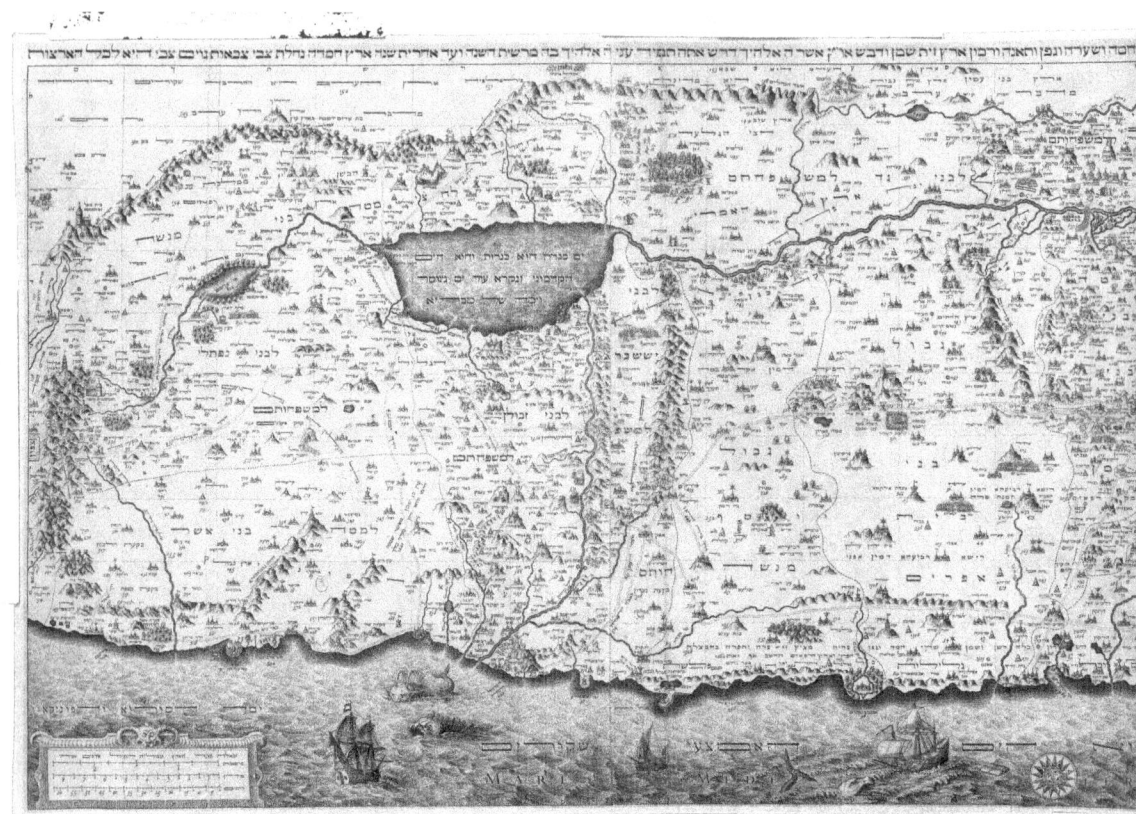

Fig. 49: Yaacov ben Avraham Zaddik, "A Drawing of the Situation of the Lands of Canaan," Bibliothèque Nationale Française, 1626 x 523 mm.

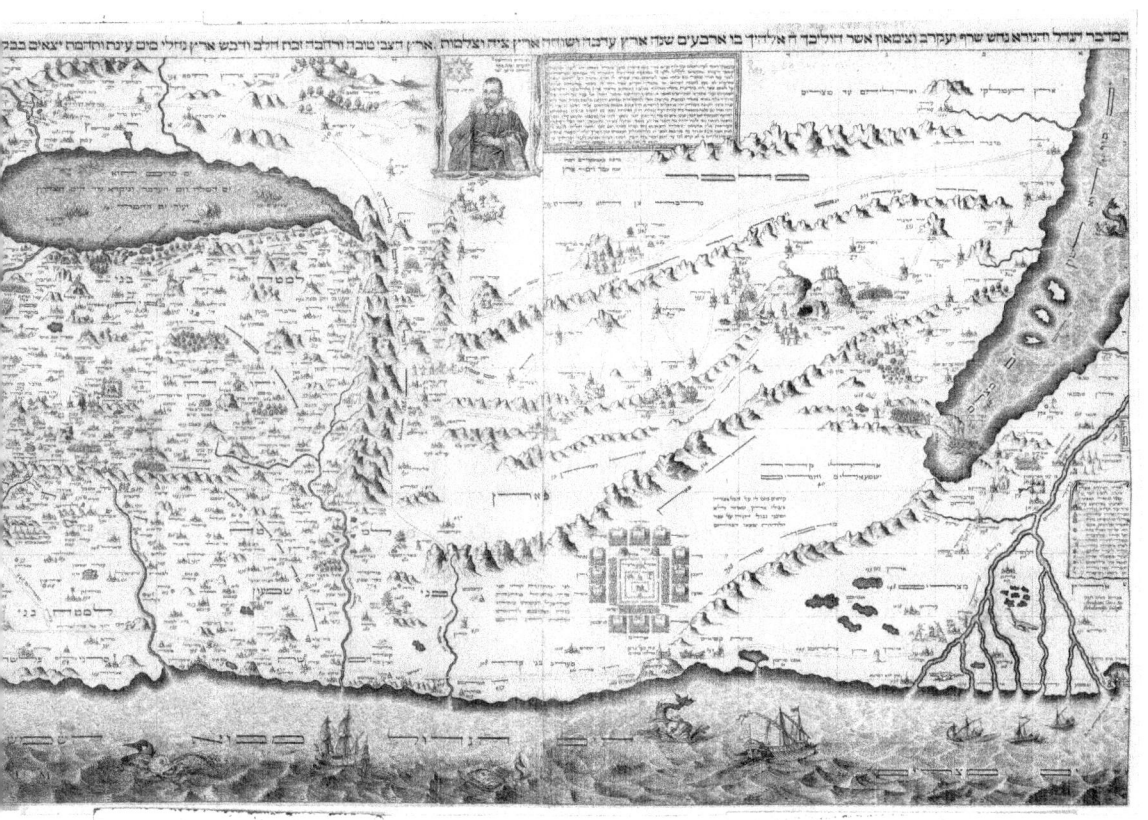

ation of an ambitious map of Eretz Yisrael. More specifically, he took Adrichom's map, copied it, and translated the text into Hebrew. With the help of Abraham Goos, a Dutch master engraver, he then fashioned a large and detailed map, which was printed in 1620–1621.[9] Two copies of this map are housed in the Bibliothèque nationale de France.[10] Although known to Hebrew bibliographers since the inception of their field, little has been written about this compelling document.[11]

Zaddik's elongated rectangular map is 1626 x 523 mm. Like Adrichom's version, the map faces east; and its borders are the Lebanon Mountains to the north, the Nile Delta and Red Sea to the south, the Mediterranean Sea to the west, and Transjordan to the East. At the top of the map, outside the frame, is a long and convoluted passage that paraphrases different verses from the Hebrew Bible. The opening words, which are situated above the Sinai Desert, read thus: "That great and terrible wilderness, wherein were fiery serpents, and scorpions, and drought [Deuteronomy 8:15] a land of deserts and of pits, in which God led you for forty years through a land of drought and of the shadow of death [Jeremiah 2:6]." The second part runs parallel to and is awash with praise for Eretz Yisrael: "The good land is fine and expansive—[one] of milk and honey[,] a land of brooks of water, of fountains and depths that spring out of valleys and hills[.] A land of wheat and barley and vines and fig trees and pomegranates[;] a land of oil[,] olives[,] and honey [Deuteronomy 8:7–8] A land which the LORD thy God careth for: the eyes of the LORD thy God are always upon it, from the beginning of the year even unto the end of the year [Deuteronomy 11:12] and [he has] given thee a pleasant land, a goodly heritage of the hosts of nations [Jeremiah 3:19] which is the glory of all lands [Ezekiel 20:6]."

On the upper right, is a "picture of the face of the copyist[,] Yaacov son of the honorable Rabbi Avraham Zaddik[,] may his Rock protect and preserve him." The said figure is arrayed in the courtly vestment of a seventeenth-century Spanish nobleman. Opposite his eyes are the words "To God is my labor." It seems as though Zaddik placed much significance on including this distinguished portrait of himself on the map.[12] Beneath the adjacent cartouche, the date of the map (1620) is embedded into a fitting verse from Habakkuk: "He stood and surveyed the earth" (3:6). Enclosed within the cartouche is a lengthy preamble offering a grandiose account of the map's production:

---

[9] While lacking a title, the map does have two signatories. Next to the portrait is the name Yaacov ben Avraham Zaddik, who is identified as "the copyist" (*ha'ma'atik*), namely the translator and editor. As opposed to the first signatory, the name and contribution of Avraham Goos—"the engraver" (*ḥakak*) who wrought the copper plate—is noted in both Hebrew and Latin.

[10] Bibliothèque nationale de France, copy 1: GE BB- 246 (17– 43/44 RES); copy 2: GE C- 4921 (RES).

[11] Nebenzahl, "Zaddiq's Canaan;" idem, *Maps of the Holy Land*, pp. 110–113; Garel, "La premier Carte de Terre Sainte en Hebreu (Amsterdam, 1620/21)." See the detailed references therein about the map in early Hebrew bibliographies.

[12] Cohen, *Jewish Icons: Art and Society in Modern Europe*, pp. 31–32.

Fig. 50: The portrait and cartouche in Zaddik's map.

I found [and] saw that the assembly of the nations have and among the sons of the gentiles there is a drawing of the situation in the lands of Canaan[,] which God has bequeathed unto the Israelites. And upon ensuring that everything concurs with the words of the rabbis of blessed memory and Rashi of blessed memory in places within his interpretations, these truths dawned upon me[.] And out of love for my fellow Jews and brethren[,] my soul deeply yearned to share with them what was pleasant to me and whose fruit were sweet to my palate [,] as they [i.e., the map's insights] were of considerable use to me for understanding certain things from the Holy Scriptures[,] for the spirit of God percolated into their [i.e., the map's] writers and His word [was] on their tongue[.] In light of the above, with great diligence and steadfast persistence[,] I assembled them [i.e., the map's content] from the outer regions of the land and copied them into the holy tongue[.] I availed myself of our sacred books and the Book of Josippon and the like[.] And since they illuminated my face to guide me in this way[,] I have taken a level path[.] And with respect to the change of names . . .[,] I replaced them with names that refer to it [i.e., the Land of Israel] in a manner that seems to be more appropriate to me[.] And also some of what came up short in them [i.e., Adrichom's map], I remedied[.] In addition, I made a keys section in which I placed marks with numbers for denoting the places[.] And if someone were to doubt any part of this[,] here in this book or a portion thereof[,] everything will be reconciled for him as per his wishes and he will see how and why it is like this[.] And I hereby duly announce that I did everything with much reflection and all the extra time I exerted on this indispensable endeavor were done with utter devotion[.] And if by virtue of the divine grace I will be fortunate enough to go to the good Land[,] may it be built and instituted swiftly in our times[,] I will further expand on the book's publication both in its plates and in its essence[.] And what is this particular Land [for] the soul [–] may it be like paradise[.] And given the fact that until today we have yet to have a work on this matter[,] I wanted to draw my picture[.] May it warrant me a good legacy before and among my people. These are the words of Yaacov Zaddik[,] may his Rock protect and preserve him.

This preamble recounts the events that prompted the author to undertake this project: To begin with, Zaddik came across the map of a non-Jew, undoubtedly Christian van Adrichom. Upon realizing the insights it possesses on the Hebrew Scriptures,

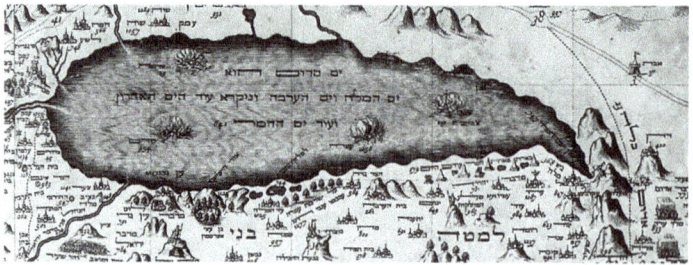

Fig. 51: The Dead Sea in Adrichom and Zaddik's maps.

he copied and translated the map. Furthermore, Zaddik proudly corrected whatever he deemed to contravene the interpretation of the Sages and Rashi, such as a handful of toponyms. The text also notes that the compiler availed himself of the Book of Josippon. This detail is rather intriguing because Adrichom evidently grounded his own map on the Latin version of Josephus' writings, which was accepted by Christian scholars. Conversely, the Jewish author preferred a version that was commensurate with his own community's sensibilities.[13] Furthermore, the text refers to a map key and book that Zaddik penned. The book may be referring to an adapted translation of *Theatrum Terra Sanctae*. However, there is no extant copy of this work, and we can assume that it was never completed. On the other hand, two copies of the key, which were printed in 1633, have survived. One is dedicated to King Christian IV of Denmark and the other to Frederick Henry, the prince of Orange.[14] These tributes bear witness to Zaddik's relations with non-Jewish society. Zaddik concludes by expressing his dream of travelling to Eretz Yisrael and learning about the Land on a firsthand basis. In sum, the most noteworthy parts of this text are the author's testimony regarding his embracement of a Christian map and the painstaking efforts to adapt it to the Jewish tradition.

As in the original, the Jewish version depicts Eretz Yisrael during the time of the Bible, without any references to the author's own era. From right to left, the stress is

---

[13] For more on the Book of Josippon and its significance, see Flusser, *The Josippon (Josephus Gorionides)*, vol. 1 and 2 [Hebrew].

[14] Garel, "La premier Carte de Terre Sainte en Hebreu (Amsterdam, 1620/21)," esp. the bibliographical references pertaining to the location of these works. One can assume that Zaddik's keys were also predicated on those in Adrichom's book.

on Egypt, the Sinai Desert, the route of the peregrinations through the wilderness, and the tribal allotments on both sides of the Jordan. Toponyms are peppered throughout the map. Alongside many of these sites is a number, which obviously directs the reader to the key. Most of the places that Zaddik incorporated in his map appear in the source, but there are fewer miniatures in the adaptation. The Dutch Jew created his own set of legends, which appears on the right side of the map. Among the categories are "royal city," "city of refuge," "Levite town," and the like.

Zaddik's detailed map barely skews from its source. This affinity is most striking in the layout of the land, namely the mountain ranges, rivers, and other bodies of water. Furthermore, the majority of site locations are identical in both maps. There are a few instances, though, where Zaddik changed his predecessor's wording. For example, Adrichom included captions of the Dead Sea's appellations in the Biblical and classical sources, whereas Zaddik listed its accepted Hebrew names along with the "Sea of Sodom," "the Last Sea," and "the Sea of Asphalt"—a Hebrew translation of *Lacus Asphaltidis* in the source. Furthermore, the Jewish author omitted the caption "Pentapolis" (the five cities) from the sea.[15] A similar revision pertains to a cavernous mountain in Samaria. Adrichom dubbed it *Mons Prophetarum* (the mountain of the prophets), while Zaddik went with "fifty people, fifty in a cave"—an illusion to the fifty prophets that Obadiah hid in a cave during the reign of Ahab (1 Kings 18:4).

In several places, Zaddik deleted Christian content. For instance, next to Gergesa (east of the Sea of Galilee), Adrichom drew a detailed miniature of Jesus exorcising demons from two people and casting the spirits into a herd of pigs (Matthew 8:28–33). Needless to say, the Jewish cartographer erased this scene, while leaving the hill and toponym. With respect to Mount Tabor, Zaddik naturally left the mountain, but omitted the scene of Jesus' transfiguration. Likewise, next to Nazareth, he replaced the caption "Saltus Domini," which refers to Mount Precipice (the site of Jesus' rejection, Luke 4:29–30), with a general description—"high mountain." To the north of the Sea of Galilee, "fertile mountain" comes instead of "Mons Christi." In Qarantal (near Jericho), where according to tradition Jesus fasted for forty days and bested the devil, he switched "Mons Quarantana" with plain "mountain." On several occasions, Zaddik apparently did not realize the source of Adrichom's toponyms. As a result, several unequivocally Christian sites that do not come up in the Hebrew Bible, such as Nazareth, Bethany, and Bethphage, were included in the Jewish version.[16]

There are a few instances where the Jewish mapmaker exhibits his independence. For example, he added the passage "the Sharon is bursting with fecundity and oil [as well as] delightful fertile fields[,] which are budding with flowers that will blossom like the Pancratium." The sea-faring vessels—a common theme in the

---

15 "Pentapolis" refers to the five kingdoms that formed an ill-fated alliance against Chedorlaomer (Genesis 14).
16 Bethany and Bethphage are villages in the vicinity of Jerusalem.

era's maps—are a secondary item in Adrichom and Zaddik's accounts, but the latter introduced a couple of symbolic changes. Of all the ships on his map, two have flags with unusual symbols. A boat sailing opposite the shores of the Nile Delta flies a flag bearing a crescent, which symbolizes Muslim rule in Egypt. Secondly, a flag decorated with the Star of David is hoisted atop a ship off the north Sinai coast. This Jewish element perhaps gives voice to the author's desire to visit Eretz Yisrael or his deep yearning for the reestablishment of the Judean kingdom.

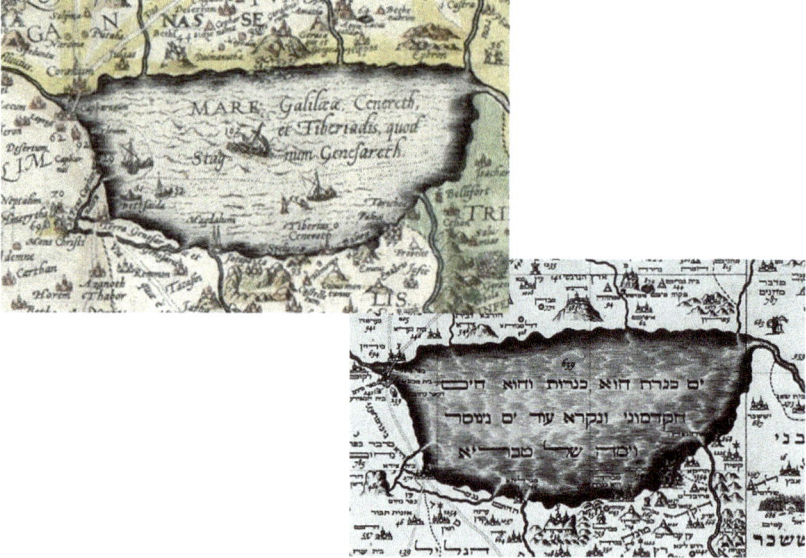

Fig. 52: The Sea of Galilee in Adrichom and Zaddik's maps.

Fig. 53: A ship with the Star of David on Zaddik's map.

The Map of Abraham bar Yaacov

The second Jewish-authored cartographic adaptation of a Christian map to be produced in Amsterdam was that of one Abraham bar Yaacob. A German-born Protestant priest, bar Yaacov converted to Judaism and became a master printer. It stands to reason that his knowledge of Latin and Christianity came in handy while preparing this map,[17] which came out in 1695 as part of the Amsterdam Haggadah.[18] This marked the first time that a map was included in a Haggadah. Although the Haggadah has been elaborated on in the literature, the map has garnered very little of this attention.[19] The relation between the two is pertinent to the subject at hand. By including the map therein, bar Yaacov visually magnified three topics that underpin the Haggadah's content: the exodus from Egypt; the Israelite's voyage through the wilderness; and the Promised Land.

The dimensions of bar Yaacov's map are 480 x 262 mm. Its preamble, which runs along the upper portion of the sheet, reads thus: "This is for the knowledge of every knowledgeable person concerning the route of the forty years of sojourning in the wilderness and the width and length of the Holy Land from the river of Egypt to the city of Damascus and from the Arnon River to the Great Sea and within it each and every tribe's part of the allotment[.] When your eyes behold [its] virtues and the wise shall understand them." The last sentence was printed in bigger letters and merges two Biblical passages: "May your eyes behold what is right" (Psalms 17:2); and "those who are wise will understand" (Daniel 12:10). The increased font size raised Brodsky's suspicion that a message was encrypted into these words, but refrained from offering an interpretation. More recently, David Stern has suggested that the meaning lies in the broader context of Daniel 12. While the entire chapter describes the End of the Days, the particular verse draws a distinction between the wicked who will not understand and the wise who will. In light of the above, Stern contends that the map's preamble alludes to the redemption at the End of the Days, whose origins go back to the exodus.[20]

A close look at this map demonstrates that bar Yaacov did not copy from Adrichom's original map, but availed himself of a later edition. The southern coastline spanning the border between Eretz Yisrael and Egypt indicates that the Jewish con-

---

[17] "Abraham bar Yaacov," *Encyclopaedia Judaica*.
[18] Abraham bar Yaacov, The map was integrated into idem, *Passover Haggadah: with a Beautiful Interpretation and Drawings [...] on Copper Plates* / by the lad Abraham bar Yaacov from the family of Abraham our Patriarch, Amsterdam: the printing press of Asher Anshel ben Eliezer and Issachar Ber ben Abraham Eliezer, 1695.
[19] Yudlov, *The Haggadah Thesaurus*, p. 11; Schubert, *Die Jüdische Buchkunst*, pp. 45–51 [Hebrew]; Nebenzahl, *Maps of the Holy Land*, pp. 138–139; Brodsky, "The Seventeenth-Century Haggada Map of Avraham Bar Yaacov."
[20] Brodsky, "The Hebrew Holy Land Map of Avraham bar Yaacov, Amsterdam, 1695;" Stern, "Mapping the Redemption: Messianic Cartography in the 1695 Amsterdam Haggadah."

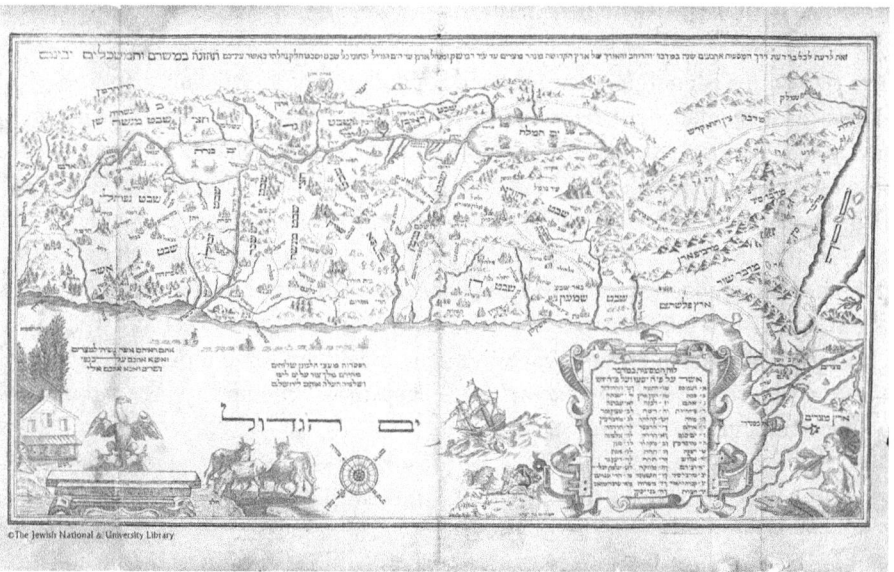

Fig. 54: Abraham bar Yaacov's map in the Amsterdam Haggadah, 480 x 262 mm.

vert's source was probably one of Hondius and Mercator' maps.²¹ Unlike Adrichom's representation, the shore does not continue in a straight line, but better approximates reality by curving west. In addition, the images of Moses and Aaron on the cover page of the Amsterdam Haggadah, which were also designed by bar Yaacov, resemble those on the bottom left hand corner of the Hondius-Mercator map. However, the figures of Aaron share a greater affinity than those of Moses. For example, only bar Yaacov's version of Moses has rays or horns emanating from the lawgiver's head. This element is consistent with Exodus 34:29: "that the skin of his [i.e., Moses'] face shone;" or according to the medieval Christian tradition, "that his face was horned" (the Vulgate uses a derivative of *keren*, the Hebrew word for both radiance and horn). In this sense, the Jewish portrayal of Moses paradoxically gives expression to a patently Christian interpretation of this verse.²² In fact, Samuel ben Meir (the RaShBaM), a prominent Jewish exegete and a grandson of Rashi, interpreted "horn in the sense of splendor, and imagining his horns to be that of the oryx's horns is nothing but drivel."²³ The disparities between bar Yaacov and Adrichom's maps also serve to disconnect the former work from that of Zaddik; put differently,

---

21 Given the similarities between all the editions, there is no way of determining which one bar Yaacov availed himself of. The earliest edition is Gerard Mercator and Hondius, Jodocus, "Situs Terrae Promissionis S.S. Bibliorum intelligentiam exacte aperiens," between pp. 649–651.
22 Mellinkoff, *The Horned Moses in Medieval Art and Thought*; idem, "More about Horned Moses."
23 Samuel ben Meir's commentary on Exodus 34:29, statement beginning "That horn" [Hebrew].

Fig. 55: The Hondius-Mercator map, 494 x 370 mm; Gerardus Mercator and Jodocus Hondius, "Situs Terrae Promissionis."

the latter was not the source of bar Yaacov's map. Be that as it may, there is nothing to suggest that the bar Yaacov was unfamiliar with his predecessor's work.

Like its source, bar Yaacov's map is oriented eastwards; and its outer boundaries are the Mediterranean Sea to the west, the Transjordan to the east, the Nile Delta and Sinai Desert to the south, and the Hermon and Lebanon Mountains to the north. Adrichom's basic geographical layout, such as the aforementioned "mistakes" concerning the Jordan River's route, the relative size of the Sea of Galilee and the Dead Sea, and the course of the Kishon, also inform the bar Yaacov and Hondius-Mercator maps; and the same could be said for the emphasis on the tribal allotments. In contrast, the number of sites and the density level of the cartographic elements are greater in the source map. Similarly, bar Yaacov provides only a handful of miniatures describing Biblical events. Among the illustrations that he does offer are the manna falling out of the sky, the twelve-stone altar to the east of Jericho, and a wagon approaching Egypt from the Land of Israel.

In the lower, most barren part of the map, bar Yaacov provided a series of drawings that was not included in the Hondius-Mercator map. At the bottom-right corner is a personification of Egypt—a practically naked woman riding a crocodile and holding a parasol. The source for this image is most likely a personification of the entire

Fig. 56: Figures of Moses and Aaron on the Amsterdam Haggadah's cover and in the Hondius-Mercator map.

African continent in Petrus Plancius' map of the world from 1594.[24] The resemblance between these two figures, especially the odd parasol, substantiates the link between the Plancius and bar Yaacov maps. This same figure was also a common personification of Africa in earlier sixteenth-century works. Alternatively, it was the accepted symbol of lust (*lussuria*) during the Italian Renaissance, most notably in Ripa's thick and highly influential tome.[25] Incidentally, the crocodile (*sans* rider) symbolized Egypt in Antiquity and the Middle Ages (see the ensuing discussion on the Ebstorf Map[26] and Roman coins, respectively). As early as the Book of Ezekiel, "O Pharaoh, king of Egypt" was described as a "great crocodile lurking in the streams of the Nile" (29:3).[27]

To the left of the Egyptian woman, a cartouche frames a key of the Israelites' forty-one stations in the wilderness as per Numbers 33. On the other side of the

---

24 Plancius, "Orbis terrarum typus de integro multis in locis emendates." Also see Shirley, *The Mapping of the World: Early Printed World Maps*, p. 207, pl. 187.

25 Ripa, *Iconologia*. The first edition of this work came out in Rome in 1593, *sans* illustrations. However, beginning in 1603, many illustrated editions were published in Italian, Dutch, and other European languages. For more on this book and its myriad editions, see, e. g., Maser, *Baroque and Rococo Pictorial Imagery: the 1758–60 Hertel Edition of Ripa's 'Iconologia'*, pp. viii-xi; Praz, *Studies in Seventeenth-Century Imagery*.

26 George, *Animals and Maps*, p. 30.

27 Pliny even describes an island in the Nile where the inhabitants rode on crocodiles and caught these beasts by jumping on their backs. Pliny the Elder, *Historia Naturalis*, pp. 66–69.

key is a pair of miniatures featuring Jonah the Prophet: one depicts the fugitive being tossed off the boat; the second portrays the fish ejecting him onto the coast. The first scene, which was quite prevalent in the era's Christian cartography, also turns up in the Hondius-Mercator map. A comparison between the two renderings of this ship reveals that bar Yaacov's version is indeed a mirror image of its source. It also bears noting that the Jewish cartographer signed his name beneath the image of Jonas sprawled on the shore.

Fig. 57: Hondius-Mercator as the source of bar Yaacov's map: Jonah's boat in the two versions.

On the left side of the Jewish-authored map are a handful of other noteworthy elements. The ornamental boats, which grace many of the period's maps, are towing cedar rafts, namely the vessels King Hiram of Tyre used to transport building supplies for Solomon's Temple. Precedents for this illustration are the Visscher family's mid seventeenth-century maps from Amsterdam and Frederick de Wit's Holy Land map, which was printed in 1680.[28]

Below the rafts are several cows and an adjacent label reading "milk." In front of the house on the bottom-left corner is an apiary with three hives. The cows and apiary embody the Biblical phrase "the land of milk and honey." In addition, the house is shaded by a few trees, which are labeled Sukkot (the Feast of the Tabernacles). Therefore, the trees are probably alluding to *skhakh* (the natural covering of a sukkah) or *aravot* (willow branches).[29] Between the cows and the house is a griffon vulture, which is capped by the following verse: "Ye have seen what I did unto the Egyp-

---

28 See, e.g., Visscher family, "Peregrinatie ofte Veertich-iarige Reyse, Der Kinderen Israels uyt Egipten [...] Door Claes Ianss Visscher;" idem, "Tabula Geographica in qua Israelitarum, ab Aegypto ad Kenahanaeam usqve profectiones en rystplaetsen der Kindren Israels;" de Wit, "Terra Sancta, sive Promissionis, olim Palestina recens delineata, et in lucem."
29 *Aravot* are one of the "four species" that make up an integral part of the Sukkot rite.

tians and *how* I bare you on the wings of griffon vultures and brought you unto myself" (Exodus 19:4). The majestic bird is hovering over an ornate pedestal. Lettering can be found on both ends of the pedestal's upper side-panel: the letter *aleph* to the right; and the letters final *kaf* and *heth* to the left. Moreover, the panel is etched in with alternating sequences of vertical and diagonal lining, which form a quasi-ruler consisting of twenty-eight units (the numerical equivalent of the three letters). On the next side-panel is a caption reading "Parasangs each of which is 2 notches." The top rung thus constitutes a linear scale with twenty-eight half-parasang units. The incorporation of this scale should come as no surprise, for it was an accepted component of early modern maps.

Fig. 58: The personification of Egypt in both the bar Yaacov and Plancius map; Petrus Plancius, "Orbis Terrarum Typus de Integro Multis in Locis Emendatus."

Bar Yaacov provides a unique, two-route account of the Israelites' crossing of the Red Sea. The first route traverses the sea from one side to the next. According to the second, the Israelites entered the Red Sea, but did not proceed to the other side; instead, they backtracked and circumnavigated the entire bay by land. As discussed below, the source of this description is an *aggadah* (non-halachic part of the Talmud). It also bears reiterating that this duplicity is alluded to in the text of the Mantua Map (see chapter 3). At any rate, there is little doubt that generations of Jews equivocated over this issue. Surprisingly, the two routes turn up even earlier, in a small map that graces a book that was penned by Quaresmius (an Italian Franciscan monk who served in Jerusalem for many years).[30]

Given the fact that bar Yaacov's map was part of the Amsterdam Haggadah, comparing its illustrations to those throughout the rest of the book promises to shed light on the map's sources. The Haggadah's drawings of Jerusalem and the Temple were taken from Matthaeus Merian's mid seventeenth-century book *Figures of the Bible*,

---

30 See the discussion on Quaresmius' map later on in this chapter.

which came out in four languages and enjoyed a wide circulation.³¹ During the copying process, the artisan who prepared the plates for the Haggadah forgot to erase a small cross atop the Jewish Temple. Since the Amsterdam Haggadah and its illustrations merited numerous replicas over the centuries, this symbol of Christ found its way into many Haggadahs.³²

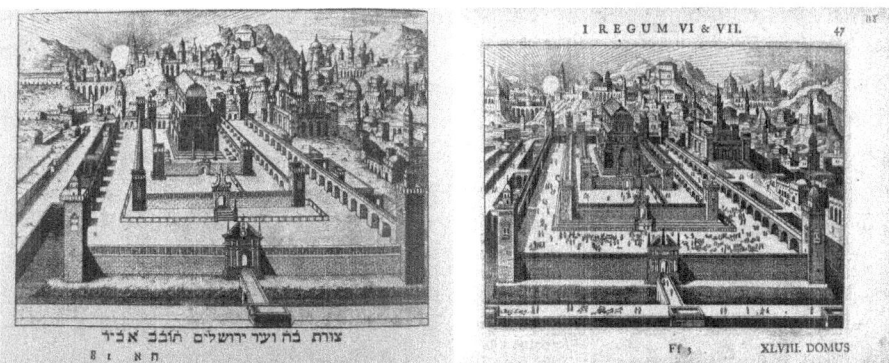

Fig. 59: The Temple in the Amsterdam Haggadah and its source: Matthaeus Merian, *Figures de la Bible*.

It bears noting that Adrichom's map was the source of both Hebrew maps from Amsterdam, but the level of adherence substantially differs. Whereas Zaddik barely deviated from the source, bar Yaacov introduced quite a few changes, such as elements from other maps. Moreover, the convert's map was based on a later adaptation of Adrichom's original version.

In the two Hebrew maps in question, as well as their later counterparts, there are detailed references to the exodus from Egypt and the Israelites' route through the wilderness. These topics are quite pronounced in the Christian Holy Land maps as well, for the Christian allegorical tradition of Biblical exegesis viewed the exodus to be a precursor of the salvation. Adrichom was not the first cartographer to accentuate the peregrinations. For instance, this theme looms large in the above-mentioned Protestant maps from the mid 1500s (see chapter 3). Additionally, the exodus and wanderings are depicted in the Hereford map of the world, which was drafted back in the

---

31 Merian, *Figures de la Bible, demonstrans les principales histoires de la Saincte Escriture, Biblische Figuren, Figures of the Bible, Icones Bibliacae*.
32 See, for example, the Haggadah that was published in Pas, Morocco in 1960; Yerushalmi, *Haggadah and History: A Panorama in Facsimile of Five Centuries of the Printed Haggadah from the Collections of Harvard University and the Jewish Theological Seminary of America*, pl. 192. The Amsterdam Haggadah continued to have quite an impact on world Jewry for hundreds of years; see Sabar, "From Amsterdam to Bombay, Baghdad, and Casablanca: The Influence of the Amsterdam Haggadah on the Haggadah Illustration among the Jews in India and the Lands of Islam," esp. fig. 19. The article includes a photo of a cross-topped Jewish Temple from the Casablanca Haggadah (1940).

thirteenth century.³³ Insofar as Amsterdam's Portuguese-Jewish community was concerned, one must bear in mind that a large percentage of its members had personally escaped the Inquisition.³⁴ Against this backdrop, it is only logical that the Amsterdam Haggadah and its attendant map featured the exodus and Eretz Yisrael as potent symbols of the redemption.

In sum, Zaddik and bar Yaacov's maps were the first direct Jewish adaptations of Christian cartographic sources. The fact that these maps were compiled in seventeenth-century Amsterdam is only fitting. To begin with, during the 1600s, Dutch Jewry blossomed into a strong, vibrant, and lettered community. Owing to the economic and educational freedom in Holland, there was also a lively dialogue between the country's Jews and Christians.³⁵ What is more, Amsterdam was an important center of cartography throughout this same period. The printing press of Jan Janssonius and his son Gerard exemplify the city's atmosphere. More specifically, the Jewish clientele and Hebrew projects of this Christian-owned establishment attest to the close interfaith relations and the intellectual discourse in this port town.³⁶ The nexus between cartography, the printing industry, and the openness of Amsterdam Jewry to general scholarship were among the factors that paved the way for these two maps.

What is more, there is reason to believe that another Hebrew map of Eretz Yisrael came out in Amsterdam during the years between the publication of Zaddik and bar Yaacov's works. The lone hint to the third map's existence is the testimony of its presumed author, Joseph Shalit ben Eliezer Riqueti. The preface of his book *Ḥokhmat ha'mishkan* (the Wisdom of the Tabernacle), which came out in 1676, includes the following passage: "I also took the trouble to create a form [i.e., a map] in the shape of the Land of Israel that has been in the making for what is now a couple of years in the holy community of Amsterdam[,] may God protect them[,] and I am indeed making it ornate and fetching with several drawings that rejoice in their variety."³⁷ That said, we lack an extant copy of this map nor is there any evidence that it was ever completed.

---

33 Harvey, *Mappa Mundi: The Hereford World Map*.
34 Mintz-Manor, "Symbols and Images in the Spanish-Portuguese Congregation of Amsterdam in the 17th and 18ᵗʰ Centuries" [Hebrew].
35 Kaplan, *From New Christians to New Jews*, pp. 32–37; van den Berg and van der Wall, *Jewish-Christian Relations in the Seventeenth Century, Studies and Documents*.
36 Friedberg, *History of Hebrew Typography in Italy, Spain-Portugal, Turkey and the Orient from its Beginning and Formation about the Year 1472*.
37 Riqueti, *Ḥokhmat ha'mishkan: panim ḥadashot ba'u le-khan be-ferush na'eh*, Mantovah: 1675/6.

Map from the Herlingen Haggadah

Held in the Braginsky Collection, Zurich, is the manuscript of a Haggadah by Aaron Wolff Shreiber Herlingen (ca. 1700 – 1760)[38]—a Jewish illustrator known for the unparalleled quality of his ornate works[39] and arguably the most prominent manuscript designer of his era. After launching his career in Pressburg (present-day Bratislava), Herlingen moved to Vienna where he embellished Hebrew manuscripts, including a few Haggadahs, for court Jews and other wealthy coreligionists. He also catered to Christians, especially in his capacity as an illustrator at Vienna's Hof-Bibliothek (Imperial Library).

The Haggadah under review was fashioned in Vienna. Like bar Yaacov's Amsterdam Haggadah, this one ends with a map that was drawn on a folded folio. Measuring 505 x 292 mm, it is a bit larger than that of bar Yaacov. The map depicts the exodus from Egypt, the route of the peregrinations, and Eretz Yisrael, which is divided into the tribal allotments on both sides of the Jordan.[40] The entire work, map included, is dated to the year *pdut* ("redemption")—gematria for 1730. Although similar Haggadahs lack cartographic content, it is evident that this map is an integral part of its framework. Herlingen's inclusion of the map calls to mind the Amsterdam Haggadah and thus raises the possibility that he emulated the latter and its attendant map. Other resemblances between the two maps include a list of the Israelites' forty-two stations in the wilderness and the opening words of each preamble: "This is for the knowledge of every knowledgeable person." These affinities notwithstanding, the two maps sharply diverge.

To begin with, Herlingen's map stands apart from that of his predecessor in many of its geographical details, such as the shape of the Dead Sea and the Sea of Galilee, the course of the Kishon River, the appearance of the Nile Delta and the Red Sea, and even the direction of the compass rose (Herlingen's faces north rather than east). Save for the opening words, the two texts are far from similar. The famed illustrator even criticizes bar Yaacov's content: "I saw in the template of the Holy Land in the Amsterdam print Haggadah[,] in the copper plate[,] and [there are] some mistake[s] I did not understand[,] and there is no one to solve [them for] me." Thereafter, the text describes how, upon arriving in Pressburg, Herlingen met "the infallible scholar R. Avraham Falklish of Prague." Falklish informed him that "I already made the drawing [i.e., a corrected map] earlier due to the entrea-

---

[38] The photograph herein comes courtesy of the collector, who kindly gave me permission to exhibit it. I would like to thank Dr. Aviad Stollman for bringing this map to my attention.
[39] Sabar, "Herlingen, Aaron Wolff (Schreiber) of Gewitsch;" idem, Seder Birkat Ha Mazon ..., in Shmuel Glick (ed.), *Zekhor Davar le-Avdekha: Essays and Studies in Memory of Dov Rappel* [Hebrew].
[40] See the catalog *Magnificent Judaica and Manuscripts: Including Property Formerly in the Furman Collection*, Tuesday, Dec. 12, 2000, New York: Sotheby's, lot 125, pp. 74 – 78.

Fig. 60: Map in the manuscript of the Herlingen Haggadah, 1730, 505 x 292 mm. Braginsky Collection, Zurich. Photography by Ardon Bar-Hama, Ra'anana, Israel.

ties of some students."[41] The mapmaker then contends that bar Yaacov erred with respect to the cities of refuge and the tribal allotments: "And the artist in Amsterdam showed that he does not understand the words of the sage[s] and their subtleties." In conclusion, the text notes that Herlingen meticulously drafted the map on the basis of his inquiries. As we can see, the objective behind this endeavor was not to emulate bar Yaacov's map, but to remedy its flaws. Moreover, the preamble suggests that Herlingen essentially adapted a map or diagram by one Avraham Falklish, which is probably no longer extant.

The map itself was crafted with a skillful hand. Unlike many of the period's maps, it faces north. Though abundant with bays, the Mediterranean coast darts from south to north in what is roughly a straight line. However, at Habesor Stream, it suddenly veers precipitously to the southwest in the direction of the Nile Delta. Running from north to south, the Jordan River is even more linear than the Mediterranean shore. As in a host of other maps, the river's source is the Ior and Dan. These two tributaries also feed "the Ain" (Numbers 34:11), namely the Hula Sea, whereupon the Jordan continues via the Sea of Galilee to the Dead Sea. The last two bodies of water are oval-shaped and laden with inlets. Like numerous other maps, the Cities of the Plain are arrayed within the Dead Sea. While also spilling into the Dead Sea, the Arnon River derives from a large lake to the East, which is identified as the Sea of Jaazer (a prevalent feature in seventeenth-century European cartography).

Scattered throughout Herlingen's map are numerous Biblical toponyms. Though critical of bar Yaacov, he too erred with respect to the location of many cities. For example, Ashkelon is situated to the north of Mount Carmel, Succoth is west of the Jordan River, and Gaza is north of Jaffa. Only a few settlements—Jerusalem, Hebron, Shechem (present-day Nablus), Shiloh, Jericho, and Sidon—are rendered as full-fledged cities, and pictograms grace several other sites. The names of the tribal allotments are highlighted with large font, and their borders are marked with double lines. As opposed to most known maps, the territory of Zebulun is located along the north Mediterranean coast. In so doing, Herlingen displays a preference for the account in Jacob's blessing: "Zebulun will dwell at the seashore. And he shall be a haven for ships. And his flank will extend to Sidon" (Genesis 39:13). Conversely, the Book of Joshua (19:10–16) situates Zebulun inland. The serpentine route of the Israelites' journey, which is also signified by a double line, circles back over its own tracks in a unique fashion. Along the length of this route are numbers denoting the stations that are enumerated in the Bible. On the far right of the map, Moses can be seen receiving the Torah atop Mount Sinai. An unconventional caption describes the Mediterranean as "the Great Sea that is the Spanish Sea." This toponym is apparently taken from ibn Ezra and the Abrabanel's commentaries on Exodus 10:19.

---

[41] I did not manage to locate any information on Avraham Falklish. That said, he is probably a relative of Eliezer Falklish (1745–1826)—a student of Yechezkel Landau (the author of *Noda be'yehudah*) and among the leaders of the anti-Frankist struggle in Prague. See "Falklish, Eliezer," *Encyclopaedia Hebraica* [Hebrew].

The map's style and some of its characteristics indeed tie it to European Christian maps, but I could not identify one particular main source from which it was copied. That said, a few of its aforementioned features can be traced back to certain maps. As per the European cartographic tradition, which dates back to the Sanudo-Vesconti map, the Kishon River flows from the Sea of Galilee and spills into the Mediterranean. While in most of the European maps, the source of the Kishon is the southern coast of the Sea of Galilee, Herlingen tethers it to the northern shore. In this respect, he adheres to the maps of Ptolemy (Vienna 1541), Ortelius and Tilemanno Stella, and the Visscher family.[42]

The shape of the Sea of Galilee and the Dead Sea closely resemble their appearance in Ortelius' map, which follows in the footsteps of Laicstain and Schrott[43] as well as Thomas Fuller. In the Fuller map, which was incorporated into his popular book on the Land of Israel,[44] a non-existent river also emanates from the Sea of Galilee's northern shore; and the Kishon begins on its southern coast. Getting back to the Herlingen map, we already noted that the Arnon River flows from the imaginary Sea of Jaazer. This route is presented in a large number of maps, foremost among them the Fuller map and that of Adrichom—the indirect source of bar Yaacov's map.

In conclusion, the evidence suggests that Herlingen's map was fashioned along the lines of the period's Christian European maps. What is more, he may have leaned on a cartographic work by Avraham Falklish (a Jewish contemporary of the illustrator). That said, Herlingen arranged his map in an independent manner and unique style that borrowed from several of his forerunners. Last but not least, Herlingen was motivated by the faults that he discovered, or at least believed so, in bar Yaacov's map.

Shlomo of Chelm and His Book Ḥug Ha'Aaretṣ

Born in Zamosc, a town in Poland, Shlomo ben Moshe of Chelm (1717–1781) was declared a prodigy at a tender age. In 1751, while serving in the rabbinate of Chelm (a Polish city), he published *Merkevet ha'mishneh* (Chariot of the Mishneh)—a commentary on Maimonides' *Mishneh torah* (Repetition of the Torah) that propelled Shlomo to fame and glory. The scholar subsequently filled rabbinic posts back in his hometown (1767–1771) and in Lemberg (1771–1777), whereupon he immigrated to Eretz Yisrael via Istanbul and Izmir. Following a brief and apparently tumultuous stay in Ti-

---

42 Ptolemaeus, *Geographia*, map no. 18; Ortelius, "Palestinae sive totivs Terrae Promissionis nova descriptio avctore Tilemanno Stella Sigenens;" idem, *Theatrum orbis terrarum*, p. 51; Visscher family, "Tabula Geographica in qua Israelitarum, ab Aegypto ad Kenahanaeam usqve profectiones en rystplaetsen der Kindren Israels" (from a Dutch Bible).
43 Ortelius, "Terra Sancta, A Petro Laicstain perlustrata, et ab eius ore et schedis à Christiano Schrot in tabulam redacta;" idem, *Theatrum orbis terrarum*, p. 97.
44 Fuller, *A Pisgah-Sight of Palestine and the Confines Thereof [...]*, facing p. 1.

berias, Shlomo left the Land of Israel—perhaps as a rabbinic emissary—and reached Salonica. It was in this Mediterranean port city that he succumbed to an epidemic in 1781 and was laid to rest.[45]

While stressing in a couple of works that his emphasis was on Torah study, Shlomo of Chelm also delved into secular topics. Among his fields of interest were algebra, geometry, astronomy, philosophy, grammar, and logic. The preface to *Merkevet ha'mishneh* states that "and I came to Elim [one of the Israelites' stations in the wilderness] where there are twelve springs of water and seventy palm trees; [there I delved] into the wisdom of whole numbers and fractions, and the wisdom of algebra and the wonders of geometry; it is tried and true, and a standard and a cornerstone, in the wisdom of astronomy, to run its course like a strongman, and I touched with my small finger the wisdom of nature and what is behind nature." Moreover, the preface calls upon others to embrace general studies: "And the Bible stands and screams [;] and hewn into it are the seven pillars [of the nations' wisdom]; and so why should this not depart from your [i.e., Jews'] mouths, for it is your wisdom and your intellect, in the eyes of the nations."[46] It bears noting that the study of "profane" disciplines was far from prevalent among eighteenth-century Jews. That said, Shlomo of Chelm was not the only Torah scholar that pursued such knowledge, as even some of his Jewish colleagues in Zamosc embarked on this path.[47] Against this backdrop, the author of *Merkevet ha'mishneh* is among the figures who have been classified in the literature as "the forerunners of the Jewish enlightenment" or members of the "early Haskalah" movement.[48]

At any rate, Shlomo was a prolific writer. One of his least known works is *Ḥug ha'areṣ* (Caliper of the Land), the majority of which has been preserved in a manuscript that survived the Holocaust and was published in Jerusalem in 1988.[49] The manuscript itself, which was bound together with another work called *Shulḥan aṣai shitin* (Table of Acacia Trees), was recently acquired by the National Library of

---

45 Berik, "Rabbi Shlomo of Chelm, the Author of Merkevet ha'Mishneh" [Hebrew]; Itzhak Alfassi, "Chelm, Solomon ben Moses."
46 Shlomo of Chelm, *Merkevet ha'mishneh*, p. 3a [Hebrew]. The cited excerpt is a paraphrase of verses from Proverbs (9:1), Joshua (1:8), and Deuteronomy (4:6). Also see Shmuel Feiner, *The Jewish Enlightenment in the Eighteenth Century*, p. 67 [Hebrew].
47 In all likelihood, studying the "wisdom of the world," mathematics and astronomy included, was acceptable among Poland's Jewish scholars during the 1500s, but this activity waned by the turn of the century. Elbaum, *Openness and Insularity*, pp. 77–79, 249–260 [Hebrew]. For more on Jews who took up the sciences in eighteenth-century Zamosc, see Freudenthal, "Hebrew Medieval Science in Zamosc ca. 1730. The Early Years of Rabbi Israel ben Moses Halevy of Zamosc."
48 Etkes, "On the Question of the Haskalah's Forerunners in Europe" [Hebrew]. As it now stands, the term "the early Haskalah" is in the ascendancy. See Feiner, *The Jewish Enlightenment*, p. 46–56. For a historiographical discussion on the beginning of the Haskalah, see Bartal, "On Periodization, Mysticism, and Enlightenment—the Case of Moses Hayyim Luzzatto."
49 Shlomo of Chelm, *Ḥug ha'areṣ ha-Shalem*, ed. S. D. Rosenthal, Jerusalem: Frank Institute, 1988 [Hebrew]. The last part of this manuscript has been lost. I am indebted to Dr. Gad Freudenthal for turning my attention to this important work.

Isreal.[50] Despite the fact that the printed edition failed to adhere to scientific standards, there is no doubt that it is an authentic text by Shlomo, for the author himself made note of and quoted from *Ḥug ha'areṣ* on several occasions in the last (Salonica) edition of the *Merkevet ha'mishneh*.[51] Furthermore, in an article that came out before the above-noted publication, Berik demonstrates that the book was cited in a few nineteenth-century sources and was known to various authors and bibliographers.[52] The fact that the work is only cited in the last edition of *Merkevet ha'mishneh*, which was printed shortly after Shlomo's death, suggests that it was compiled towards the end of his life. If so, we can assume that *Ḥug ha'areṣ* was motivated, at least in part, by its writer's aspiration to immigrate to Eretz Yisrael. As we shall see, there is also reason to believe that he revised the manuscript and added a few details on the basis of his personal experiences in the Land itself.

The extant book is comprised of several parts. The first is the preface, which was written in a lofty style and in rhymes, is an expression of Shlomo's love of the Land and includes the following prayer: "If heaven forbid we shall not merit to see it [i. e., Eretz Yisrael] in its built-up state then I will ask but this of God and beg to see it in its ruin. . . .[,] to walk 4 cubits there before I perish." Comparing himself to no less than Moses the Lawgiver, the Polish rabbi then wrote that if he is prevented from reaching the Land, "I will stand from afar on the observatory and look at its appearance on a map." After unfurling the difficulties that identifying various places in Eretz Yisrael entails, he reported that "all its borders and cities" were drawn "with a *meḥugah* [caliper]"—hence the title *Ḥug ha'areṣ*.[53]

The opening and principal chapter of this work constitutes an alphabetical list of the toponyms in Eretz Yisrael, from Avela to Tirzah (from *aleph* to *tav*). Each entry provides no more than a few lines about the site. Chapter 2 consists of an exegetical discussion on *shalosh artzot le'biur* (literally three lands for removal, or zones for transferring offerings to their designated beneficiaries). Shlomo of Chelm's third chapter is a detailed list of the cities in each tribal allotment as per the Book of Joshua. It also includes six maps of the tribal allotments and one of the entire Land, which is also arranged by tribe. The topic of the fourth chapter is the Land's borders. In the next chapter, Shlomo takes stock of the Israelites' peregrinations and provides an alphabetical list of the cities in each tribal area. The layout of the final surviving chapter, "The System of the Second Map," resembles that of the first chapter, as the

---

50 The National Library of Israel Ms. Heb. 28°8310; I would like to thank Mr. Avishai Galer for allowing me to examine the manuscript before its acquisition by the Israel National Library.
51 Shlomo of Chelm, *Merkevet ha'mishneh*, Laws of Forbidden Foods, vol. 1, 6; Laws of Offerings, vol. 1, 5 [Hebrew].
52 Berik, "The Fate of a Manuscript by R. Shlomo of Chelm" [Hebrew]. The work is indeed mentioned in the extensive bibliographical survey of Zunz, "On the Geographical Literature of the Jews from the Remotest Times to the Year 1841," p. 290.
53 Shlomo of Chelm, *Ḥug ha'areṣ*, p. 47. Henceforth, every mention of this work refers to the print edition.

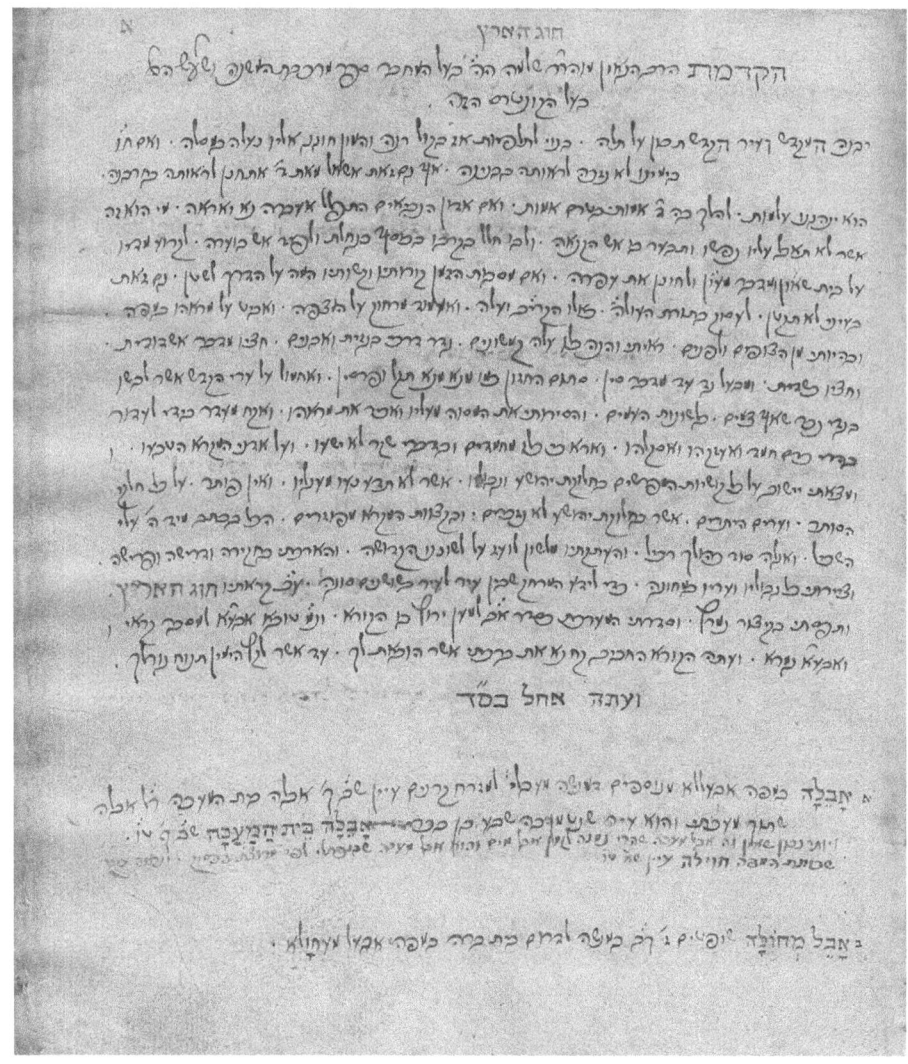

Fig. 61: The cover page of Ḥug ha'areṣ in the manuscript at the National Library of Israel.

toponyms are arranged in alphabetical order, from Avigdor to Peor. At this point, though, the manuscript comes to a premature end.

The Sources of Ḥug ha'Areṣ

Ḥug ha'areṣ contains no express hints of the sources that Shlomo of Chelm leaned on while producing this work. Given his attitude to the Bible, it is evident that the primary source was the lists of settlements in the Book of Joshua's account of the division into tribal allotments. However, the rabbi's book also discusses "the remaining cities that are not mentioned in Joshua's division. And in the extremities of the Bible they are scattered."[54] In other words, he complemented these lists by drawing on other parts of the Bible. Moreover, he refers to the settlements from these texts using the self-contrived term "addees" (*menusafim*). An analysis of the lists in the first and last chapters of *Ḥug ha'areṣ* demonstrates that Shlomo also availed himself of Latin texts. For instance, the first page contains the following passage: "Abel Shittim in Numbers [33:49] [is referred to] in the map [as] Abe'al Sata'im, and it is south of Zereth Shahar in Reuben."[55] As is the case throughout the length of *Ḥug ha'areṣ*, it is clear from this excerpt that he was copying from a non-Hebrew text. More specifically, he transcribed these place names using an accepted Yiddish method of transliteration in which the letters *aleph* and *ayin* serve as *matres lectionis* (consonants employed as vowels). Further examples of Latin derivatives are the Hebrew transliteration of *Saltum Ephraim*, vis. the Forest of Ephraim, where Absalom son of David was killed (2 Samuel 18:6). At times, Shlomo directly referred, albeit in Hebrew transliteration, to a term from the Latin map that he had access to: "the well of living water [is referred to] in the map [as] *Puteus Aquarum Vivantium*;" "Joseph's well [–] the mouth of the pit in which his brothers threw him into [is referred to] in the maps [as] the *Cisterna Joseph*;" "the rock of disagreements or the rock of the wilderness [is referred to] in the map [as] *Petra devisionis* and also the *Petra Deserti*;" "Tabor [is referred to] on the map [as] the *Mons Thabor* [–] a name of a mountain in Zebulun to the south of Sippori, and more."[56]

A wide range of Christian works could have stood at Shlomo's disposal, for the preceding generations included many European scholars who explored the Hebrew Bible. Foremost among them were the so-called "Hebraists," who occupied themselves with Judaism, the Holy Scriptures, and the Land of Israel.[57] A comparative inquiry demonstrates that the source of the first list in *Ḥug ha'areṣ* is Christian van Adrichom's *Theatrum Terrae Sanctae*, which played a similar role in Zaddik's map.[58] There is little doubt that Shlomo copied and translated whole entries from this thick Latin tome into Hebrew with virtually no changes.

Shlomo of Chelm's reliance on *Theatrum Terrae Sanctae* notwithstanding, the first list also draws on other sources. Adrichom considered Hazar Susah (Joshua 19:5)

---

54 Ibid, author's foreword, p. 47.
55 Ibid, p. 48.
56 "Forest of Ephraim," ibid, p. 52; "the well of living water" and "Joseph's well," ibid, p. 54; "the boulder of disagreements," ibid, p. 84; and "Tabor," ibid, p. 96.
57 The topic of the Hebraists is beyond the limited purview of this study. That said, I would like to point to a couple of Hebraist books: Reland, *Palaestina ex monumentis veteribus illustrata*; and Fuller, *A Pisgah-Sight of Palestine*. Both of these works are arranged in alphabetical order. An extensive bibliography of such texts can be found in Röhricht, *Bibliotheca Geographica Palaestinae*. It is also worth noting that the ranks of the Hebraists were comprised of both Catholics and Protestants alike.
58 Adrichom, *Theatrum Terrae Sanctae*.

Fig. 62: The tents of Kedar in the Adrichom map.

The text in *Ḥug ha'aretz* (my italics in this column)	The text in Adrichom's ***Theatrum Terrae Sanctae***
Cana [is referred to] in the map [as] *Chana Maior* [,] that is to say Cana Major.[a]	Cana sive *Chana* Maior, est in tribus Aser. . . .
Cana that is called Galilean Cana or [Cana] Minor [is referred to as] *Chana Galilaeae* or *Minor* [in the] map.[b]	Cana, alias *Chana Galilaeae* civitas est quae ob in Evangelica historia.
Dothan [is referred to] in the map [as] *Dotaim* [–] addees to Zebulun.[c]	Dothain, alias *Dothaim, Dotaim*, [...] (in Zabulon).
[In] the tabernacle of Kedar in the map are drawn 3 tents called *tabernacula Cedar*[,] that is to say sukkot or a tabernacle [sic] or tents of Kedar.[d]	Tabernacula Cedar, Inter *Suetam & Cedar* civitates
Iscariot [Kerioth] [is referred to] in the map [as] *Iscariotes* and it is Ephraim's addees to the east of the Dan Mountain. And this city is mentioned in the EG [Evangelists][,] that it is the city of Judah the informer from Kerioth [i.e., Judas Iscariot].[e]	Iscariot, vicus, in quo natus est *Iudas* Apostolorum duodecimos, *Christi* proditor; unde dictus est *Iscariotes* (in Ephraim).

[a] Shlomo of Chelm, *Ḥug ha'aretz*, entry 15, p. 92.
[b] Ibid, entry 16, p. 92. This entry lacks Adrichom's reference to the New Testament.
[c] Ibid, entry 9, p. 65.
[d] Ibid, entry 46, p. 79. In Adrichom's map, there is an illustration of three tents.
[e] Ibid, entry 23, p. 50.

and Sansannah (ibid 15:31) to be two different places, whereas the rabbi preferred to identify them as one: "Hazar Susah or Hazar Susim or Sinsnot, and in the map *Sa'asana* or *Hatzer Susa*." Moreover, Shlomo's entry for "Sansannah" directs the reader to the entry "Hazar Susah."[59] The entry for "Rhinocorurah" (al-Arish) reads thus: "*Rhinocorura* is indeed in the map, outside of the Land belonging to Judah, to the west of Yavne'el; and in another map I found that it is Gerar[, the kingdom of] Abimelech[;] see what we said [under] Gerara."[60] This express reference to "another map" leaves no doubt that the author consulted multiple cartographic works.

Two of the entries perhaps allude to his short-lived residency in Tiberias. Taricheae—a city along the shore of the Sea of Galilee, which is mentioned by Josephus[61]—was included in both *Theatrum Terrae Sanctae* and *Ḥug ha'areṣ*. In his own entry, Adrichom cites from Suetonius and Josephus concerning Titus' activity therein.[62] Shlomo leaves out this material, but makes note of the city twice. Under the letter *teth*, he writes that "Taricheae is [among] the addees of Issachar[;] see what we said under Ta'arichi and it may be [named for] a *tarit* fish [sardine] and this is a species of fish called sard'ala that is" also known as "taricheae." Under the letter *taw* is the following entry: "Ta'arichi [is referred to] in the map [as] *Tarichea* and it is on the coast of the Sea of Galilee in Issachar [....,] and the main [spelling is] Taricheae and it is a type of salty, undersized fish."[63] In both instances, Shlomo ties the city's name to a fish that, in his estimation, is endemic to the Sea of Galilee. Seven maps accompany the third chapter:

- The allotments of Judah, Simeon, Benjamin, and Dan.
- The allotments of Ephraim, half of Manasseh, Issachar, and Zebulun.
- The allotments of Naphtali and Asher.
- The allotments of half the Tribe of Manasseh in Transjordan.
- The allotment of Gad.
- The allotment of Reuben.
- A small map of the entire Land, which is also divided into the tribal allotments.

In drafting these maps, Shlomo of Chelm apparently relied on Adrichom's book as well. To begin with, *Theatrum Terrae Sanctae* offers detailed maps covering multiple tribal areas. For example, Benjamin, Ephraim, and Dan are included in one of his maps and half of Manasseh, Issachar, and Zebulun on another. Like Adrichom, the Polish rabbi places the allotments of Judah, Simeon, Dan, and Benjamin on a single map, which he called "The Tribe of Judah." What is more, certain details in the rabbi's maps bear a marked resemblance to those in the Catholic priest's work. Among the numerous examples are "Mount of the Prophets" in the allotment of half the tribe of Manasseh west of the Jordan, which is identical to Adrichom's

---

59 Ibid, entry 23, p. 69; entry 13, p. 84.
60 Ibid, entry 9, p. 94.
61 Josephus, *War of the Jews*, book II chap. 20; book III, chap 10; and other places therein.
62 "Tarichea, sive *Tarichae*, Suetonius in Tito *Tarrachaim* apellat; urbs munitissima, sita sub monte in littore maris Galilaeae distat a Tiberiade [...];" Adrichom, *Theatrum Terrae Sanctae* (1628 edition), pp. 37–38 (in the description of Issachar's allotment).
63 Solomon of Chelm, *Ḥug ha'aretz*, pp. 71, 96.

Mons Prophetarum; both incorporate the four Cities of the Plain in the Dead Sea; and the caption "the Sea of *kinneret* [the Hebrew name] or the Galilean or of Tiberias" is analogous to the caption "Mare Galilaeae, Cenerth et Tiberiadis" in *Theatrum Terrae Sanctae*. Shlomo also copied at least one glaring mistake from Adrichom. More specifically, he situates Lydda, Ono, and Hadid between Jericho and the Jordan River—quite a distance from the cities' actual location on the inner Shephelah (the Judean foothills). This error indeed constitutes firm evidence of the replication process involving these two maps.

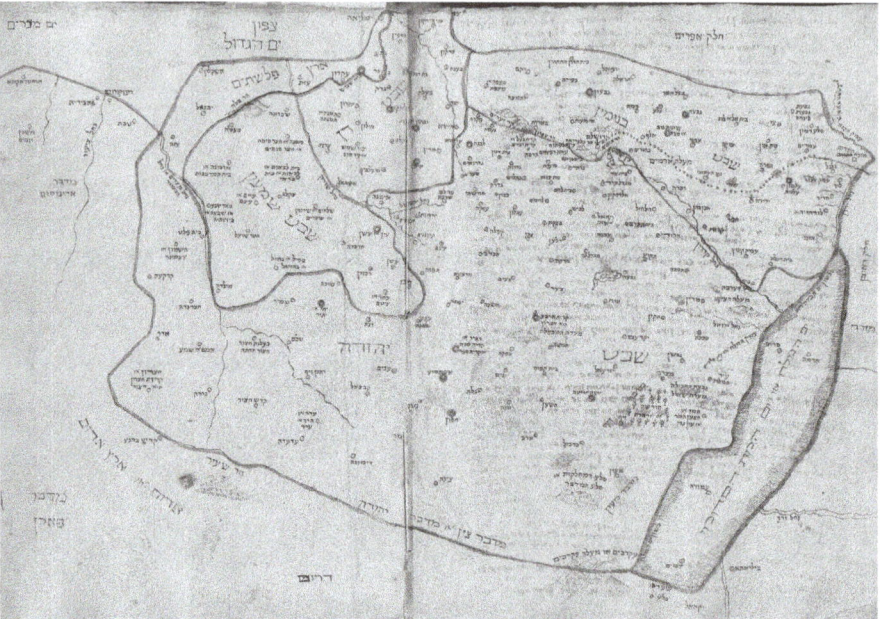

Fig. 63: Shlomo of Chelm's map of Judah, Simeon, Benjamin, and Dan.

Fig. 64: The plain of Jericho and the location of Lydda and Hadid in the Adrichom and Chelm maps.

In Shlomo of Chelm's chapter on the peregrinations, there is a medium-sized map of the 42 stations bearing the title "The Circuit of the Israelites from Egypt to

Canaan via the Wilderness." Although it is obviously not a replica of Adrichom's representation, the slithering route of the Israelites from one station to the next was influenced, like many other cartographic takes on this theme, by the graphical template of the Israelite's journey in the Catholic priest's map. Adjacent to this map in the print version of Ḥug ha'areṣ is a map of the entire Land of Israel. The book's editor states that he personally added this map, so that it is not an original component of Shlomo's work.

The final existing part of Ḥug ha'areṣ contains another list of place names that is titled "The System of the Second Map." This section was clearly influenced by a Latin source as well. For instance, in the entry for Beth She'an, Shlomo simply translates into Hebrew the Latin clause "*Bethsan olim Schythopolis*" (Beth She'an formerly Schythopolis). Since many toponyms of this sort informed the era's cornucopia of Latin maps, my identification of the source map is predicated on several entries containing unusual Hebrew transcriptions of complicated and relatively long entries. The Hebrew formulations in Shlomo's entries make no sense unless we assume that the text is a synthesis of replication and translation from Latin. Examples of these sort of passages include the following: "The new Be'er Sheva on the border of Judah, in the map [is] Beersaba Novachabar Saba versabium to the southeast of Antipater;" or "Achshaph on the border of Asher[,] namely Acre, in the map [is] Aco, Acci, Ptolemais, Arce, Achschaph to the south of Achzib."[64] Comparing these entries to various Latin maps from this period leads to the conclusion that Shlomo of Chelm's source is Matthaeus Seutter's map, which was printed in Augsburg, Germany in 1745.[65] To wit: this map includes long captions that are absolutely identical to the entries on the Polish rabbi's Hebrew list.[66]

Like numerous other maps from the sixteenth to the nineteenth centuries, Seutter adopted a broad chronological approach to the Land of Israel. He included both the tribal allotments along with the regional division that was in force during the Hellenistic and Roman eras: Judea, Samaria, and the Galilee (as well as Perea in Transjordan, Bashan in north Transjordan, and Phoenice to the north of the Galilee). What is more, Seutter's map contains the names of settlements from the Hebrew Bible, the New Testament, the classical sources, and even a couple from the literature of the Sages. This wide scope made it difficult for Shlomo of Chelm to identify later toponyms with ancient settlements, thereby spawning a handful of interesting crossbreeds: Even though none of these names surface in the Hebrew Bible, Seutter's Herodium turned into *ḥarudim*, Apolonia became *haploni*, and Hippos (alias Sussita) *ḥipot*. In essence, these are neologisms that were concocted by the Jewish scholar.

---

64 Ibid, p. 139.
65 Matthaeus Seutter, "Palaestina seu Terra a Mose et Iosua occupata et inter Iudaeos distributa per XII Tribus vulgo Sancta adpellata." in Seutter, *Atlas novus*.
66 For example: the captions for Acre and Beer-Sheba in Seutter's map: Aco, Ace, Ptolemais, Arce, Achschaph, Cho; Bersabe nova Chabarsaba, Versabium.

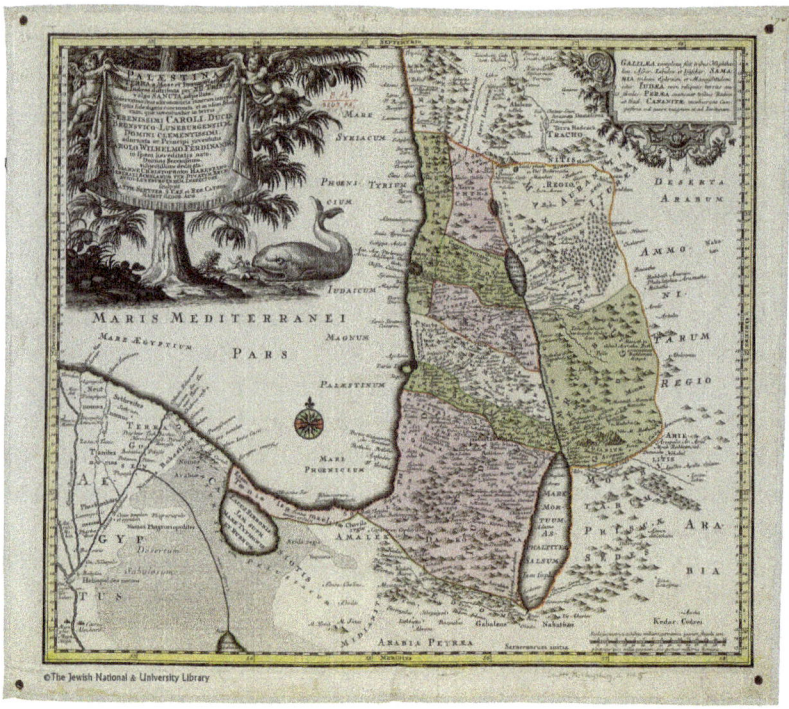

Fig. 65: Matthaeus Seutter, "Palaestina seu Terra a Mose et Iosua occupata et inter Iudaeos distributa per XII Tribus vulgo Sancta adpellata," 570 x 493 mm.

Fig. 66: Details in the Seutter map: Acre, Be'er Sheva, Rhinocorurah and Lake Bardawil.

In his text Shlomo of Chelm did indeed follow Seutter's map from 1745. However, the outline and the border-lines of his own map of the twelve tribes, followed another map by Seutter, printed in 1725,[67] which was also copied by Tobias Conrad Lotter in 1759 (see fig. 4.29).

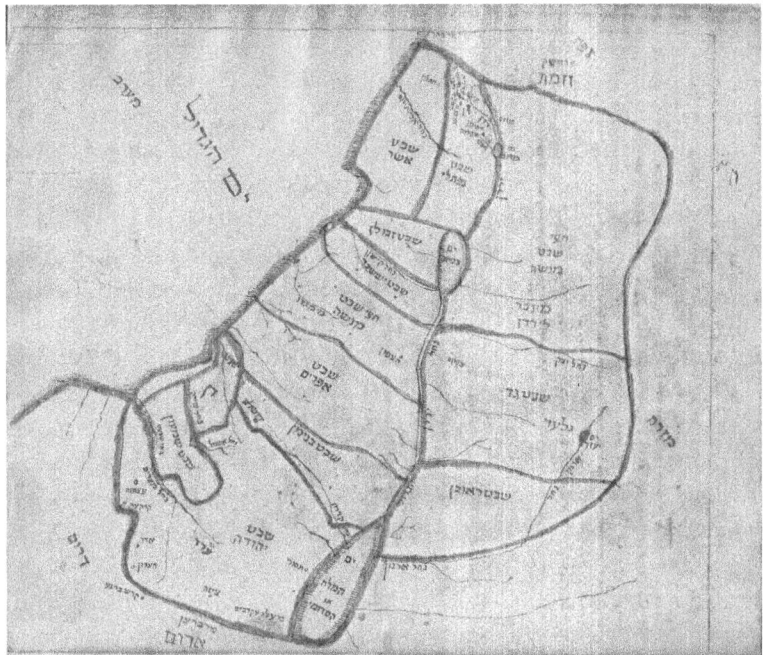

Fig. 67: Shlomo of Chelm's map of the Twelve Tribes

In relying on Christian sources, Shlomo's greatest challenge was not the identification of later toponyms, but how to contend with non-Jewish content. His objective was to write a book on Eretz Yisrael for the Jewish reader. As we have seen, Jewish copyists tried as best as they could to "expurgate" maps for their target audience and even openly stated as such in their preambles. That said, Shlomo's exposure to the European Enlightenment naturally led him to rest heavily on Adrichom's comprehensive book, Seutter's map, and in all likelihood other works as well. In balancing these competing impulses, the Polish rabbi completely erased a significant number of sites that are linked to Christian traditions and omitted or papered over the Christian meaning of other sites. Additionally, he changed the meaning of certain Second Temple period toponyms that pertain to Jesus and the Apostles. For example, these sort of names were identified as places that are ostensibly mentioned in the Hebrew Bible. These efforts notwithstanding, a few unequivocally Christian sites remained

---

[67] Matthaeus Seutter, *Regio Canaan seu Terra Promissionis postea Iudaea vel Palaestina nominata hodie Terra Sancta vocata. Quae olim XII Tribus hac autem aevo in VI Provincias Distincta est iuxta recentissimas et accuratissimas descriptionis adornata*, from: his *Atlas Geographicus oder accuratae Vorstellung der ganze Welt*, Augsburg 1725.

on his map. Nazareth and Kfar Cana in the Galilee are simply listed as is, without any background information. Alternatively, he described Gergesa, the place where Jesus exorcised sprits, thus: "Gergesi [is referred to] in the map [as] Garasa or Gargasa [–] addees to Manasseh."[68] On the other hand, there were entries like "Calvari [or] M[ount] Calavaria on the map[,] that is to say the name of a mountain in Benjamin;" and "Nephtoah, the name of a spring or well in Benjamin[,] Phonis Nephto in the map [,] that is to say the fountain of Nephtoah or the spring of the waters of Nephtoah in the verse and it is to the south of Mount Calavariah."[69] Both passages indeed contain an explicit reference to Calvary, the site of the Crucifixion.

The most problematic Christian site is what Shlomo describes as "Iscariot, [or] *Iscariotes* in the map, and it is [one of] Ephraim's addees to the east of the Dan Mountain. And this city is mentioned in the EG [Evangelists] that it is [sic] the city of Judah the informer from Kerioth." This entry undoubtedly refers to the birthplace of Judas Iscariot—the disciple that turned Jesus in to the Roman authorities (Matthew 26; Mark 14).[70] What is more, the abbreviation EG constitutes an overt reference to the New Testament. Therefore, it must be assumed that Shlomo of Chelm realized the significance of including an entry for "Iscariot" in his book. If so, the following questions beg asking: Why did he leave this reference in his book? Did his Jewish audience understand who this "Judah" is? And how did those Jewish readers who identified the disciple react to this entry? In any event, we lack the source material to answer these freighted questions.

In sum, *Ḥug ha'areṣ* is a geographical work that was nourished by Shlomo of Chelm's love of Eretz Yisrael. These feelings are epitomized by his attempt to settle the Land. The book is predicated on Shlomo's broad general knowledge, not least his ability to read Latin and adapt Christian sources. Owing to the rabbi's attentiveness to his readership and the fact that he too was wedded to the Jewish tradition, steps were taken to suit this book to its targeted Jewish audience. To this end, *Ḥug ha'areṣ* was almost entirely cleansed of hints to its Christian sources. In fact, these submerged influences most likely failed to register in the minds of the vast majority of his readers. Only someone that was capable of tying between these two worlds—Jewish and secular learning—could have compiled such a book. At any rate, the fact that this work was never printed attests to its limited distribution and influence. On the other hand, references to *Ḥug ha'areṣ* in other Jewish works demonstrate that it did gain some traction.

---

[68] Shlomo of Chelm, *Ḥug ha'aretz*, entry 36, p. 63.
[69] Ibid, entry 12, p. 92; entry 41, p. 83.
[70] In all likelihood, Iscariot is not a "genuine" toponym, but rather a distortion stemming from Greek and Latin derivatives of an ancient term. There are doubts as to whether the name Iscariot should be rendered *ish krayot* (the man from Kerioth) as in Mandelkern's Hebrew translation of the New Testament. Whatever the case, there is no way of determining if Judas was indeed from a place called Kerioth.

The Map of Yehonatan ben Yaacov

In 1784, one Yehonatan ben Yaacov printed a map of Eretz Yisrael and the Sinai Desert in Nowy Dwor (a town outside of Warsaw). Just before the Second World War, a copy of the map reached Jerusalem from the Polish capital as part of a shipment from the "Well-Wishers of The Hebrew University." Printed on a single folio, the map was subsequently published by Abraham Ya'ari (1942)[71] and Zev Vilnay (1968).[72] The fact that it had not hitherto been listed in any bibliography suggests that it was an extremely rare map. Since the present location of the original is unknown, I have sufficed with a copy from Ya'ari's article.

The 41 x 30 cm map was printed from a woodcut. At the top is a crown-shaped cartouche flanked by two lions. The short text inside presents the author and his objectives: "Upon consulting my inner thoughts [//] I etched from my stronghold [//] a drawing of the picture of the Israelites' voyages and their borders [//] that were in their allotments [//] I the meager and young Yehonatan son of my tea[cher] Yaacov [,] a righteous person of blessed memory from Veyaloish[, am] the proofreader from the New Printing Press [of] Nowy Dwor." On the left margin is a Hebrew text with the following sentence: "Yehonaton said[,] I neither relaxed nor rested until I published my mysteries and thoughts of copying a *landkarte* from Ashkenazic [German] to Hebrew as an image and its picture[;] and for a year I made [print] marks so that the reader can race through it." Further down, he provided a legend with the various symbols for towns, mountains, rivers, and the like. The date was provided at the end of the text: "Executed and completed today[,] Sunday, the head of the month of Nisan[.] In the Hebrew year[:] and may you mercifully bring us to Zion your city. Here in Nowy Dwor."[73]

Interestingly enough, the printing press in Nowy Dwor was under the ownership of Anton Krieger, a German Christian who specialized in printing Hebrew books. Founded in 1777, Krieger's establishment was a bustling center for printing, proofreading, and commerce in Hebrew books until at least 1819.[74] Among its printed books were an edition of the MaHaRShA's *Sefer ḥiddushei halakhot*, which was mentioned in chapter 2, and Alexander Ziskind's *Yesod ve'shoresh ha'avodah* (the Foun-

---

[71] Ya'ari, "Miscellaneous Bibliographical Notes 41: Hebrew Map of Palestine Printed at Nowy Dwor 1784" [Hebrew].
[72] Vilnay, *The Hebrew Maps of the Holy Land* [Hebrew].
[73] The date (1784) is ciphered by marks above certain letters in the Hebrew words for merciful, your city and Zion.
[74] A discussion on the history and output of Krieger's establishment can be found in Friedberg, *History of Hebrew Typography in Poland*, 75–82 [Hebrew]. Friedberg lists the books that were printed therein through 1816, a task that is completed by Ya'ari, "Miscellaneous Bibliographical Notes 1: On the History of the Hebrew Printing at Nowy Dwor [Hebrew]. It also bears noting that the Israel National Library's catalog totals 98 entries printed in Krieger's press from 1777 to 1819. Also see Pilarczyk, "Hebrew Printing Houses in Poland against the Background of their History in the World."

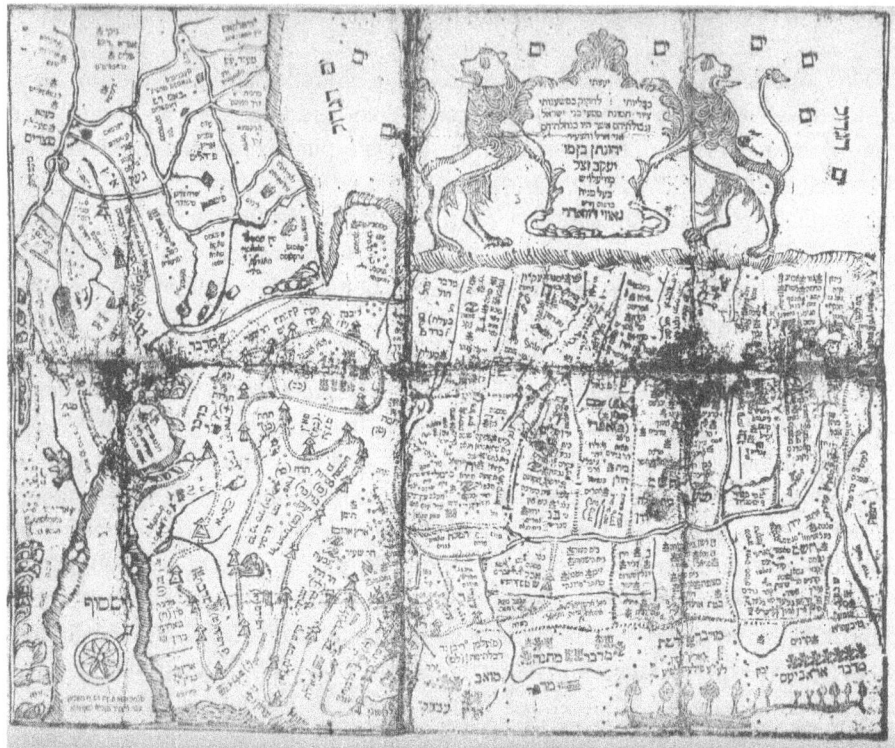

Fig. 68: The map of Yehonatan ben Yaacov, 300 x 410 mm, in Ya'ari, "Miscellaneous Bibliographical Notes 41."

dation and Root of Worship), which will be discussed in the next chapter.[75] Yehonatan ben Yaacov was not only Krieger's proofreader, but served as the supervisor and manager of his Warsaw book trading house. In 1794, ben Yaacov would open his own printing house in Dubno.[76]

To the best of our knowledge, the map at hand was not part of a book. Given its release date in Nisan (the Hebrew month in which Passover falls), Ya'ari raised the possibility that the map was slated for a Passover Haggadah. However, he ruled out this thesis on the grounds that no Haggadah is known to have been printed in Nowy Dwor during these years. Alternatively, Ya'ari proposed that the map was a decoration for the walls of a house or sukkah—a suggestion that is indeed commensurate with its dimensions.

From a design standpoint, the map is far from beautiful. The Mediterranean Sea covers two-thirds of the upper portion, so that the map faces west. The remainder of

---

75 Liberman, *Rachel's Tent*, [Hebrew].
76 Ya'ari, "Miscellaneous Bibliographical Notes 41," p. 215.

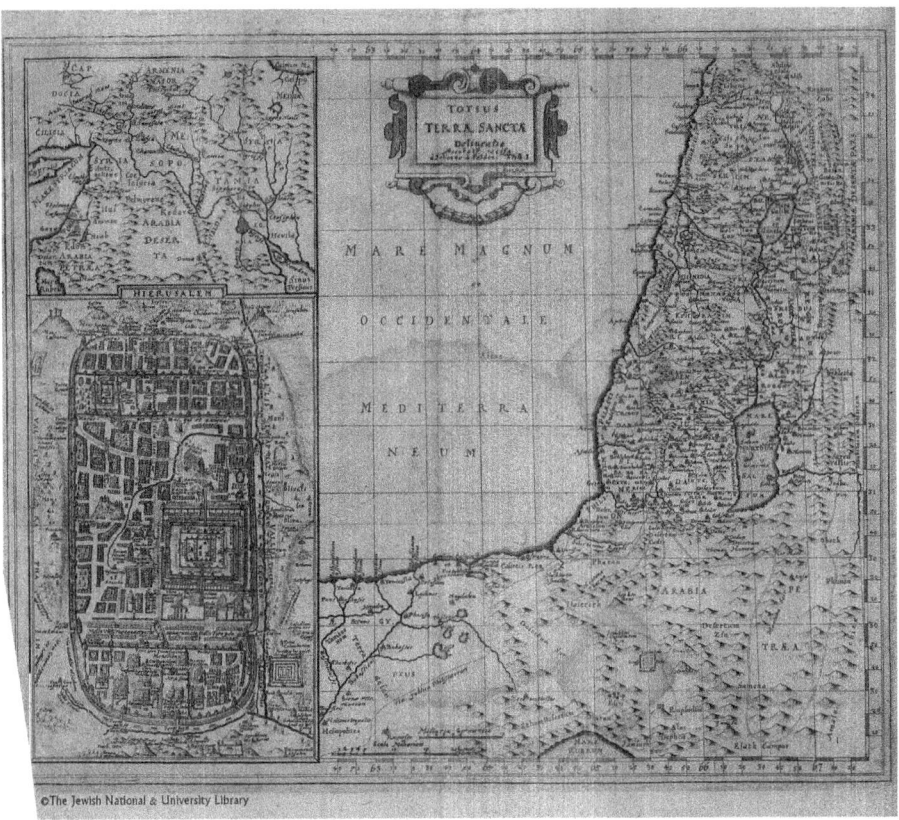

Fig. 69: Is Heidmann the source of Yehonatan ben Yaacov's map? Christoffer Heidmann, "Totius Terrae Sanctae Delineatio," 320 x 268 mm.

this section is taken up by Egypt. On the bottom-left are the Red Sea and the Sinai Desert, while the right half consists of the Land of Israel and Transjordan. Most of the land mass is peppered with toponyms from the Hebrew Bible. However, some of the sites derive from Latin texts that were transcribed into Hebrew. Adjacent to most of the map's toponyms is one of the symbols from the above-mentioned legend. As per the era's accepted format, the Nile Delta is conceived as a network of the river's arms, which eventually spill into the Mediterranean Sea. The route of the Israelites' peregrinations is signified by a double-dotted line. Along the length of this winding route are the stations that are enumerated in the portion of *Mas'ei*, up to and including "Abel Shittim" (acacia meadow)—the last stop before the Israelites traversed the Jordan. Eretz Yisrael is divided into the tribal allotments, which are signified by acronyms, such as HTM = half the tribe of Manasseh or by the first few letters of each tribe: Be = Benjamin, Ephr = Ephraim, G = Gad, and the like. Each allotment contains a host of settlements from the Bible.

Oddly enough, trees line the area classified as the "Arabias Desert," that is the Arabian Desert (most likely a translation of *Arabia deserta*). Yehonatan ben Yaacov indeed wrote that standing before him was a *"landkarte"* that he rendered "from Ashkenazic to Hebrew" over the course of an entire year. In other words, he translated a German map into the holy tongue, edited its content, and devised symbols for its entries. The Latin names that were transcribed into Hebrew indeed point to a non-Jewish source map. As alluded to earlier, the area of Eretz Yisrael is devoid of Latin toponyms. However, such names dot the Egyptian expanse (particularly the Delta region) and parts of Transjordan. Other hints of the European cartographic tradition are the names of four Cities of the Plain inside the Dead Sea;[77] the inclusion of a river that connects the Sea of Galilee to the Mediterranean; and the three names given for the former—"the Sea of *kinneret*" (the accepted Hebrew name), "the Sea of Tiberias," and "the Sea of Galilee." Artistically speaking, the map is hardly a masterpiece and fails to live up to the standards of eighteenth-century European cartography. Moreover, its western orientation stood out during this period. These unusual elements hinder our efforts to definitively identify the source map. That said, quite a few of ben Yaacov's elements are reminiscent of those in Christoffer Heidmann's map,[78] such as the following: the shape of the Nile's branches in the Delta region; the lakes to the east of the Delta; and the route of the Arnon River, especially its upper portion's northerly orientation; the head of the Yarmouk, twisting down from the north; and parts of the Israelites' route through the wilderness. Despite these similarities, it is not entirely certain that the Heidmann map constitutes the Jewish proofreader's source.

In sum, the circumstances behind Yehonatan ben Yaacov's map are riveting. Upon coming across a "German" map, a Jew who was heavily involved in the printing and book industry decided to copy and suit it to the tastes of a Jewish audience. Although he considered this undertaking a personal project, he operated within the framework of a Christian-owned printing house in Nowy Dwor that specialized in Hebrew books. These facts shed light on the importance of the cultural encounters that transpired in European printing houses at around the turn of the nineteenth century.

The Maps of Yaacov Auspitz in *Be'er ha'Luḥot*

In 1817, Yaacov Auspitz published a booklet that was accompanied by five annotated maps. The first edition came out in "Old Oven" or "Ofen" (the German moniker for Buda—the western part of Budapest), and the second edition was printed a year later in Vienna.[79] At the very outset of the work, Auspitz defined his goal:

---

77 Rubin, "Mapping a Myth: The Cartographic Image of the Overthrowing of Sodom and Gomorrah."
78 Heidmann, "Totius Terrae Sanctae Delineatio."
79 Auspitz, *Be'er ha'luḥot*, the Ofen edition [Hebrew]; ibid, the Vienna edition.

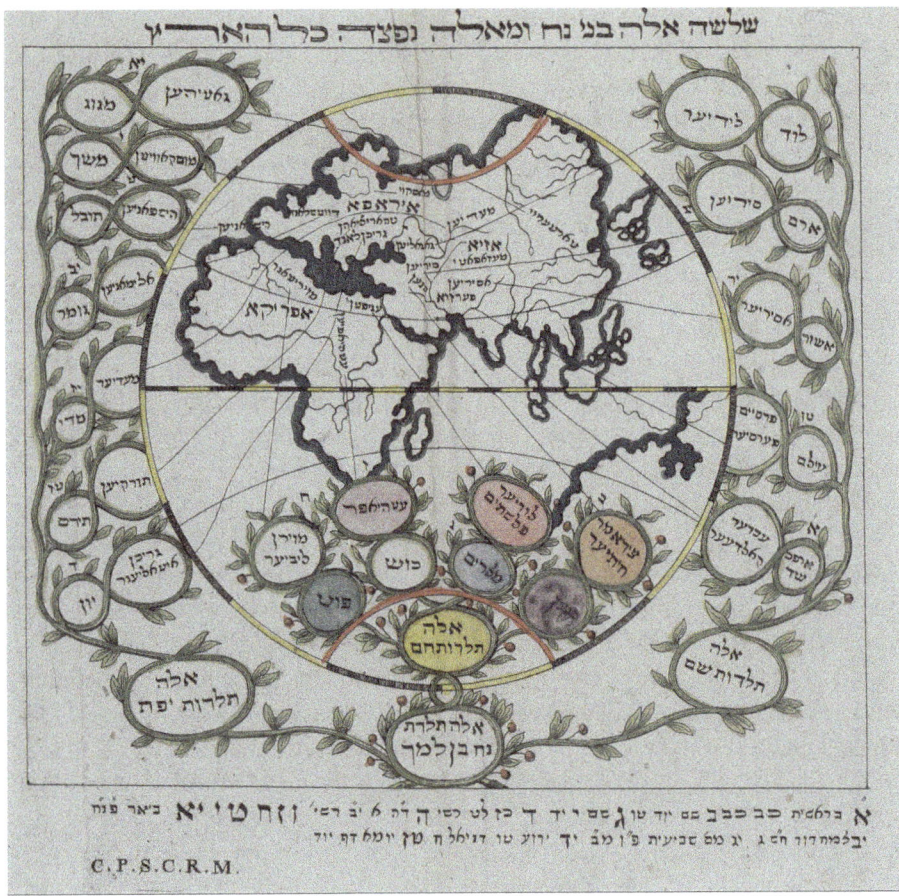

Fig. 70: Yaacov Auspitz's the sons of Noah map, 158 x 138 mm, from *Be'er ha'luḥot*.

> [A] comprehensive explanation of the five plates that are found at the end of the book, which I copied into the holy tongue with much change from what I found in the Roman tongue. With clear proofs and faithful views as per the views of sages and those wise in science from among the sages of the nations, […] and I always toiled to find vitality for myself in the mountains of holiness and in places scattered throughout the Hebrew Bible and the Talmud and the exegesis of our Sages of blessed memory[;] may your eyes behold what is upright."

Auspitz then describes the developments that led to the adaptation of these maps: "And it was a full year ago that some plates in the Roman [i.e., Latin] language came into my hands, and I saw that drawings were etched on them which I was interested in, but were nevertheless like a sealed book to me because I did not understand the Roman language." As a result, he turned to others for help with the Latin. The preface then explains that the maps had to be modified for Jewish consumption because "even the drawings themselves were not in line with our holy Torah and the exegesis of our Sages of blessed memory. And every time I saw that the authors of these Roman plates veered away from this and did not go down the path of the exegesis of our Sages of blessed memory, I interpreted and duly inquired and modified [the content] here and there."[80]

The five maps, or what Auspitz referred to as "plates," are on the following topics: the division of the world between the sons of Noah; the encampment of the twelve tribes in the wilderness; the Israelites' peregrinations; Eretz Yisrael and the tribal allotments; and the land's future apportionment as per Ezekiel.

The Sons of Noah Map

Printed onto a square folio, Auspitz's map of the dispersion of Noah's sons is crowned with a fitting verse from Genesis 9:19: "These three are the sons of Noah, and from them the whole Earth was populated." At the center of the map is a circle encompassing a modern template of the three known continents of the ancient world. The content reflects the tradition whereby the offspring of Shem dwelled in Asia, Japheth's sons in Europe, and those of Ham in Africa.[81] It bears noting that this tradition appears in the Christian exegesis from as early as the Middle Ages and is depicted on classic T-O maps. Surrounding the continents in the Auspitz version is a genealogy chart comprised of medallions, which are formed by running vines. Every pair of medallions bears the name of a scion of Noah and a Yiddish transliteration of the nation that he sired. For example, Tubal is the forefather of "*hishpanin*" (the Spanish); Gomer is linked to "*allmanien*" (the Germans); Tiras with "*turki'an*" (the Turks); Elam to "Persians *parsiar*;" and Ashur to "*assiriar*" (the Assyrians).

The dispersion of Noah's sons in accordance to Genesis 9–10 was a prevalent topic among European cartographers. Some limited their purview to the Mediterranean basin, while others spread the nations throughout the three continents of the ancient world.[82] I did not manage to definitively identify the source of Auspitz's map. However, the most likely candidate is Bendecitus Arias Montanus' map,[83] which indeed merited numerous replicas and adaptations.

---

80 Ibid, the author's introduction.
81 Wajntraub and Wajntraub, "Hebrew Map Showing the Dispersion of the Sons of Noah."
82 Le Jay, *Sedes Filiorum Nohha vt legunt hebr. Genes*; Moxon, *A Map of All the Earth and How After the Flood it was Divided among the Sons of Noah*; Kircher, *Tabula geographica divisionia gentium et populorum per tres filios Noe, Sem, Cham, Japhet...*; Calmet, "A Geographical Map of the Old World according to the Division of it among Noah's Sons after the Dispersion of them which Happened at Babel," between pp. 384 and 385.
83 For a discussion on Montanus and his work, including the map under review, see Shalev, "Sacred Geography, Antiquarianism and Visual Erudition: Benito Arias Monatno and the Maps of the Antwerp Polyglot Bible."

The Map of the Israelites' Peregrinations

The title of Auspitz's second map is taken from Numbers 33:1: "These are the Israelites' Journeys: Until They Crossed the Jordan." Its purview ranges from the Nile Delta to the west to the Land of Moab to the east and the Sinai Desert to the south to an invisible line connecting Jericho and Ashkelon up north. A windrose is situated in the "Great Sea." The roomy cartouche on the lower righthand corner is adorned with an illustration of the two spies delivering the cluster of grapes; it also includes the date of publication, 1817, and the artisans' signatures: "A. Günther" and "sc. Pestini." The former may have been the painter; and the letters "sc." before "Pestini" credit him with fashioning the printing copperplate. The Israelites' route through the wilderness—from the Land of Goshen to the Plains of Moab—is signified by a twisting contoured line. Over the length of the route are the 42 stations that are mentioned in Numbers 33.

Figure 71: Yaacov Auspitz, "These are the Israelites' Journeys" map, 158 x 205 mm.

The Israelites' peregrinations were an extremely popular theme among European cartographers. Following its appearance in the Geneva Bible's maps, this topic became a near fixture in Bible cartography. In all likelihood, the map at hand draws heavily on Augustin Calmet's map on the same topic.⁸⁴ This is evident from the geographic purview of both maps; the overall topology of the bodies of water and land mass; the shape of the Nile Delta and Lake Bardawil; the inclusion of two islands in the Red Sea, to the south of Eilat; and the route of the Israelite's journey. Like Calmet's other maps, this one merited countless editions as well as close and loose imitations. Consequently, it is impossible to ascertain which one of the numerous versions was the direct source of the Auspitz map. However, given the Jewish cartographer's professed use of maps "in the Roman tongue," it stands to reason that his own was predicated on the Latin edition of the Calmet map.⁸⁵

Fig. 72: Augustin Calmet's map of the Exodus from Egypt, 430 x 310 mm.

His reliance on Calmet's map notwithstanding, Auspitz preserved a modicum of independence. This desire to plot his own course is epitomized by two details that

---

84 Calmet, "Carte du Voïage des Israëlites dans le desert, depuis leur Sortie de l'Egipte (sic!) jusqu'au passage du Jourdain, Dressee par l'Auteur du Commentaire de l'Exode, et executée par Lièbaux Geografe," facing p. 381.
85 Calmet, "Tabula Itineris, et Stationum Israelitarum in Deserto, delineata et incisa juxta Auctoris systema a P. Starck-man."

are faithful to the Hebrew sources. First, the map depicts "the way through the Land of the Philistines," namely the shortest possible route from Egypt to the Land of Canaan—a reference to Exodus 13:17: "God led them not through the way of the Land of the Philistines though that way was shorter." Second, with respect to the crossing of the Red Sea, Auspitz offered two different routes. One efficiently traverses the sea widthwise in a straight line. The other route also enters the water, but soon loops back to the southern coast before circuiting the head of the bay via land. Abutting the second route is the following caption: "As per the view of the Tos[fot] Arakhin [p.] 16." The circumnavigation of the Red Sea is indeed based on the introduction to tractate Arakhin (estimations) in the Babylonian Talmud and the accompanying commentary in the Tosfot. The latter cites an aggadah (legend) according to which the Israelites vacillated over whether to enter the sea upon finding themselves hemmed in between the sea in front of them and the Egyptian army closing in from behind.[86] According to this commentary, which was widely accepted by later exegetes, the Israelites entered the sea, headed north, backtracked in a semi-circular fashion, left the sea, and then circuited the eastern bay.[87] The basis for this discussion on the

---

[86] "It was taught: R. Judah said, With ten trials did our forefathers try the Holy One blessed be He: two at sea, two over water, two over the manna, two over the quails, one in connection with the golden calf, and one in the Wilderness of Paran. 'Two at sea': one during the descent, the other during the ascent. . . . This teaches that the Israelites were rebellious at that very hour, saying: Just as we go up from this side, so will the Egyptians go up from the other side. The Holy One blessed be He said to the Prince of the Sea: Cast them out onto the dry land! He answered: Sovereign of the Universe, is there a slave to whom his master gives a gift and then takes it away from him? He said to him: I shall give you one and a half times as many of them. He said before Him: Sovereign of the Universe, is there any slave who can claim anything against his master? He said: The brook of Kishon shall be a surety. At once he cast them onto dry land."

[87] "'Just as we go up from this side (they said) so will the Egyptians go up from the other side'—one must wonder how the Israelites of that generation were so short on faith that they reasoned the Holy One blessed be He would do miracles for Egypt in this fashion transferring them [i.e., the troops] from their land to EY [Eretz Yisrael]. And R. says in the name of his father Rabbi Shmuel that the Israelites did not cross the sea widthwise from this side to that, for if so they would have rushed off to EY. Instead, they advanced a single stretch in the sea within the sea before turning to the wilderness to one side, for they would not have been able to go from any side if they had not passed through the sea. And they had Egyptians to their right and their left; therefore, they passed one stretch of the sea before turning to the wilderness from one side. And thus [they said:] 'Just as we go up from this side and came [sic] to the wilderness so will the Egyptians go up from another side in the wilderness and will chase after us and catch up to us,' for they assume that the Egyptians would not pass through the sea, but would come from another side. And this, then, reconciles what is written: 'And they set out from Marah' etc. (Numbers 33:9) 'and camped beside the Red Sea' (ibid 10); and if they had crossed the width of the sea until the [other] side from a different side[,] how did they still hit the sea[?] . . . [D]id they turn back? But learn from this[:] After setting out into the sea and coming out[,] once again along the sea to the south [toward] Egypt, for Egypt is also to the southern direction of EY next to [its] southern end to the west side and the Nile ceases between Egypt and EY and the [Great] Sea is to the south side of the Land of Egypt from east to west and they set out to the seashore from north of the sea which is south of Egypt, whereupon they got out after proceeding like

Talmudic aggadah in Arakhin and the Tosafists' interpretive acuity may well be Psalms 106:7: "And they rebelled by the sea, the Red Sea."

The two alternative routes are alluded to in the text of the Mantua Map and depicted in bar Yaacov's map, which was discussed in the first half of this chapter. As we shall see, it also turns up in other maps. Surprisingly, this popular "Jewish" narrative already surfaced in a map by Franciscus Quaresmius—a Franciscan monk who was active in the Holy Land for many years. The map was part of his expansive book[88], which was published in 1639.[89] A slightly different version of the two routes informs Binder's late eighteenth-century map[90] and perhaps those of other authors as well. Most interestingly, the fact that an explanation of Talmudic provenance was incorporated into a Catholic map long before its adoption by Hebrew cartographers also bears witness to a theoretical and cartographic discourse between Jews and Christians.

Map of the Israelites' Encampment

Titled "Each Under their Standard and Holding the Banners of their Family" (Numbers 2:2), the third plate depicts the layout of the Israelites' camp around the roving Tabernacle—a topic that was quite prevalent in both Christian and Jewish art. We already touched on the Israelites' encampment in the discussion on a similar map in the *Me'am lo'ez* (chapter 2). However, Auspitz's version appreciably differs from a graphical standpoint. At the center of the camp stands the Tabernacle. Due east of (below) the walled compound are Moses and Aaron's tents. The Levite families, which were entrusted with the work in the Tabernacle, enclose the structure from the other directions: the sons of Merari to the north; Gershon to the west; and Kohath to the south. Pursuant to the Biblical account, the camp's inner core is surrounded by the twelve tribes—three to each side. Filling the bottom of the page are the census results from Numbers 2 concerning the number of military-fit men in each tribe. The information is duly crowned with the words "These are the Israelites aged twenty and over who were counted."

The Israelites' encampment was indeed a highly popular theme as both standalone drawings and as part of cartographic accounts of their journey to the Promised

---

a round semi-circle and got out on their northern side which is south of the Land of Edom and east of the Land of Egypt. And thereafter they walked a few days until Marah and then they were still by the sea. And when they had to enter EY[,] they did not have to pass through the sea, but circuited the Land of Edom and the Land Moab which is south of EY and east of the Land of Egypt" (Tosafot Arakhin 15a, statement beginning "Just as we go up").

88 Quaresmius, *Historica theologica et moralis Terrae Sanctae elvcidatio*.
89 Quaresmius, "Imago transitvs filiorvm Israel per Mare Rvbrvm.", there, vol. 2 p. 973.
90 Binder, "Charte worauf das iuedische Land vornehmlich, wie es zur Christi und der Apostel Zeiten gewesen ist, vorgestellet wird, Sebast. Dorn sculps."

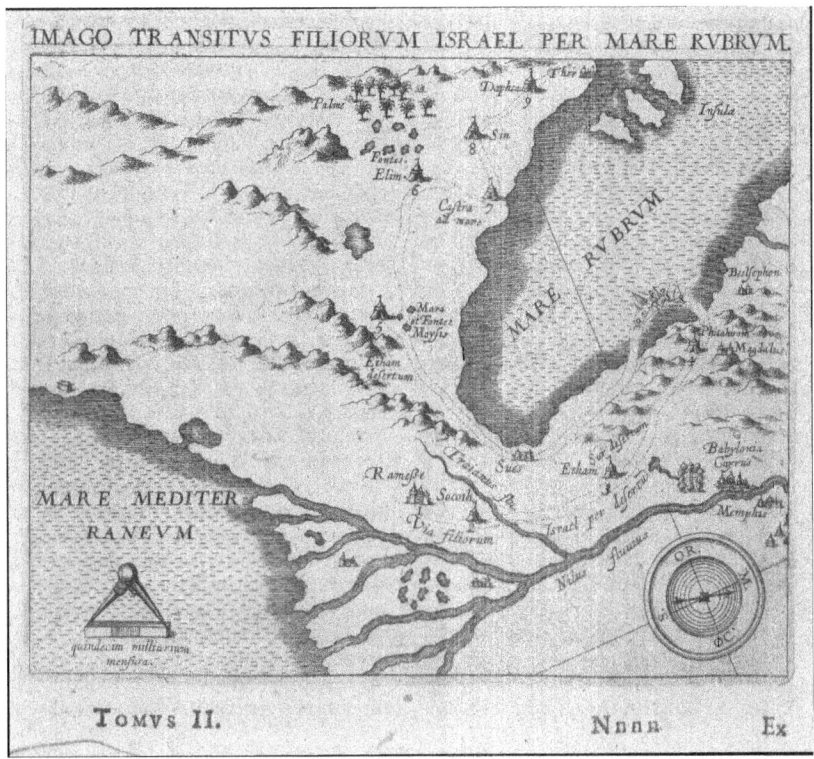

Fig. 73: The circumnavigation of the Red Sea in Franciscus Quaresmius' map, "Imago transitvs filiorvm Israel per Mare Rvbrvm," 150 x 118 mm.

Land. What is more, there are myriad variations on this theme. Unlike the depiction in the *Me'am loez*, the one under review lacks so much as a hint of Miriam's well. In a few of the maps, the tribal camps are signified by tents that are arranged in either geometric patterns or less rigid alignments. The squares in Auspitz's rendering seem to emulate a model that was prevalent in the Visscher family's maps of the Israelites' wanderings, which merited a large audience between the mid-seventeenth century and the mid-eighteenth century.

The Holy Land Map

The fourth map in Auspitz's booklet depicts Eretz Yisrael divided into the tribal allotments. The Land is delimited by the Paran Desert and the Wilderness of the Peoples (i.e., the eastern Sinai) to the southwest, the edge of the Dead Sea to the south, the lands of Moab, Ammon, and Bashan to the east, Mount Hermon and Sidon to the north, and the Great Sea to the west. Dotting the map's surface are many of the Land of Israel's cities that are mentioned in the Bible. The legend on the right side

Fig. 74: Yaacov Auspitz's map of the Israelites' encampment, 191 x 140 mm.

consists of four urban classifications: royal city, city of refuge, Levite town, and Israelite town. Salient, multi-colored lines divide the Land into the tribal areas.

The map's upper-left portion features an embellished cartouche. Seated on the outer side of this frame and donning the priestly vestments and breastplate is Eleazar the Priest. The warrior to the right is none other than Joshua ben Nun. His colorful flag is adorned with the names and symbols of the twelve tribes. Behind his outstretched left hand are the sun and a small crescent moon. The former is crowned by the verse "Sun stand still over Gibeon" (Joshua 10:12). These drawings aside, no miniature illustrations of Biblical events were incorporated into the main body of the map.

Perched above the cartouche is the beginning of Auspitz's preamble "A map of the Holy Land [–] explained as it is in the Bible[,] and interpreted are the locations as per the words of our rabbis of blessed memory." Within the cartouche itself is the rest of the introduction: "Copied from the Latin language by the h[onorable] Yaacov Auspitz of Old Oven: with the approval of our lord and teacher[,] the true genius[,] a presiding court judge[,] and head of our community's yeshiva[,] his honor our teacher and rabbi[,] Moshe Mintz[,] may his candle illuminate forever! And he laid down the law that no one should dare move back the border of the copyist[. And] the listener will obtain [as much] joy and happiness as his soul desires[,] amen: The year 1816..." Under the cartouche is the government's permission to print the map: "Cum Priv. S.C. Majest." Etched in miniscule font by Eleazar's left foot is the name of the

Fig. 75: Yaacov Auspitz's map of the Holy Land, 445 x 326.

copper engraver: "Prixner sc." In sum, the cartouche includes the map's title, the printer's name, the year of publication, the fact that the map was copied, the authorization from both the rabbinical and government authorities, and a quasi-copyright declaration.

Holy Land maps depicting the tribal allotments were indeed quite popular from the early 1500s until the emergence of the modern map in the nineteenth century. In all likelihood, the source of this particular sub-genre is the maps of Nicolas Sanson (1600–1667), which first came out in the seventeenth century. His son, Guillaume Sanson, subsequently printed a few new editions and numerous copies were made as well.[91] The distinguishing feature of Sanson's maps are the inlet-heavy coastline; the diagonal, northeasterly slant of the Land; an elongated and swirly Dead Sea; and the Jordan's arching into Transjordan between the Hula and the Sea of Galilee. Among all the maps that fit under this category, the figures on either side of the cartouche link Auspitz's version directly to that of Tobias Conrad Lotter.[92] It is worth noting that Lotter grounded his own map on one that was produced by his father-in-law,

---

[91] Nicolas Sanson, *Geographiae Sacrae ex Veteri et Novo Testamento desumptae, Tabula Secunda in qua Terra Promissa, sive Iudaea in suas Tribus Partesq, Autore N. Sanson Abbavillaeo.*
[92] Lotter, *Terra Sancta sive Palaestina*, Augsburg 1759.

Matthaeus Seutter. The latter also embellished the cartouche with figures, but not those of Joshua and Eleazar.[93]

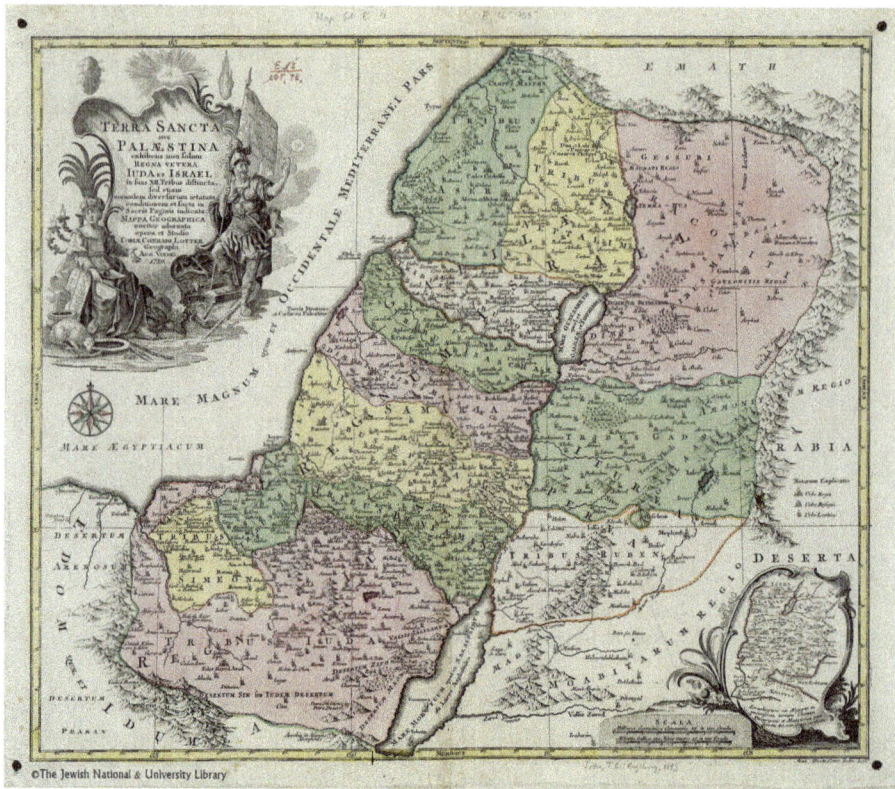

Fig. 76: Tobias Conrad Lotter, *Terra Sancta sive Palaestina*, 480 X 570 mm.

A comparison between Lotter and Auspitz's maps indicates that the latter moderately revised the cartouche. The Jewish author replaced Eleazar's feminine visage with a more masculine, hirsute look and identified the priest with a caption. Moreover, the Lotter version does not include Joshua's name, the moon, or the attendant Biblical verse. Needless to say, Auspitz erased the holy trinity from above his cartouche and filled the vacant space with the map's title. A few changes were also introduced to the body of the map. Lotter's small map on the bottom-right corner was removed, and the same can be said for the toponyms that derive from the Second Temple period, the New Testament, and the classical sources. For example, the Jewish cartographer omitted Judea, Samaria, Galilea, Rhinocorurah (al Arish), Caesarea,

---

93 Seutter, "Palaestinae accurata descriptio geographica, ita adornata ut diversarum aetatum regna conditio et fata, in sacris oraculis indicata intelligi possint," Augsburg 1741 and another version from 1744.

and Ptolemais (Acre). Likewise, the captions in the Sea of Galilee and the Dead Sea were pared down considerably.

Fig. 77: The figure of Joshua ben Nun in the Lotter, Auspitz, and Aharon ben Haim maps (right to left).

In sum, the map in question is by and large a translated replica of Tobias Lotter's map, which drew on earlier works in its own right. More specifically, Auspitz suited his map of Eretz Yisrael to the sensibilities of the Jewish reader by, among other things, adapting its content to the spirit of the Hebrew Bible and "the words of our rabbis of blessed memory."

A Futuristic Map of Eretz Yisrael

Auspitz's fifth map, "the Division of Eretz Yisrael in the Future," offers a schematic, rectangular representation that is oriented to the west. Undulating across the top side is "the Great Sea;" to the left is "the River of Egypt;" to the east are the Jordan River, the Dead Sea, and further down Gilead. The right-hand side is stacked with boxes containing sites that comprise the northern border as per the Hebrew Bible. The area of the Promised Land is comprised, *inter alia*, of twelve parallel strips running from east to west, each of which signifies one of the tribal allotments. Furthermore, an appreciable portion of the Land's center is taken up by Jerusalem and the Temple, which are apportioned to the "priests," "Levites," and "the city's workers." This diagram makes for an idyllic graphical representation of Ezekiel's apportionment of the Land in the End of the Days (45:1–8; 48:1–29).

As far as can be seen, Auspitz copied this particular map from Yehiel Hillel Altschuler's cartographic take on the same theme. Both maps will be elaborated

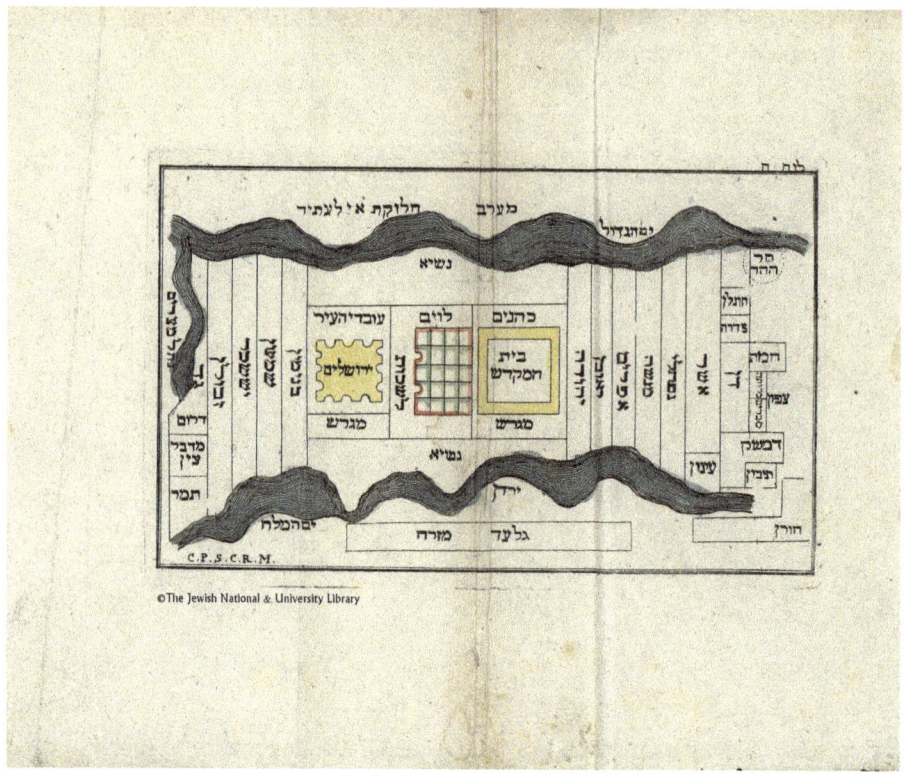

Fig. 78: Yaacov Auspitz, "The Division of Eretz Yisrael in the Future (according to Ezekiel)" Map, 180 x 106 mm.

on in the next chapter. At this juncture, I will merely reiterate two points: the prototype of this cartographic template for Ezekiel 45 and 48 are maps that are attributed to Rashi; and earlier Christian adaptations of Rashi's diagrams would subsequently inspire Jewish authors to re-embrace this subject matter.

Dov Baer (ben Joseph) Yozpa

The two main topics of Dov Baer (ben Joseph) Yozpa's map are the route of the exodus from Egypt and the tribal allotments. In 1820, the first of this map's three editions came out in Vilnius and Grodno as part of a Hebrew Bible with the Rashi commentary.[94] Printed off a different block the same year as the first, the second edition adds

---

94 *The Books of the Prophets and Hagiographa from the Twenty Four [Books of the Bible]: [...]the Translation [...] Corrections of the Mistake in Rashi's Commentary [...] ha'Toldot Aharon [...] We Also Added the Commentary of [...] the Vilna Gaon [...] the Minḥat Shai Co[mmentary] [...] Davar Ḥadash [...] in the Name of Mivaseret Ṣion [...] to Note Every Verse to a Verse that Resembles it From the Stand-*

the following information on the lower-right margin: "Lithography by Peretz Feinroit in Warsaw."[95] Although the design in both these editions is identical, the artistic quality of the latter is a cut above its counterpart.[96] Another version with minor changes is preserved in manuscript form (ink on canvas) at the Moldovan Family Collection in New York. Its cartouche was left blank, so that there is no way of ascertaining the circumstances behind its creation. We cannot, therefore, establish whether this manuscript served as a draft for the printed versions or was a copy of either one.[97] The third print edition was published, with negligible changes, by Moshe Danzigerkron in Warsaw over sixty years after the first two.[98]

All told, the geographic layouts in all the versions of Yozpa's map are practically the same. The map faces east and presents the expanse between the Red Sea to the south and the Euphrates River and Mount Hermon to the north. The Euphrates warrants special attention. Its mouth taps into the sources of the Jordan River to the east. From there, the Euphrates flows west before spilling into the Mediterranean Sea. Not only is the direction of the river's flow illogical from a geographic standpoint, but it belies the Euphrates' location relative to Aram-Naharaim (see the discussion on this topic in chapter 2).[99] Returning to the purview of Yozpa's map, the western border of Eretz Yisrael is the Great Sea. The Red Sea constitutes its outer limits to the south and southeast, and the eastern border runs through the allotments of Gad, Reuben, and half of Manasseh. As alluded to earlier, the Euphrates connects Mount Hermon up in the northeast with Mount Hor to the northwest.

The verse "And I will set your border from the Red Sea to the Sea of the Philistines" (Exodus 23:31) arches across the map's upper righthand corner. Lodged into the upper-left corner are the following words: "Mount Hermon that is Senir that is Sirion that is Sion[,] for these are [its] 4 names." The aforementioned personification of Egypt (the woman riding a crocodile) graces the lower-right corner. On the bottom left is a rectangular cartouche, which is capped by a griffon vulture and a passage from Exodus 19:4: "Ye have seen what I did unto the Egyptians and how I bare

---

point of Language or Subject Matter[...] Grodno: Vilnius: Simha Zymel ben Menahem [Press], 1820, 6 volumes, at the end of *Minḥat Shai*, first part.

95 The name of the printing house, Feinroit Lithography, points to the fact that the new technology of lithography had penetrated the world of Hebrew printing. The lithography is noted beneath the apiary in the Warsaw edition.

96 Unfortunately, the copy at the Israel National Library's Laor Collection has several tears on the margins.

97 I would like to thank the late Dr. Alfred Moldovan for showing me this map.

98 The title and preamble of this edition is as follows: "A List of the Surrounding Borders of Eretz Yisrael [–] to a precise degree and in the correct order as per the order of the genius[,] rabbi of all the Diaspora's sons[,] our teacher the rabbi [–] Elijah of Vilnius of blessed memory[.] And he built from it the order of the Israelites' journey and the Land's borders in the portion of *Mas'ei* and the Book of Joshua, Moses Danzigerkron Lith[ography,] Warsaw, 1881 [Hebrew]."

99 This depiction of the Euphrates may be connected to the fact that it is one of the four rivers said to flow through the Garden of Eden, but this is not self-evident in the map.

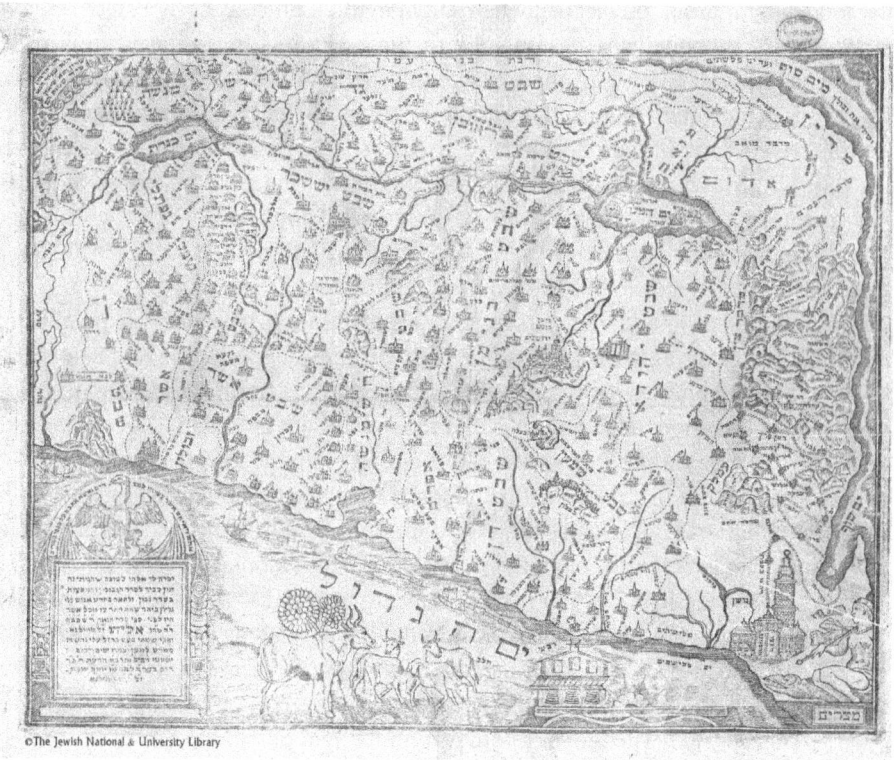

Fig. 79: The map of Dov Baer Yozpa, 388 x 235 mm.

you on the wings of griffon vultures and brought you unto Myself." The text inside the cartouche reads thus:

> Remember me my God for good on account of [my] pondering for an extended [period of] time [over how] to arrange the borders and the voyages in the right order. And for describing in common script on a folio with greater vigor and greater intensity than all my predecessors. As per the layout of the genius[,] the rabbi of all the Diaspora's sons[,]the one and only[,] our teacher the rabbi[,] Elijah of blessed memory from Vilnius. And I placed my imprint with an iron pen on gleaming copper so that it shall last for many days: May many people run to and fro and knowledge proliferate [Daniel 12:4.] These are indeed my words Dov Baer son of my master my father my teacher and my rabbi[,] Joseph Yozpa of blessed memory [from] the holy community of Vilnius.

In the verse "May many people run to and fro and knowledge proliferate" there are dots above the letters Shin, Resh, Vav, He and Ain, encoding the date of the map to 1820/1821. Chronograms, i.e., encoding the date by marking specific letters and calculating their numerical values, are extremely rare in the world of European cartography, but were rather prevalent in Hebrew books.[100]

Fig. 80: Source and adaptation: details from the Bar Yaacov and Yozpa maps.

Scattered throughout the map are scores of Biblical settlements. The vast majority are denoted by a small cartographic symbol: a building with a tower in the middle. Conversely, Jerusalem and Hebron are signified with relatively large cityscapes. In the case of Hebron, the illustration may be a depiction of the Cave of the Patriarchs. For its part, Egypt merits an even larger urban scene. The Israelites' journey through the wilderness is marked by a double dotted line that twists and turns from Egypt to Gilgal and Jericho. The route passes through all the stations, each of which is also denoted by a variation of the above-mentioned tower-studded building. An appreciable portion of this route slithers through a mountainous region. Eretz Yisrael is divided into the tribal allotments, but the map does not contain any renderings of Biblical events.

Needless to say, the accompanying illustrations—the crocodile rider, the rafts transporting cedar wood, the cows, the apiary house, and the griffon vulture atop the cartouche—directly tie this map to that of bar Yaacov in the Amsterdam Haggadah. However, there are also significant discrepancies between the two. A case in

---

[100] For more on chronograms in maps, including their rarity, see Rubin, "A Chronogram Dated Map of Jerusalem."

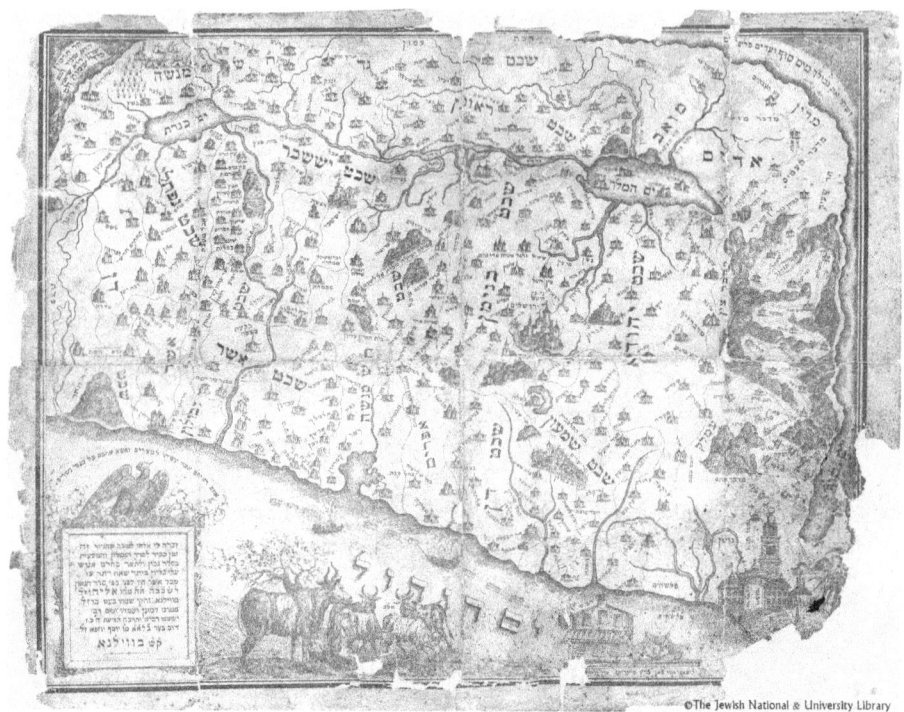

Fig. 81: Warsaw edition of the Yozpa map, 388 x 235 mm

point is Yozpa's signification of *"antiokhi"* (Antioch) on the Land's northern border, which clearly alludes to Rashi's identification of this site with Hamath in Numbers 34:8.

Compared to the Vilnius-Grodno version, there are a few changes to the cartouche's text and the secondary cartographic details in the first Warsaw edition. Above all, the artisanship and the quality of the plate are superior. Prime examples of this difference are the images of the cows, the griffon vulture, the mountains, and the ships at sea. Interestingly, one of the boats towing the above-noted rafts erroneously faces north.

The third print edition, which was published many years later, also contains some minor revisions. A title, "The Map of Eretz Yisrael's Borders," was added to the top of the folio. Moreover, the cartouche and all its accompanying illustrations that graced the first two editions were erased. That said, the printer added his own, unframed preamble in its place, but refrained from mentioning the original version. It also bears noting that all three editions sought to ride on the coattails of the *Vilna gaon* (Genius from Vilnius), namely R. Elijah ben Shlomo Zalman. However, none of them bear even the faintest resemblance to the maps that are attributed to the famed savant, which are discussed in the following chapter.

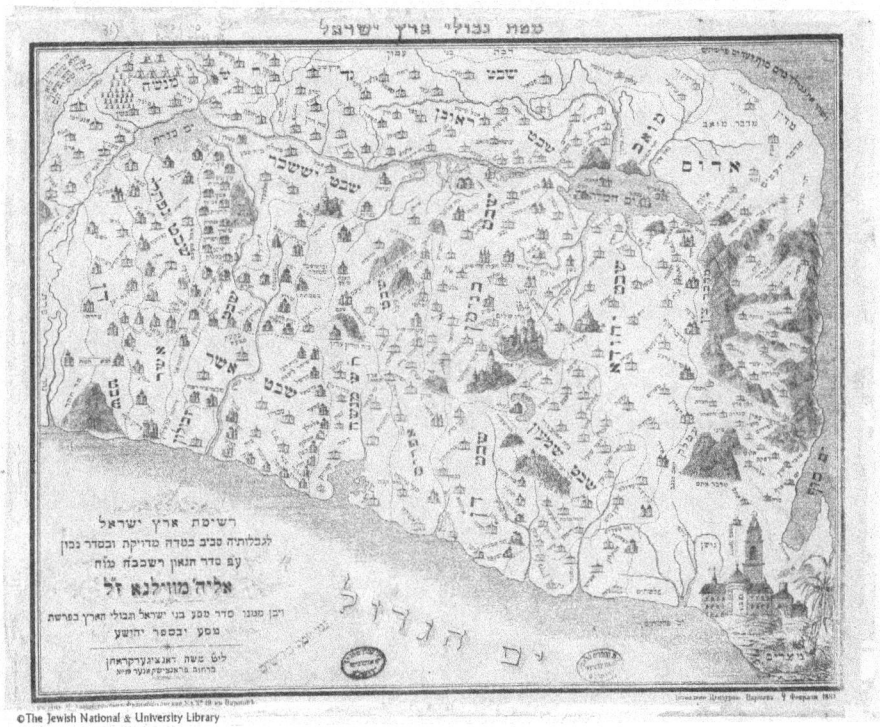

Fig. 82: Moshe Danzigerkron's edition of the Yozpa map, 380 x 300 mm.

The Map of Aharon ben Haim

In 1836, Aharon ben Haim published the booklet *Sefer moreh derekh* (the Book of Guidance for the Road). As noted in the preamble, the work's objective was to shed light on three geographical issues that pertain to the Hebrew Bible: the Israelites' peregrinations through the wilderness; Eretz Yisrael's borders as per the portion of *Mas'ei* (Numbers 33–36); and the tribal allotments according to the Book of Joshua.[101] The author also notes therein that he followed in the footsteps of "our rabbi the great explorer Rashi of blessed memory." Part of his goal, ben Haim stressed, was to clear up contradictions that rear up in various interpretations that he found in "drawings of the borders," that is earlier maps. The first half of *Moreh derekh* is an exegesis on *Mas'ei*, and the second half covers Joshua 15–19. Two more ed-

---

101 Aharon ben Haim, *Sefer moreh derekh* (Book of the Travel Guide), Grodno: Yaacov Yechezkel and Yechezkel Press, Hebrew date [encrypted in the words]: To conceive a paved road (1836). The book's cover includes a title in Russian and the same date of publication as the map.

itions of this booklet came out in Warsaw in 1879 and 1883, at the initiative of one Meir Isaac Hirsch—"the grandson of the author's brother."[102]

Ben Haim's work is accompanied by a map depicting the route of the exodus and the division of Eretz Yisrael into tribal areas. A table on the final page locates different sites on the map using their longitudinal and latitudinal coordinates. This key indicates that the map was an integral part of *Moreh derekh*.

Fig. 83: The map in Aharon ben Haim, *Sefer moreh derekh*, 370 x 282 mm.

Facing east, the map ranges from the Red Sea to the south and Mount Hermon to the north and from the Great Sea to the west and the Jordan to the east. Its basic cartographic design is somewhat reminiscent of the Yozpa map. This affinity lies in the map's territorial purview, the shape, location (north), and form of the Euphrates River, and the Kishon linking the Sea of Galilee to the Mediterranean.[103]

These common attributes notwithstanding, there are many differences between the two maps. Pithom and Raamses were added to ben Haim's depiction of Egypt

---

102 Aharon ben Haim, *Sefer moreh derekh*, Warsaw: Izaak Goldman Press, 1879; idem, 1883.
103 The first cartographer to depict the Kishon flowing into the Mediterranean was Marino Sanudo, and numerous Christian mapmakers followed in his footsteps.

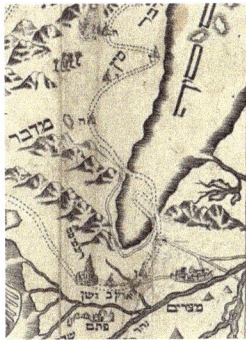 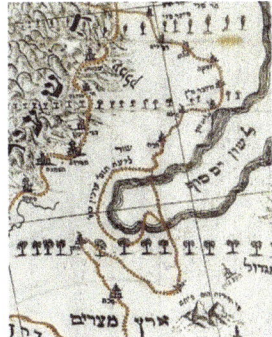 

Fig. 84: The crossing of the Red Sea in maps of bar Yaacov, ben Haim, and Auspitz (left to right).

in the form of large set of buildings. Below these illustrations is a buffalo grazing in "the fields of Zoan." The mountain ranges of Seir and Gilead are drawn in a figurative style. Save for a few vessels and aquatic creatures in the Mediterranean Sea, the map's lower portion is devoid of accompanying illustrations. Likewise, his floral cartouche stands apart from Yozpa's and contains the following text: "In the area of the map is included the [Israelites'] voyages and the borders of the good Lan[d] and the Carmel as per Rashi of blessed memory[;] I compiled [the map] and made [it] fine with thirteen sieve[s]. Here are my words, Aharon son of my teacher Haim." In the corner of the map, next to the cartouche, is an oval frame reading "Copyright made[104] by the above-mentioned in Grodno may his Rock protect and preserve him." On the upper-right edge is the figure of Joshua ben Nun, which was undoubtedly taken from Auspitz's map. Other components, such as the Cities of the Plain within the Dead Sea, turn up in many earlier sources. In light of the above, it is impossible to determine where all the elements of this map derive from. Conversely, ben Haim was the first to give cartographic expression to a topic in Joshua 17:9. As per this verse, Ephraim had cities in the territory of Manasseh. Accordingly, these settlements are depicted as a strip within the area of Manasseh that is labelled "From the land of Tappuah to Kanah Ravine."

Like the maps of bar Yaacov and Auspitz, this one also provides two alternatives for the crossing of the Red Sea. As above-noted, the provenance of this interpretation is a Talmudic aggadah. I already mentioned that the tradition of the partial crossing of the Red Sea appears in the maps of Quaresmius' and Binder. Ben Haim's cartographic account of this event bears a resemblance to that of the latter.

A similar version of this map was incorporated into the second and fourth editions of ben Haim's book, which came out in Warsaw in 1879 and 1883, respectively.

---

104 Perhaps the Hebrew acronym of *qoph* and *shin* in the original stands for *kinyan shalem*, which is basically a warning against copyright infringement.

Fig. 85: Binder's map, 317 x 404, in "Charte Worauf das Iüdische Land vornehmlich, wie es zur Christi und der Apostel Zeiten gewesen ist, vorgestellet wird. Nach dem Entwurff des Herrn W. A. Bachiene von neuem gezeichnet von M. Binder aus Schæssburg in Siebenbürgen, Sebast."

Its cartouche includes the map's title and a short note concerning the printer.[105] In the folio's lower-right corner is a Russian inscription that informs the reader of the

---

105 The text of this particular cartouche reads thus: "'Map of the borders of the good land, and the Israelites' journeys in the wilderness upon leaving the Land of Egypt, until their arrival in the Land of

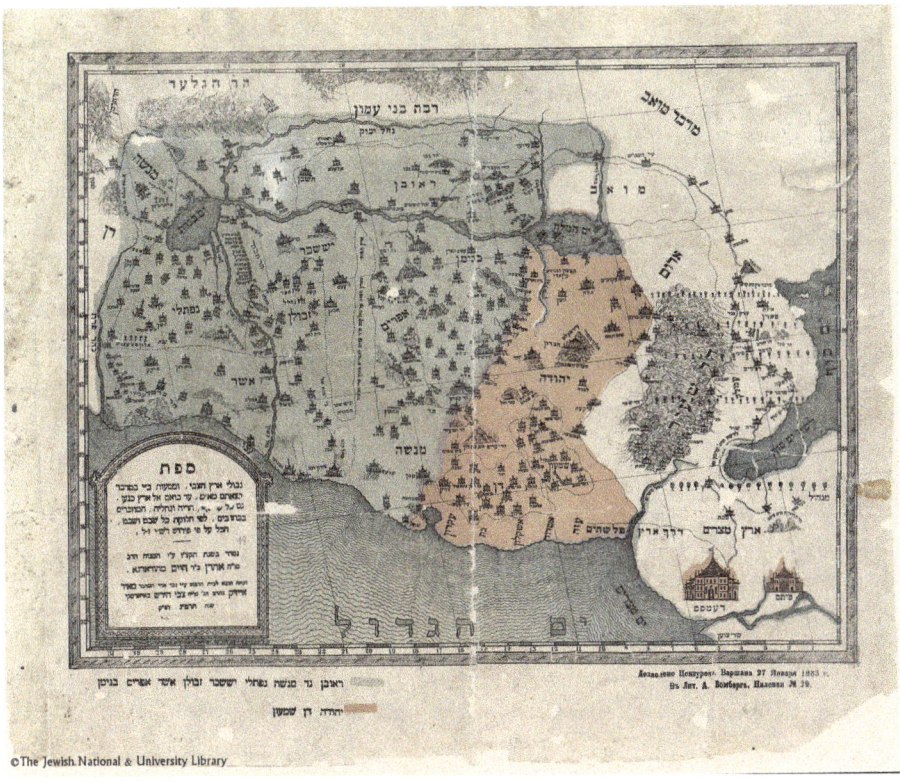

Fig. 86: The Meir Isaac Hirsch edition of the Aharon ben Haim map, 375 x 285 mm.

printing house in Warsaw, the lithographer's name, and the year of publication. While close in form, territorial scope, and content to the original map, this edition is less impressive from an artistic standpoint. For example, the images of Joshua and the buffalo were removed, the cartouche was toned down, and the quality of detail pales in comparison. However, its geographical coverage and content, including the toponyms, are exactly the same.

Another edition of this map was printed in 1882 by one Meir Yekutiel Galgor.[106] This version is furnished with headings in Hebrew (above) and Russian (below) as

---

Canaan, along with all the cities of EY, its mountains, rivers, which are noted in the scriptures, as per the allotment of each and every tribe, all in accordance with the commentary of Rashi of blessed memory.' Arranged in the year 1836 by the late rabbi[,] our teacher the rabbi[,] Aharon ben R. Haim of Grodno. And currently brought to press by the grandson of the author's brother[,] Meir Isaac son of the g[reat] rabbi[,] our teacher the rabbi[,] Tzvi Hirsch Boiarsky. The year 1883[...]"

106 At the top of the map is its Hebrew title and preamble: "Map of the Land's Borders: An explanation of our forefathers' journeys from the time of their exodus from the Land of Egypt until reaching inhabited land. And also the apportionment of the land of Sihon and Og . . . on the eastern side of the

Fig. 87: The Galgor edition of the Aharon ben Haim map, 360 x 280 mm.

well as sizable texts on either side of the map. In the first two lines of the Russian heading is the censor's permission to publish the map as well as the name and address of the printing house.[107] The Hebrew date is encrypted just above the Russian title in the large letters of the passage "If I forget thee, O Jerusalem, let my right hand forget its cunning." The Hebrew block to the right bears the title "Explanation on the sojourns of our forefathers who sojourned for forty years from the time of their exodus from the Land of Egypt until their arrival in an inhabited land. As per the commentary of our great rabbi Rashi of blessed memory." Below these words is the text

---

Jordan[,] which Moses our Rabbi may he rest in peace bequeathed and allotted to two tribes and a half [sic]. And also the Land of Canaan's apportionment among the nine tribes and a half [sic] as per our great rabbi Rashi of blessed memory. This was done by the land-surveying yeshiva student Meir son of our teacher and rabbi Yekutiel Galgor from Grodno. It was arranged in the year 1836 by the deceased[,] the rabbi[,] our teacher and rabbi[,] Aharon ben Haim of blessed memory from Grodno." Below the map is the following Russian title: "Маппа гевулот го-орецъ, т.е. карта Палестины. Пособіе при изученіи св. писанія. Составилъ и копировалъ землемѣръ – таксаторъ Мейеръ Гельгоръ, Вильна: тип. Маца, 1882" [Map of the Borders of the Land, i.e., map of Palestine, to help the study of the Holy Scriptures, copied and edited by land-surveyor Meir Galgor, Vilna, the Matz Printing House, 1882].

107 "Permission to print granted on May 7, 1882; the N. Matz Lithography [Studio], Vilnius."

from Numbers 33–34 enumerating the Israelites' stations in the wilderness and the borders of the Promised Land. Underneath is a key providing the coordinates for all the sites on this map. The text to the left consists of chapters 15 to 19 from the Book of Joshua (the end of chapter 19 is located under the map). This block opens with the words "An explanation of the borders of the nine tribes and a half that took their allotments in the Land of Canaan [...]." Cartographically speaking, the map's geographical purview and content are identical to the original, but the former was crafted by the hand of a master. More specifically, the illustrations were improved upon, the figure of Joshua, the Egyptian cities, and the grazing buffalo assume a new style, and the introduction of color to this map is aesthetically pleasing.

The Map of Gershon Chanoch Leiner

Following in the path and under the tutelage of his father Yaacov and grandfather Mordechai Yosef, one Gershon Chanoch (Henich) Leiner (1839–1891) was crowned the third rebbe[108] of the Izhbitza-Radzin Hasidic court in 1878. Together with his father, Leiner moved from Izhbitza to Radzin where he would spend the rest of his life.[109] According to relevant biographical sources, the rebbe was an educated person who wrote on a variety of topics. His activity hints to a multifaceted individual. On the one hand, Leiner was an opponent of modernity and the Jewish Enlightenment who delved into Hasidism and the kabbalah. On the other hand, he was knowledgeable in many and manifold general fields, particularly the natural sciences, chemistry, and medicine, and mastered a handful of languages.[110] Accordingly, he was best known for his Halakhic-cum-scientific research on the origins of the color Tyrian or imperial purple. It also bears noting that he frequently polemicized with other rabbis —both Hasids and Mitnagdim alike.[111] In any event, Leiner's broad horizons did not come to expression in his Hasidic court leadership or halakhic writing.[112]

Among the rebbe's works is *Sefer sidrei tohorot* (the Orders of Purities Book). This treatise gleans information on *Keilim* (Utensils) and *Oholot* (Tents)—two tractates from the Mishnah that lack orderly Talmudic counterparts—from throughout the six orders of the Talmud. *Sidrei tohorot* is comprised of two volumes, which came

---

108 The Hebrew term for rebbe is ADMOR—an acronym for our master, teacher, and rabbi.
109 Ben-Dor, "Leiner, Gershon Hanokh (Henik) ben Jacob."
110 Magid, *Hasidism in Transition: the Hasidic Ideology of Rabbi Gershon Henoch of Radzin in Light of Medieval Jewish Philosophy and Kabala*; idem, *Hasidism on the Margin: Reconciliation, Antinomianism, and Messianism in Izbica/Radzin Hasidism*.
111 Marcus, *Hasidism*, 243–246 [Hebrew]. Literally "opponents," the Mitnagdim resisted the Hasids' spiritual-cum-pietist innovations.
112 Magid, *Hasidism in Transition*, p. 58, note 34.

out thirty years apart.[113] At the end of the first volume is a map that bears the title "This shall be Your Land as Defined by its Borders All Around" (Numbers 34:12). The map faces north and is arranged on a grid of coordinates ranging from 27 to 34 degrees longitude and 49 to 56 degrees latitude.[114] These features attest to the fact that it was predicated on a modern map.

Leiner's map depicts the Land of Israel and a portion of the Sinai Desert. More specifically, its purview stretches from Sidon and Mount Hor to the north to the Israelites' stations in the wilderness to the south, and from the Egyptian border to the southwest and the Mediterranean Sea to the west to Damascus and the hills of Transjordan to the east. The western shoreline of Eretz Yisrael is practically rectilinear. Accurately situated below Acre, the Bay of Haifa is recessed much deeper than it actually is. Kanah Ravine empties into the Mediterranean at a small inlet by the coast of Dor. Between the latter and Ekron, the seacoast juts out to the west (!). The most salient feature on the northern Sinai coast is Lake Bardawil, which is identified by a caption as *yam suf* (literally: Sea of Reeds—see the discussion in the next paragraph). The Jordan emanates from two tributaries and flows, via "the Sivkhi Lake (i.e., Sumchi or Hula), to the body of water labelled "the Sea of Galilee and Gennesaret." At this point, it continues on to the Dead Sea, which is elongated and somewhat pointy. The majority of the map's surface is devoted to the allotments of the tribes, each of which is denoted in large font. Within each tribal territory, the settlements are printed in small letters. With the exception of Benjamin and Judah's border areas, the author evidently sought to keep the number of toponyms to a minimum.

A close look at the map's features reveals that it was arranged along the lines of Matthaeus Seutter's corresponding map.[115] To begin with, both authors identify Lake Bardawil with the Red Sea. In fact, Seutter labels it *"Lacus Sirbonis Jam suph Mare Typhonis et Rubrum."* Put differently, he also mistook the lake for the Biblical *yam suf* (Sea of Reeds) or the "Red Sea" in the classical sources. In the two maps, the Jordan proceeds from north to south in nearly a straight line. Unlike many of the period's other maps, the Jordan does not go off on a looping tangent between the Hula and the Sea of Galilee. The Bay of Acre is recessed and round. Skewing wildly from reality, the Yabok (Zarqa) River empties into the Yarmouk from the north. Both maps refer to "the Abarim Mountains" (Latin: *Abarenorum montes*) in plural, rather than the single form of the Torah—"Mount Abarim" (e.g., Numbers 27:12).[116] Furthermore,

---

113 Leiner, *Sefer sidrei tohorot*, vol. 1, tractate Keilim, Józefów: Shlomo and Baruch Zetser and Yechezkel Raner Press, 1873; vol. 2, tractate Oholot, Piotrków: Shlomo Belchatovski Press, 1908 [Hebrew].
114 Whereas the longitudinal lines are similar to the current standard, the latitudinal lines differ. Leiner's coordinates derive from his source, Seutter's analogous map. It bears noting that the latter came out well before Greenwich became the universal prime meridian.
115 Seutter, "Palestina sev Terra a Mose et Iosua occupata et inter Iudaeos distributa per XII Tribus vulgo Sancta adpellata, a Ioanne Christophoro Hernberg [...]," Augsburg 1745.
116 "Abarim Mountains" occur also in plural in the Hebrew bible in Numbers 33, 47–48.

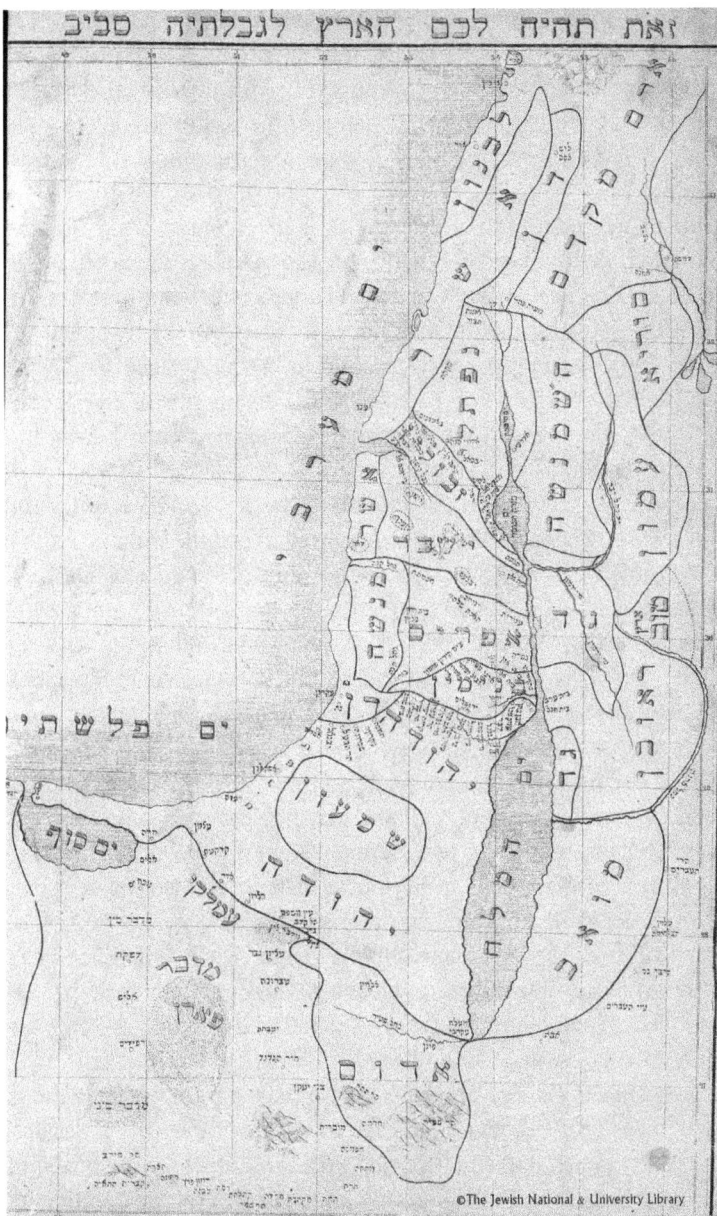

Fig. 88: Gershon Chanoch (Henich) Leiner, The "This shall be Your Land as Defined by its Borders All Around" Map, 224 x 360 mm. The map is located at the end of his book, *Sidrei tohorot*, vol. 1, tractate Keilim.

they locate these mountains far to the east of the Jordan, on the border of Moab and past the allotment of Reuben. These points of resemblance suffice to tie the two maps and differentiate Leiner's from other late nineteenth-century maps. Especially noteworthy is the identification of Lake Bardawil with *yam suf.* Although quite a few European maps refer to Lake Bardawil by its classical name of Lacus Sirbonis, Leiner and Seutter are the only cartographers who identified it with the Sea of Reeds or the Red Sea. In contrast, most other mapmakers felt that the Sea of Reeds was synonymous with the Red Sea (i.e., the major body of water to the south, east, and west of the Sinai Peninsula).

In comparing the source map to that of Leiner, it must be emphasized that the former is larger and much richer in detail—toponyms and drawings included. Moreover, the Hasid did not mechanically copy/translate Seutter's map word for word. Instead, he adapted certain elements and pursued his own course. An interesting detail that bears witness to Leiner's independence, as well as his preference for Jewish sources, is his addition of "the cave of Pameas" near the mouth of the Jordan River. According to the Talmud, the "Jordan emanates from the Pamyas Cave and proceeds into the Sivkhi Lake, and the Lake of Tiberias, and the Lake of Sodom."[117] In other words, a cave over the mouth of the Banyas constitutes the source of the Jordan's water. Similarly, the rebbe preferred to call the Hula by its Talmudic name of Sivkhi rather than its Biblical name of Waters of Merom (Joshua 11:7), which is found on many maps.

As opposed to Seutter, Leiner considered Laish-Leshem and Dan separate places and even situated them at quite a distance from each other. With respect to the Yabok River, the Hasid added a segment that flows from north to south, thereby deviating from both the source map and reality. Whereas Seutter leaves out Kanah Ravine, Leiner made note of it on the Ephraim-Manasseh and the Ephraim-Asher borders. The relation between the territories of Ephraim and Manasseh on this map is unconventional. On most Holy Land maps, the two allotments are parallel strips between the Jordan and the Sea, with Manasseh sitting north of Ephraim. However, Leiner dovetailed these areas from north to south. In other words, Manasseh is to west of Ephraim and blocks it off from the Sea. Manasseh barely reaches the shoreline itself on account of Dan's narrow corridor hugging the coast. This representation of Ephraim and Manasseh contravenes the straightforward interpretation of the Hebrew Bible whereby both territories juxtapose the Mediterranean Sea (Joshua 16:8; 17:10). Leiner's depiction of Gad is also unusual, as it is broken into non-contiguous parts: an enclave southwest of Reuben; and a swath of land between the latter and half of Manasseh. In the vicinity, one can also find "the land of Tob"—the haven of Jephthah the Gileadite (Judges 11:3). To the best of our knowledge, yet another unique feature of this map is a river that flows from the south of Damascus to three fictional lakes. Leiner also deviates from the literal reading of the Hebrew Bible on the matter of Ezion-Geber. Instead of placing it on the coast, near Eilat, the site is in the middle of the desert. The source of this error is indeed the Seutter map. However, one would have expected a student of the Bible not only to identify Ezion-Geber as one of the

---

[117] Babylonian Talmud, tractate Bekhorot, 55a.

Israelites' stations, but to remember its maritime role as one of King Solomon's shipyards (1 Kings 9:26; 2 Chronicles 8:17).

Leiner's map thus reflects a trend among rabbis and Talmud scholars from the seventeenth to the nineteenth centuries to take a map from a Christian source, copy it, and adapt it for a Jewish audience. The rebbe did not suffice with mere replication, but added components of his own that derive from the Hebrew Bible and Mishnah. What is more, he introduced a few elements that appear to be inexplicable errors.

The Map of Benjamin Lichtenstaedter

Amtaḥat binyamin (Benjamin's Satchel), a book by one Benjamin Lichtenstaedter, was printed in 1846. The lion's share of this work consists of explanations for various matters that come up in the Hebrew Bible.[118] According to the preface, the writer was born in Baiersdorf, Germany in 1794. Moreover, Lichtenstaedter himself testifies that he was not a Torah scholar, but a simple person. As alluded to in "the song of thanks" at the end of *Amtaḥat binyamin*, he went through childhood as an impoverished orphan. Upon attaining a measure of success and comfort, Lichtenstaedter decided to publish his thoughts on Biblical exegesis. On several occasions, the book even takes issue with famous exegetes.

At the outset of *Amtaḥat binyamin* is a map of the Near East. According to the map's preamble, it is a "reference for understanding several verses in the Book of Genesis and the rest of . . . the Hebrew Bible." The map's geographical purview extends from Persia in the east to Egypt in the west and from Mount Ararat to the north to the Red Sea and Arabia down south. It provides an in-depth account of various rivers, settlements, and regions, all of which are labelled in Hebrew. Even at first glance, it is obvious that many of the toponyms, such as "rocky Arabia" (*Arabia Petrea*) or "the felicitous Kingdom of Arabia (*Arabia Felix*), derive from Latin sources, so that the map is clearly based on a non-Hebrew source. Similarly, a few captions—"*ṣiliṣia*" (Cilicia) and "*capadoṣia*" (Cappadocia) included—are Hebraicized versions of Latin place names in which the derivative replaces the Latin "c" with the Hebrew *sadhe*.

Alongside toponyms from classical sources and the author's own era, the map contains full-fledged Biblical topics and sites. For example, it includes an illustration of Noah's ark balanced atop the "Ararat Mountains" and toponyms like Migdol, Raamses, Pathros, and Zoan. On the other hand, there are non-Biblical sites like Cairo and Memphis in Egypt and Sura—the seat of a famous yeshiva in Babylon. Lichtenstaedter also offers a unique rendering of the Garden of Eden. Two of Paradis-

---

[118] Lichtenstaedter, *Sefer amtaḥat binyamin*, Fürth: Zürendorffer & Sommer, 1846. Another edition came out in 1848.

Fig. 89: Benjamin Lichtenstaedter's "The Lands of the Orient" Map, 395 x 251 mm, in *Amtaḥat Binyamin*.

e's rivers, the Tigris and the Euphrates, basically adhere to their real-world courses, and the confluence between them (presently known by the Arabic name of Shatt al-Arab) is identified as Pishon. The Gihon runs parallel to and spills into the Persian Gulf to the west of the wide estuary, but no such river actually exists. Paradise is flanked by Babylon to the west and the Land of Nod to the east. Knifing through Eden is the aforementioned confluence of the Tigris and Euphrates. An illustration of Adam and Eve's expulsion from Paradise adorns the Garden's eastern outskirts.

An in-depth analysis points to the Visscher family's much-emulated "Map of the East" as the likely source of Lichtenstaedter's map. Netting myriad editions in various sizes, the source was rather commonplace in Dutch and German Bibles. For this reason, scholars would be hard-pressed to identify the exact version that the Jewish writer availed himself of.[119] The similarities between the two is also pronounced in the above-noted details of the Lichtenstaedter's map, including its translated Latin names as well as the mountain ranges and rivers for all their twists and turns. Par-

---

[119] The earliest version of this map was, in all likelihood, the Visscher family, "Die Gelegentheyt van t'Paradys ende t'Landt Canaan Mitsgaders de eerstbewoonde Landen der Patriarchen, door C.I. Visscher, Abraham van den Broeck fecit, C.I. Visscher Excu," in a Bible, Graven-Hage 1645. However, since Lichtenstaedter's *Amtaḥat binyamin* was written and printed in German, there is a strong possibility that its source is the German edition of this Christian map: "Die Gegend des iridischen Paradieses und des Landes Canaan […]," Erben 1736.

ticularly striking is the resemblance between the two maps' account of the Pishon and Gihon Rivers, their incorporation of a fictional river to the west of the Tigris-Euphrates estuary, their design of the Garden of Eden and its proximity to Babylon, and their imaginary lake to the south of Paradise.

Fig. 90: The Visscher family, "Die Gegend des iridischen Paradieses und des Landes Canaan," 505 x 378 mm.

Most compelling are Lichtenstaedter's revisions *vis-à-vis* the original map, which he incorporated on the basis of his own knowledge of and outlook toward the Hebrew Bible. For instance, he added "Edrei[,] land of the Rephaites [giants]" (the site of the Israelites' battle against Og)[120] on the eastern side of the Jordan and replaced Palmyra with the Hebrew name Tadmor. Moreover, to the east of Eretz Yisrael and Syria, he integrated the following reference to the Arab nomads: "The sons of Keturah are the *zaraṣanan* foreigners." In other words, he deems the peripatetic Saracens to be Abraham's progeny from Keturah (Genesis 25:2–4).

Lichtenstaedter's "Lands of the East" Map stands out among those that Jews copied from Christian European cartographers. Like many of the other maps under review, it was published within the framework of a traditional Jewish exegetical work. That said, it is probably the only known Hebrew map that displays all the

---

120 See Numbers 21:33; Deuteronomy 3:1.

lands of the Orient. In so doing, the author vastly expanded the accepted cartographic purview of Hebrew maps from Eretz Yisrael and Sinai to the entire ancient East.

Conclusion

In this chapter, I surveyed Jewish-authored Hebrew maps that were the fruit of the emulation and replication of maps by Christian-European cartographers. The Jewish compilers imbued their own works with geographic knowledge and representations of the Holy Land from the legions of contemporaneous Christian maps. Put differently, they adopted their Christian counterparts' graphical image of the Land's topography, borders of the tribal allotments, site locations, and settlement types (e.g., cities of refuge or Levite towns). Moreover, the Jewish authors frequently included elements from their source material's artistically rich illustrations—from both the maps themselves and the accompanying marginalia—in their own work. Content-wise, the two most prominent characteristics of these Hebrew maps were indeed central themes in the era's Christian Holy Land cartography: Eretz Yisrael's apportionment into tribal territories; and the Israelites' route through the wilderness. What is more, the Jewish emulators introduced their sources' encyclopedic approach to Hebrew cartography. In consequence, quite a few of their maps are bursting at the seams with information.[121]

Over the course of the replication, translation, and adaptation process, the makers of these Hebrew maps toiled to suit their work to the tastes of the Jewish reader. In fact, many of them documented these steps in their preambles. Christian traditions were by and large erased; however, traces of this lore often remained in the Hebrew map, usually due to errors or lack of knowledge. Conversely, the Jewish authors added more than a few sites and traditions that did not exist in the Christian sources. Needlessly to say, these innovations were motivated by a desire to present aspects of the Jewish tradition.

As part of my analysis, I endeavored to identify the Christian source of each Hebrew map. In several cases, I am quite confident in the accuracy of my identification. However, given the proliferation of map replication between the seventeenth and nineteenth centuries, it stands to reason that for the rest of the maps there was an additional source that I was unable to track down. In all likelihood, these "missing pieces" served as an intermediate link between the original and the adaptation. A distinction should be made between two approaches to this sort of replication. The first, which clearly guided Zaddik, bar Yaacov, ben Yaacov, and Auspitz, involves copying directly from Christian-authored maps. These Jewish authors expressly stated that they had come across a Christian map in a "foreign" language. Furthermore, they testified that they had personally copied and translated the source, while ex-

---

[121] Barber, "Visual Encyclopaedias: The Hereford and other Mappae Mundi."

punging any and all items that deviated from the Jewish tradition. The second approach informs the cartographic work of Yozpa and Aharon ben Haim, both of whom leaned on bar Yaacov's map. Unaware of the fact that bar Yaacov had drawn heavily on Christian maps, Yozpa and ben Haim were under the impression that their source was "a hundred percent kosher."

Auspitz's map of the future apportionment of Eretz Yisrael involved a more complex form of replication. As will be discussed in the next chapter, the genesis of these sort of maps is most likely Rashi's exegesis. However, the savant's insights were discussed in Christian works well before Jewish cartographers re-embraced this theme. In this respect, Quaresmius and Binder's allusions to the Talmudic tradition of the crossing of the Red Sea are intriguing, for they reveal that emulation was not a one-way street.

Beginning with the maps produced in seventeenth-century Holland, the subgenre that was featured in this chapter indicates the extent to which the copying of Christian maps was accepted in the Jewish community. In the first place, this sort of borrowing implies that there were appreciable cultural ties between Jews and their Christian neighbors. Within this framework, geographic and other types of knowledge flowed from Christian to Jew. This was not only the case in Italy and Amsterdam, but also in eighteenth- and nineteenth century Eastern Europe. As evidenced by the works of Shlomo of Chelm and Gershon Chanoch Leiner of Izhbitza, it even penetrated the very heart of rabbinic scholarship. Of course, it bears noting that the replication of maps was but one, rather minor, expression of this phenomenon.

Educated Jews in both Western and Eastern Europe were apparently quite familiar with Holy Land maps that were printed in Latin and other European languages. Besides the cartographic representations of Shlomo of Chelm and Yehonatan ben Yaacov, another example of this phenomenon is two maps that are included on the inventory of one Simhah Pinsker's library (his son published the list in a booklet called *Mazkir le'vnai reshef* [A Reminder to the Seraphim's Sons]).[122] The existence of the maps from Amsterdam or those in the possession of educated, Western European Jews in the second half of the 1800s should come as no surprise. Less expected, though, are the maps that were copied in East Europe between the eighteenth and early-nineteenth century. These replications attest to the fact that there were also Eastern European Jews who welcomed the currents blowing in from outside the Jewish community. In fact, these same individuals, most of whom were Talmud scholars, were open to the continent's languages and general studies even *before* the rise of the Jewish enlightenment.

Thanks to the publication of Robinson and Smith's seminal book, Heinrich Kiepert's research, and above all the subsequent advances in pertinent fields of study, new and accurate information about the Holy Land's geography was available by the

---

122 Bardach, *Mazkir le'vnai reshef: [...] A List of the Collections and Manuscripts that Remain from the Estate of [...] R. Simhah Pinsker [...]* [Hebrew].

mid-1800s. However, this knowledge was not integrated into the Jewish maps that we have discussed until now. As discussed in chapter seven, this sort of modern, scientific information would only penetrate the Hebrew map within the framework of the Haskalah movement.

5 The Debut of the Tribal Allotments on the Traditional Hebrew Map

The division of the Land of Israel into the tribal allotments is a quintessential geographic topic. Its display on maps feels almost natural, and Christian cartography indeed adopted this theme from the very outset. For instance, an extant part of the Madaba Map signifies the territories of Simeon, Judah, Benjamin, and Dan with captions.[1] This map likely drew on a now lost precursor that accompanied Eusebius' *The Onomasticon*. Furthermore, many place descriptions in *The Onomasticon*'s text refer to tribal affiliation.[2] Jerome's *Liber Locorum*—a fifth-century Latin translation of Eusebius' above-mentioned work—was prevalent in the West and influenced many intellectuals in the generations following its creation. In addition, it had a profound impact on medieval Christian maps.[3]

The tribal allotments surface, either expressly or implicitly, in manuscript-based maps. For example, it is clearly the topic of a map that graces a ninth-century manuscript of the Book of Joshua.[4] There are also a couple of world maps from the early Middle Ages with hints of the tribal areas.[5] In Marino Sanudo's early fourteenth-century map, the division into tribes looms large in the spatial arrangement.[6] This map deeply influenced Holy Land maps from the dawn of the print era, not least the aforementioned maps of the Catholic scholars Montanus and Adrichom. What is more, the allotments are a major topic in the early print maps of Protestants, like Jacob van Liesveldt and Jacob Ziegler.[7]

In contrast, the preferred topic of the traditional Hebrew maps—those of Rashi and his legion of emulators—was the borders of the Promised Land. As discussed in the opening chapters, their authors completely eschewed the apportionment of the Land by tribe. These maps continued to employ diagrammatic, square templates. More specifically, this genre remained faithful to the Hebrew map's traditional framework, namely a schematic design of Eretz Yisrael and its borders that is devoid of any realistic geographic representations in all that concerns the layout of the land, the

---

[1] For more on the Madaba Map, see Avi-Yonah, *The Madaba Mosaic Map*; Piccirillo and Alliata (eds.), *The Madaba Map Centenary 1897–1977*.
[2] Eusebius, *The Onomastikon of Eusebius*. The book itself is discussed in the introduction.
[3] Levy-Rubin, "From Eusebius to the Crusader Maps: The Origin of the Holy Land Maps."
[4] O'Loughlin, "Map as Text."
[5] Edson, *Mapping Time and Space*; Harvey, Paul Dean A., *Mappa Mundi, The Hereford World Map*. A fine example of this genre is the so-called "Cottonian World Map;" see Barber, *The Map Book*, p. 47.
[6] Sanudo, *The Book of the Secrets of the Faithful of the Cross*.
[7] Jacob van Liesveldt, Die ghelegentheit ende die palett des lants van Beloften, From a Dutch Bible, Antwerp: Jacob van Liesveldt, 1526; Jacob Ziegler, Quinta tabula universalis Palaestinae, continens superiores particulares tabulas; idem, *Quae intus continentur Syria, [...] Palestina, [...] Arabia, [...] Aegyptus [...]*, Argentorati, Apud Petrum Opilionem, Strassburg: M.D.XXXII [1532]), Signum Biiij, map no. 5.

shoreline, the mountains and valleys, the rivers and brooks, etc. The only Jewish authors to embrace the tribal allotments were the aforementioned adapters of Christian maps. Against this backdrop, it is evident that this popular Christian cartographic topic entered the Jewish field by way of replication and emulation.

Towards the end of the eighteenth century, the traditional Jewish mapmakers introduced, what for them, was a new wrinkle: the division of Eretz Yisrael into tribal allotments. In light of the above, the following question begs asking: Was the depiction of the tribal areas in this Jewish genre a direct result of Christian maps, or did this "Christian theme" reach traditional Jewish cartography through the mediation of the above-mentioned Hebrew adaptations? In the pages ahead, I will trace the development of the Hebrew maps that presented the tribal allotments on the traditional, square, and unillustrated templates. Besides this innovation, other components that were heretofore alien to Jewish cartography entered these maps, and these will be discussed as well. We will return to the question of the influences behind these developments in the chapter's conclusion.

The Map of Joshua Feivel ben Israel

In 1772, an all but unknown scholar by the name of Joshua Feivel ben Israel (Teomim) of Tarnogród (d. 1777)[8] published his work *Sefer kiṣvai areṣ* (Book of the Land's Edges).[9] According to its cover page, the book is "an explanation of the Land of Canaan for all its surrounding borders in the portion of These are the Voyages [i.e., *Mas'ei*] and the arrangement of the Land's division into the nine tribes and the half a tribe as is written in Joshua 15, 16, 18 [...] and the arrangement of the Land's division as is written in the end of Ezekiel." Joshua Feivel ben Israel (henceforth also Teomim) endeavored to provide a systematic account of Eretz Yisrael's borders and reconcile the contradictions between various sources in the Bible and the literature of the Sages, including different commentaries on these texts. As such, *Kiṣvai areṣ* frequently refers to both the first circle—Ammon, Moab, and Midian—and more distant circles—Persia and Media as well as India and Cush (Ethiopia)—of neighboring lands. Towards the beginning of the first edition, Teomim inserted two exceedingly plain and coarsely designed maps: the first covers the Land's borders and the tribal allotments; and the second is a map of Eretz Yisrael as per the prophecy of Ezekiel. In crafting the former, he pioneered the inclusion of the tribal areas in works of Hebrew cartography.

---

[8] There is not a single reference to Joshua Feivel ben Israel in the entry for "Tarnogrod" by Nathan Gelber in the *Encyclopaedia Judaica* (2nd ed., vol. 19, 2007, p. 516.). For a few snippets of information about Teomim and his family, see Buber, *A Lofty Town, is the City of Zholkva*, p. 32 [Hebrew].

[9] Joshua Feivel ben Israel of Tarnogród, *Sefer kiṣvai areṣ*, Zholkva: Printing House of the Grandsons of Uri Feivish Segel, 1772 [Hebrew].

The first map bears no title. Square and schematic, it faces east. The map's layout is structured with the help of straight lines. Abounding in toponyms, the map contains no artistic or realistic elements. At its core, Teomim's map bears the imprint of the Rashi cartographic tradition. This legacy is evident in the place names and the Land's boundaries: the northern border extends from Mount Hor to Hazar Enan; the eastern border runs from Riblah via Ain, the Sea of Galilee, and the Jordan to the tip of the Dead Sea; the southern border runs from Kadesh Barnea, on through Azmon, to the River of Egypt; the unlabeled Mediterranean coast is the western border. It also bears noting that the Jordan, the Sea of Galilee, and the Dead Sea are not artistically rendered, even in the form of simple lines. Put differently, the only signification that these bodies of water merit are captions.

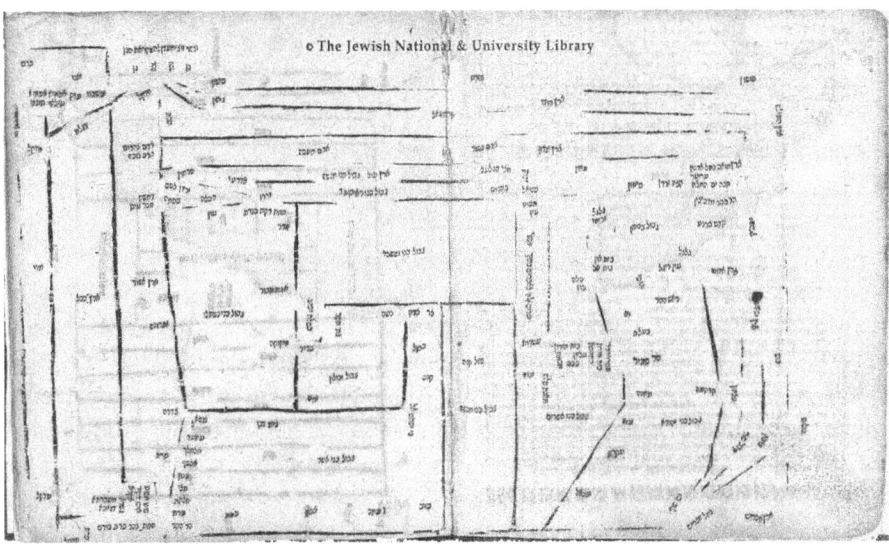

Fig. 91: Joshua Feivel ben Israel (Teomim), map of Eretz Yisrael, 310 x 200 mm.

An analysis of the many place names on Teomim's map discloses at least five toponymic categories. Leaning on Genesis 2 (11–14), the first category includes the four rivers of Paradise, "the land of Havilah," and the "river that leaves Paradise to water the Garden."[10] The Gihon and Pishon flit across the upper portion of the map, forming its eastern edge. Flowing from east to west into the Mediterranean Sea, the Tigris and Euphrates can be found on the north side. The land of Havilah and the Garden of Eden are on the southeastern and northeastern corners, respectively.

---

10 Representations of Paradise and its four rivers were quite prevalent in medieval as well as in early printed Christian maps. That said, there is no evidence that Teomim could have been familiar with these maps. He may very well have predicated his own map on the Book of Genesis.

Eretz Yisrael's borders as per the portion of *Mas'ei* constitute the second category. In this respect, the map perpetuates the cartographic tradition of Rashi and Elijah Mizrachi. The names of the border towns dot the Land's boundaries.

The third category is the tribal allotments according to the Book of Joshua. Each territory is delimited by straight lines. Some of these areas suffice with just the name of a tribe (e.g., "Border of the Issacharites"), while others contain one or more toponyms. Relatively-speaking, Judah is awash in sites. Located in the middle of this tribe, Jerusalem is exclusively referred to as "Jebusite"—its pre-Davidic name. Along the northern border of Asher, the towns appear as a quasi-list. Teomim also incorporated the strip of "These cities of Ephraim within Manass[eh]," while leaving out the territories of Simeon and Dan. The only hint to the Biblical story of the Danites' relocation up north is the signification of Leshem—a city near Mount Hermon that was conquered and occupied by the tribe.

Settlement names deriving from the Mishnah and Talmud comprise the fourth category. For instance, Hammat and Rakkath are denoted near the Sea of Galilee, and Hukkok and Gerar in the area of Naphtali.[11] By the upper-left corner are the following places in Babylon, which are mentioned in tractate *Kiddushin* (Betrothals) of the Babylonian Talmud: Moxoene, Diglath (the Aramaic name of the Tigris River), Ma'avarata, Upper Apameas, and Lower Apameas.[12]

The final category is places beyond the boundaries of Eretz Yisrael. Within this framework, Teomim introduces another circle of neighboring lands that come up in the Bible. To the Land's north are three parallel strips running from east to west: the land of Assyria; the land of Babylon; and Persia and Media. The Tigris River separates between the last two areas. Although not marked with a line, the Euphrates River appears twice along Eretz Yisrael's northern border. The Land of Edom and the Land of Ishmael are adjacent to its southern boundary. Cush is situated further right, south of Egypt. As noted, the land of Havilah is located on the top-right corner; and to its east is "the Land of India." By covering a wide swath of territory outside the Land of Israel's borders, "from India and until Cush," the map naturally alludes to the Scroll of Esther. From the standpoint of the Bible's geographical purview, the map can be said to encompass the entire world, with the Land of Israel at its center.

In sum, little is known to us about the map's compiler; and save for the book, there is no information on the circumstances behind its creation. At any rate, the map's primary importance lays in the fact that it is the earliest of the traditional Hebrew maps to depict the tribal allotments. Put differently, Joshua Feivel ben Israel should be credited with introducing a new theme into this cartographic sub-genre.

---

11 Hukkok surfaces in the Jerusalem Talmud, tractate Sanhedrin 3 9, p. 21d. Gerar is mentioned in idem, tractate Sabbath 18 2, p. 16c. In the main text of *Kiṣvai areṣ*, Teomim distinguishes between the Gerar near Tiberias and the one near Gaza.
12 Babylon Talmud, tractate Kiddushin 71b.

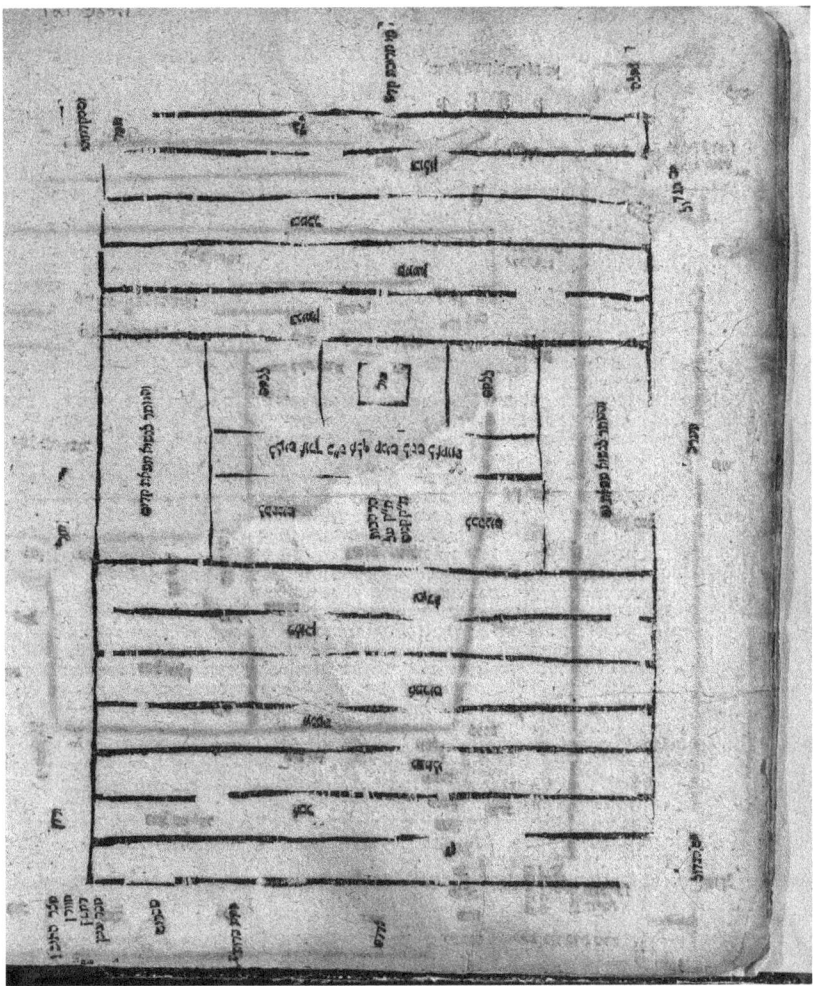

Fig. 92: Teomim's map of the tribal allotments as per the Book of Ezekiel, 200 x 310 mm.

The second map depicts the Land's apportionment at the End of the Days as per the prophecy of Ezekiel (45:1–8; 48). Interpreting these passages, Teomim underscored the Temple and portrayed the areas of Jerusalem that are to be designated for the priests, Levites, and the *nasi* (president, or political head), among others. The rest of Eretz Yisrael consists of tribal areas, which take the shape of parallel strips aligned from east to west. Seven of the tribes are positioned to the north of Jerusalem, and five to the south. The provenance of this graphic model is a map that is attributed to Rashi, which was copied by Christian authors from as far back as the Middle Ages and later appeared in print.[13]

---

13 For a discussion on the graphic model of Rashi, see chapter 1. Sed-Rajna examines the replication

Hauran, Damascus, Hazar Enan, Sibarim, Hamath, Berothah, and Zedad (Ezekiel 47:16) can be found on the northern edge of the map. Lining the southern border are the Dead Sea, Tamar, the disputed waters of Kadesh (as per Numbers 20), and the River of Egypt. Conversely, the east and west suffice with the Great Sea and the Jordan River, respectively. The rectangle at the heart of this map is divided into three parallel strips. Its northernmost unit reads "To the priests" (twice) and "The Temple Mount [is] 500 by 500 cubits." An informative sentence takes up most of the middle compartment: "For the Levites, a length of 25 thousand cubits for them for an estate." The final strip is broken into three parts: the middle one reads "City," while both of the flanking units read "For Food" (Ezekiel 48:18). To the left and right of the middle rectangle are the following captions: "And that which remains for the president to the eastern border;" "And that which remains for the president to the western border." The allotments of Dan, Asher, Naphtali, Manasseh, Ephraim, Reuben, and Judah are stacked one atop the other to the city's north, and Benjamin, Simeon, Issachar, Zebulun, and Gad on the other end. Needless to say, this content diverges sharply from the Book of Joshua's meticulously detailed geographical account of the tribal allotments. In all likelihood, Teomim was the first Jewish author to print this cartographic model of the future Land of Israel since Rashi's progeny some 750 years earlier.

The Maps of Alexander Ziskind

Alexander Ziskind of Grodno (d. 1794) was an acclaimed rabbi and scholar, especially in the fields of morality and the Jewish occult. In his writing, he often referred to Eretz Yisrael and Jerusalem. Moreover, Ziskind viewed the settling of the Land to be a great virtue, and even collected funds on behalf of its Jewish inhabitants. What is more, the Polish rabbi instructed his students to immerse themselves in the Bible's geographical accounts of the Land and familiarize themselves with its layout.[14] His book *Yesod ve'shoresh ha'avodah* (the Foundation and Root of Worship), which netted quite a few editions, first came out in 1782.[15] Among its chapters is "Corrections of

---

of these maps in idem, "Some Further Data on Rashi's Diagrams to his Commentary of the Bible." In the wake of De Lyra's emulation of this model, it became rather popular among Catholic authors. See the discussion below on Altschuler's map.
14 "Ziskind, Alexander," *Encyclopeadia Hebraica*; Klausner, "R. Alexander Ziskind of Grodno—the Hasid amongst Mitnagdim" [Hebrew]; Klausner, "Alexander Susskind ben Moses of Grodno."
15 Alexander Ziskind ben Moshe, *Yesod ve'shoresh ha'avodah: of the Prayer, and the Torah and the Commandments [...] with Simplifying Intentions [...] in All the Prayers of the Entire Year [...] and the Enormity of the Virtue of Learning [...] and Lastly [...] Corrections of the Mistakes that Befell Rashi's Commentary on the Prophets and the Hagiographa and an Explanation of Joshua's Borders and the Borders of the Future and an Explanation of Solomon's Building and the Building of the Future [...] All this My Hands have Done [...]* Nowy Dwor: the J. A. Krieger Printing House [...] [1782] [Hebrew]. Alexander Ziskind ben Moshe, *Yesod ve'shoresh ha'avodah*, Grodno, Baruch ben Yosef [Press], 1795 [Hebrew].

the Mistake that Befell Rashi's Exegesis on the Prophets and Hagiographa." Within this section, he included a four-page work titled "The Book of Joshua [–] an Explanation of the Borders." At the end of this insert are two germane maps, which were printed on either side of the same folio.

One of the maps, which appears on the first page of Ziskind's comments on the Books of Judges, is called "Drawing of the Borders of the P[ortion] of *Mas'ei*." Following in Mizrachi's footsteps, Eretz Yisrael is shaped like a rectangle. Sprinkled around the borders are a few toponyms. All told, this map is similar to the ones that were discussed in chapter 2. The Land's outer circumference is a regular straight line, while the inner one is decorative. Between these two hems are several toponyms that surface in *Mas'ei*'s account of the borders. Inside the frame are the words "Eretz Yisrael" as well as the four winds. The Jordan is essentially detached from the main rectangle. Running diagonally, the river is also signified by a double line; and the interstice is filled with dots. The Jordan feeds into the Dead Sea, which abuts the main box. Lastly, the Great Sea comprises the entire western border.

The second map is vastly more intricate. As noted in its title, the cartographic design leans on Teomim's map: "Drawing of Joshua's Borders that Begin with Ch[apter] 15 in the Book of Joshua[,] Copied from *S[efer]kiṣvai areṣ*." While citing his source, Ziskind incorporated some additions and changes. The map is oriented to the east and preserves the genre's rectangular frame. Its purview is limited to the borders of Eretz Yisrael, so that there are no references to Paradise, the four attendant rivers, or India and Cush. Along the northern edge are the border settlements as per *Mas'ei*. A narrow strip on the northeastern side includes the Gilead Mountains, the Land of Og, and "the border of Manasseh." Below these sites are Mahanaim as well as "The border of the Reubenites and Gadites." "The waters of Jericho,[16] the Jordan, and "the tongue [i.e., the northern bay] of the Dead Sea"[17] are further south. For some reason, Ziskind leaves out the Sea of Galilee and the southern border towns. The western boundary makes due with the Great Sea.

The Land itself is divided by means of straight lines into the tribal allotments, most of which run from east to west. The southernmost strip is the territory of Judah. To its left is Benjamin followed by Ephraim. On the lefthand portion of the latter is a strip that is identified as "These cities of Ephraim within Manasseh." North of Ephraim is indeed the allotment of Manasseh. Unlike the rest of the tribes

---

Alexander Ziskind ben Moshe, *Yesod ve'shoresh ha'avodah* [...] Yehuda Lima ben Aryeh Leib Segal and Simcha Zimel ben Yechezkel [Press], 1810. Alexander Ziskind ben Moshe, *Yesod ve'shoresh ha'avodah* [...], Grodno and Vilnius: Menachem Mann and Simcha Zimel Press, 1817, p. 88, side 2.

16 The waters of Jericho appear in the exegesis, but not in the Bible itself. For example, according to the *Pesikta zutra* (*Midrash lekach tov*) commentary on Exodus 15, these are the waters that cured Elisha. Moreover, the *Meṣudat david* commentary on Joshua 16 deems the water of Jericho to be a border marker. This may be the reason why this ostensible toponym was included in Ziskind's map.

17 Needless to say, the intention here is to the tip of the Dead Sea as described in Joshua 18:19. The modern geographical meaning of "tongue" had yet to be coined.

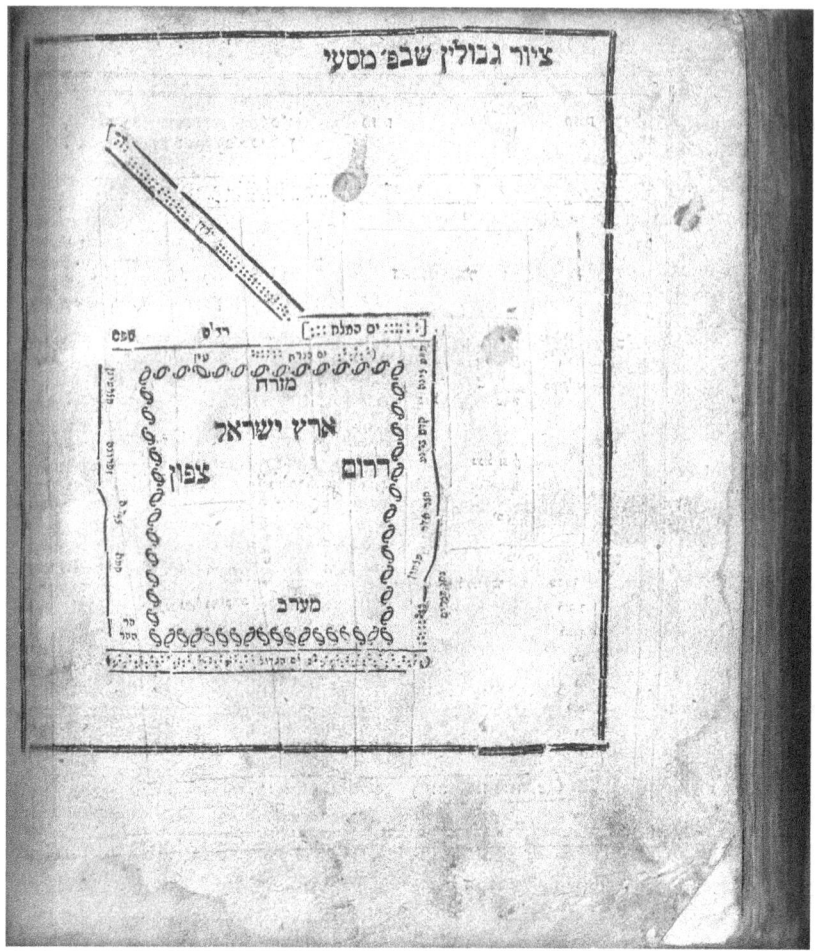

Fig. 93: Alexander Ziskind, "Drawing of the Borders of the P[ortion] of *Mas'ei*," 145 × 170 mm, in *Yesod ve'shoresh ha'avodah*, 1795, p. 102b.

which have eponymous captions, Manasseh has a text explaining that it had cities in Issachar and Asher. In back of Manasseh is the realm of Asher—the largest of all the allotments. Issachar is the only mainland territory that runs from south to north. Zebulun is landlocked by Issachar and Naphtali. Here too, no territory has been assigned to Simeon or Dan. A mere handful of the tribal areas contain a few settlements. In Judah, there is a reference to the description thereof in Joshua 15. As in Teomim's map, Asher contains a list of settlements on its west and north side. It also bears noting that the cities of Tyre and Sidon are erroneously located on Asher's eastern tip (near its border with Issachar), rather than on the coastline.

Krieger's press in Nowy Dwor printed the first edition of Ziskind's book, map included, two years before handling the map of Yehonatan ben Yaakov. Although the

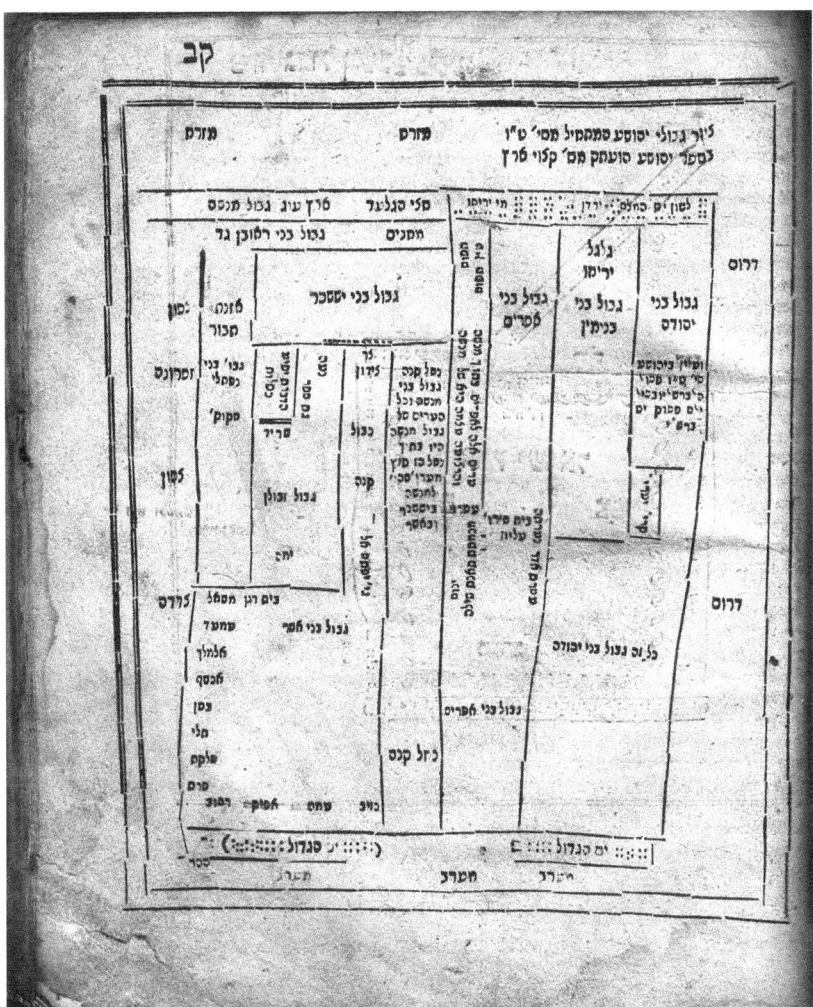

Fig. 94: Alexander Ziskind, "Drawing of Joshua's Borders," in *Yesod ve'shoresh ha'avodah*, 135 × 165 mm, p. 102a.

latter ultimately chose to create an illustrated map on the basis of a German source, it is possible that Ziskind's served as one of his inspirations.

The Maps of Yehiel Hillel Altschuler

Born in around 1720, Yehiel Hillel Altschuler was active during the mid-eighteenth century. He completed an expansive work on the Prophets and Hagiographa that his father, David ben Aryeh Leib Altschuler (active in around the 1690s), had begun. The commentary is divided into two parts: "*Miṣudat Ṣion*" (Citadel of Zion), which explains the words in the Prophets and Hagiographa; and *Miṣudat David* (Citadel of David), which provides a literal interpretation of their content.[18] Parts of the book's later versions were published in Zholkva (present-day Zhovkva, Ukraine) and Berlin, but the entire first edition was printed in Livorno.[19] A map of the tribal allotments was included in the section on Joshua.[20] The initial printing process evidently took quite some time, as the *haskamot* (rabbinical imprimaturs) of Livorno's rabbis were written in 1778/1779, the cover page of the Book of Joshua is dated to 1779/1780, and the final page did not come off the presses until 1782.

Altschuler's map takes up a full page. The upper portion is devoted to the map itself, while the lower third contains a key. Lacking any distinction from the rest of the fonts, the title is situated atop the page: "Map of the Apportionment of Eretz Yisrael and its Borders." The key is crowned by a preamble: "For the sake of knowing the Land's borders I have mounted this drawing[;] and every person that will look at it will understand in his heart that this is the most proficient road and negates the first ones[;] and here before you is a table [i.e., key] of the border markers in the order that is stated in the Book of Joshua." Below these words are six columns of 106 referenced toponyms, whose corresponding numbers are spread throughout the map.

Altschuler's map faces north. On its left or west side is a wavy rendering of the Great Sea, to which the River of Egypt feeds into from the south. On the opposite end is another body of water that amalgamates the Sea of Galilee, the Jordan, and the Dead Sea. Next to the mouth of this waterway is a caption reading "Leshem is Dan." Neither the Sea of Galilee nor the Hula is mentioned by name. Although these particular drawings were influenced by the period's cartographic style, the author did not render the layout of the Land or its geographic characteristics in a realistic manner. Instead, he preserved the schematic framework of the traditional Hebrew map with the objective of illuminating the Book of Joshua—the source of all his toponyms—and its commentaries.

The parts of the map are arranged into a series of inter-connecting lines, along the lengths of which are circles of various sizes. Most of the circles include a Hebrew number that refers to a toponym in Joshua's description of the borders. In the southwest is the allotment of Simeon; to its right is Judah, which is underneath Benjamin. Northwest of the latter is Ephraim and further right is Dan. Perched over Benjamin is the allotment of half Manasseh. This area features seven empty circles that are accompanied by the caption "Cities set apart for Ephraim." Above Manasseh is the territory of Issachar. The realm of Asher borders Dan and the Great Sea. Nestled in the upper-left corner is Zebulun. Only one settlement, Jerusalem, lacks a circle, as its

---

18 "Altschuler, R. David ben Aryeh Leib," *Encyclopaedia Judaica* [Hebrew].
19 Shisha Halevi, "Miṣudat Zion and Miṣudat David and their Author" [Hebrew].
20 Yehiel Hillel Altschuler, *Bible: Book of Joshua, with Two Commentaries [...] the Handiwork of [...] the Honorable[,] Our Teacher and Rabbi[,] the Rabbi[,] Rabbi Yehiel Hillel Altschuler [...]Miṣudat david [...] Miṣudat ṣion*, Livorno: E. Y. Kashtilo and A. Saadon Press, 1780–1782.

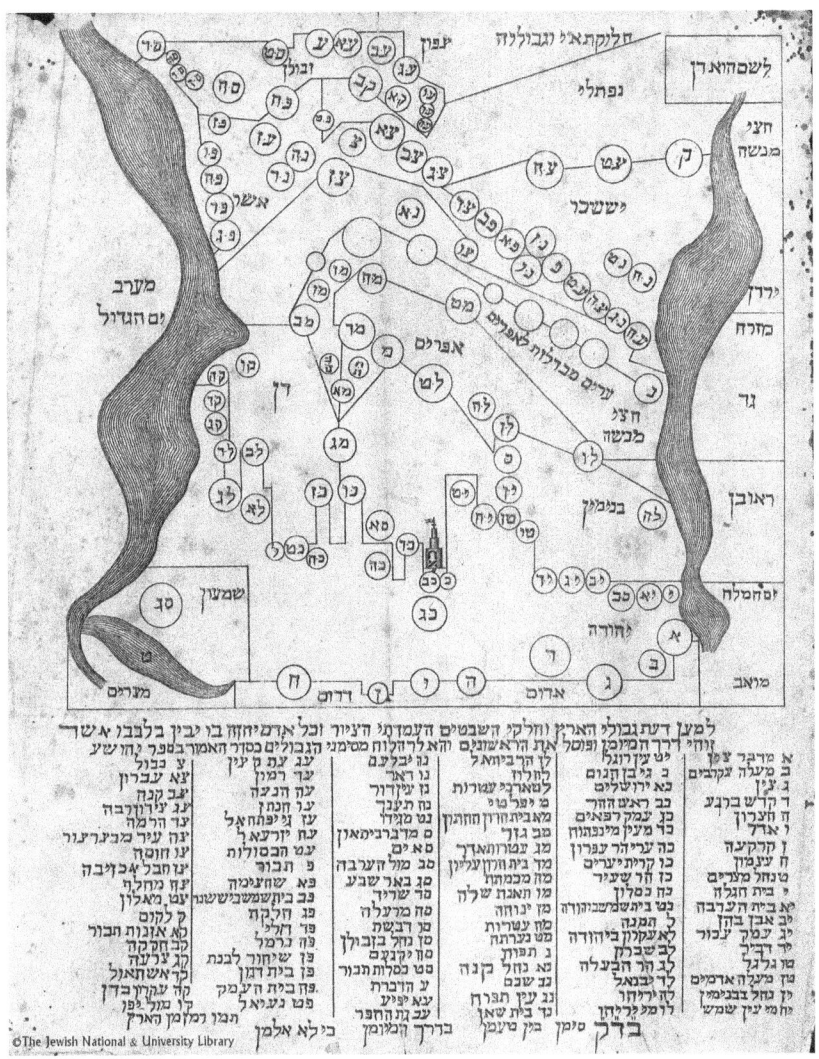

Fig. 95: Yehiel Hillel Altschuler, "Map of the Apportionment of Eretz Yisrael and its Borders," 174 x 212 mm.

number (21) is tucked sideways into a flag-capped tower. Unlike the rest of the map, Transjordan is split into clearly demarcated compartments, each of which contains a short caption: Half of Manasseh, Gad, Reuben, and Moab.

In his commentary to Ezekiel, Altschuler included a flyer that he called *Sefer binyan ha'bayit* (the Book of the House's Building), which was also disseminated on an individual basis. This booklet comes with a map of the Land's future division as per

Ezekiel.²¹ It is essentially a variation of Teomim's aforementioned map on this topic, as the two are identical from the standpoint of content. However, this rendering constitutes a marked improvement from a design and printing standpoint. For instance, the Jordan River and Mediterranean Sea resemble their figurative counterparts in Altschuler's map of the tribal allotments. What is more, Auspitz replicated Altschuler's futuristic map with practically no changes in *Be'er ha'luḥot*.²²

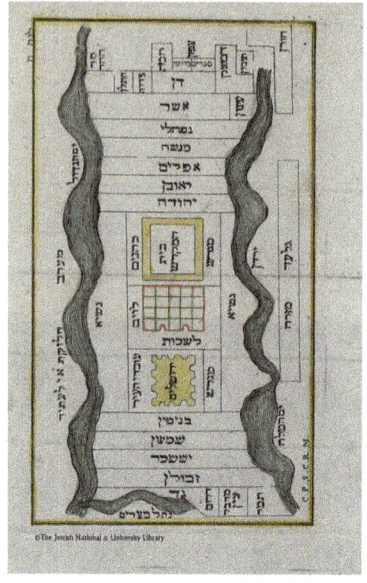 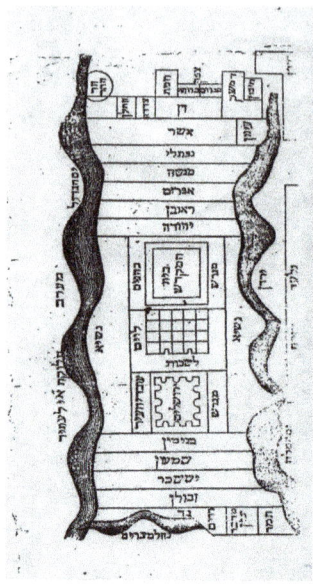

Fig. 96: Altschuler and Auspitz's maps of Eretz Yisrael's future division as per Ezekiel. (Auspitz's map 106 x 180 mm)

As already noted, the roots of this map go back to Rashi. It is difficult to say whether the map's re-emergence towards the end of the eighteenth century in Teomim and Altschuler's works and at the start of the nineteenth century in Auspitz's booklet was triggered by a manuscript or early print of Rashi's version that fell into one of their hands or by one of the Christian adaptations of the savant's map. In any event, it is worth noting that Christian versions of this map were published as early as the second half of the 1500s by Pierre Eskrich and du-Puy.²³ Moreover,

---

21 Yehiel Hillel Altschuler, *Sefer Binyan ha'bayit: Wherein shall be Described the Building of Ariel[,] its Qualities, Chambers and Offices*, Zholkva, 1775. The preamble of this edition notes that it included a drawing from the 1785 Livorno edition. For the possibility of a 1753 edition, see Preschel, "More on Yehiel Hillel Altschuler" [Hebrew].
22 Auspitz, *Be'er ha'luḥot*, Oven, 1817, plate v. Also see the discussion on this topic in chapter 4 above.
23 Delano-Smith and Ingram, *Maps in Bibles*, pp. 61–62 and fig. 36 (p. 70).

Fig. 97: Calmet's map of the division of the Land according to Ezekiel, 186 x 355 mm.

Calmet widely distributed the template of this map in the eighteenth century.[24] Even though the map's content was eschatological, the Christian authors drew their ver-

---

24 Plan et Distribution de la Terre de Chanaan, Suivant la Vision d'Esechiel Chap. XLVIII Laquelle ne

sions of the future tribal allotments on a contemporaneously accepted Holy Land map. In other words, their versions offered a geographically realistic depiction of local elements, such as the seacoast, the path of the Jordan's flow, and the Sea of Galilee. Conversely, the Jewish cartographers were predisposed to their tradition's long-standing schematic framework that eschews geographic "facts," thereby distinguishing their maps from the Christian versions. That said, Altschuler made some strides in all that concerns the rendering of bodies of water, albeit in a highly abstract manner.

The Reputed Maps of the Vilna Gaon

R. Elijah ben Shlomo Zalman, the *Vilna gaon* (Genius from Vilnius, 1720–1797), is considered one of the greatest minds in the history of the Jewish people. The revered scholar delved into all the traditional realms of Jewish erudition, while also taking an interest in geography and many other scientific fields. Although he was publically opposed to the Jewish Enlightenment, there are scholars who deem the *Vilna gaon* to be one of its forerunners. Some seventy books and other works are attributed to him, but not one of them came out in his lifetime; instead, they were all copied and published by his progeny and students posthumously.[25] R. Elijah frequently grappled with questions concerning Eretz Yisrael and its borders. Among the many books that are credited to him is *Sefer ṣurat ha'areṣ le'gvuloteha* (the Book of the Layout of the Land for All its Borders), which was printed in Shklov (present-day Shkloŭ, Belarus) in 1802. This work explicates sources and attendant commentaries that pertain to the Land of Israel.[26] As noted on the cover, the work was brought to print by R. Elijah's kith and kin. Similarly, it is quite certain that the *Vilna gaon* did not personally arrange the extant version of this book. In any case, *Ṣurat ha'areṣ le'gvuloteha* faithfully reflects, at least to some extent, his interest in various issues that concern the Promised Land. This work is by and large a Biblical exegesis[27] on the following topics: the tribal allotments in Joshua (chapters 15–19); the Jewish Temple's construction as per 1 Kings (6–7); and the description of the future Temple in Ezekiel (40–47). Throughout most of *Ṣurat ha'areṣ*, Biblical verses fill

---

fut jamais executee a la lettre, depuis le retour de la Captivite, Calmet, Augustin Antoine, *Commentiaire litteral sur tous les livres de l'Ancien et du Nouveau Testament*, Paris: Emery, Saugrain, Pierre Martin, T.VI., 1726, opposite p. 607.
25 Etkes, *The Gaon of Vilna—The Man and His Image* [Hebrew]; Katz, *Rabbanut, Hasidut and Haskalah*, vol 2; Maimon, *Annals of the Vilna Gaon* [Hebrew]; Klausner, Mirsky, Derovan, and Kaddari, "Elijah ben Solomon Zalman."
26 Elijah ben Shlomo Zalman, *The Book of the Layout of the Land For All its Borders* [*Sefer ṣurat ha'areṣ le'gvuloteha*] *Around and Around and the Plan of the House*, Aryeh Leib ben Shene'ur Feivish, Shabbatai ben Ben-Zion [, printers], Shklov 1802 [Hebrew].
27 Cohen, *From Dream to Reality: Descriptions of Eretz Israel in Haskalah Literature*.

the center of the page and are surrounded by Rashi and the author's commentaries. At the end of the book, though, a few pages consist entirely of the *Gaon*'s exegesis.

A folio that was attached to *Ṣurat ha'areṣ* contains a map on either side. The first is titled "The Layout of the House of Ezekiel from the Genius[,] Our Rabbi Elijah[,] May the Memory of the Righteous and Holy Person from Vilnius be Blessed;" and the second is "The Layout of the Land For All its Borders from the Genius[,] Our Rabbi Elijah[,] May the Memory of the Righteous and Holy Person from Vilnius be Blessed." The manner in which this page was inserted into the book suggests that it was *not* an integral part of the overall work. Moreover, it explains why the map is quite rare and indeed absent from many known copies of *Ṣurat ha'areṣ*. An identical manuscript-based version of the second map is housed at the Israel National Library's Laor Collection. Both of the maps were reprinted in 1897 on a single folio in Jerusalem, with the objective of raising money for the Elijah Beit Midrash —a yeshiva in the savant's memory that was probably based in Jerusalem.[28] "The Layout of the Land" map also appeared in *Aderet eliyahu* (Elijah's Coat)—a commentary on the Prophets and Hagiographa, which the *Vilna gaon*'s grandson published in Jerusalem at around the turn of the twentieth century. However, the map was not included in other editions of this same book.[29]

"The Layout of the Land" map is square and faces east. Accordingly, Transjordan is depicted on its upper portion. The top outer strip contains "the wilderness of Kedemoth" (Deuteronomy 2:26) and the Land of Ammon. Below this frame, the expanse is divided into the two and a half tribal allotments and the neighboring lands. Edom and Yemen take up the southeastern corner; to the north, between the Zered Brook and Arnon River, sits Moab. Further north is a compartment that is headlined by Reuben and Sihon; however, it also includes Aroer and Heshbon, which are portrayed as cities. The allotment of Gad hems in Reuben from the north and east. It includes the caption "Gad took half of the Gilead." Beyond the Jabbok River to the north is the Land of Bashan as well as the following information: "The second half of Gilead was taken by half the tribe of Manasseh."

Dotting Eretz Yisrael's northern extremities are the Hermon, which is rendered as a large mountain, and the border towns of Mount Hor, the entrance of Hamath, Zedad, Ziphron, and Hazar Enan. Next to the Hermon is a caption reading "Springing out of Bashan"—a reminder of the blessing received by the Danites (Deuteronomy 33:22). The Red Sea is aligned just inside the map's southeastern and southern edges. Raamses, Migdol, Succoth, Etham, and Pi Hahiroth lie between the Red Sea

---

[28] As stated on the folio's margins, the price of this item was "five kop" (perhaps kopecks). "Its fruit are to be consecrated to the tremendous and lofty and the like enterprise [–] the Elijah Beit Midrash." On the other side of the folio was a drawing of "The Form of the House of Ezekiel," which is attributed to the *Vilna gaon* as well.

[29] *Aderet eliyahu*, a Commentary on the Prophets and the Hagiographa, by Elijah of Vilnius[,] a Righteous Person of Blessed Memory, published by Elijah son of Rabbi Eliezer Landau[,] grandson of the Genius Rabbi Elijah, Jerusalem (1905).

Fig. 98: Two versions of Elijah ben Shlomo Zalman's "Layout of the Land" map in *Ṣurat ha'areṣ le'gvuloteha*: MS Shklov 1802, 209 x 209; and the 1905 Jerusalem edition, 210 x 215 mm.

and the Mediterranean—the obvious western boundary. From its starting point in Egypt, the route of the Israelites' voyage winds its way along the southern and southeastern parts of the map and then through Transjordan, before entering the Promised Land near Jericho. North of the Israelites' route is a thick line marking the Land's southern border as per *Mas'ei* (Numbers 33–36). Along this line are six border

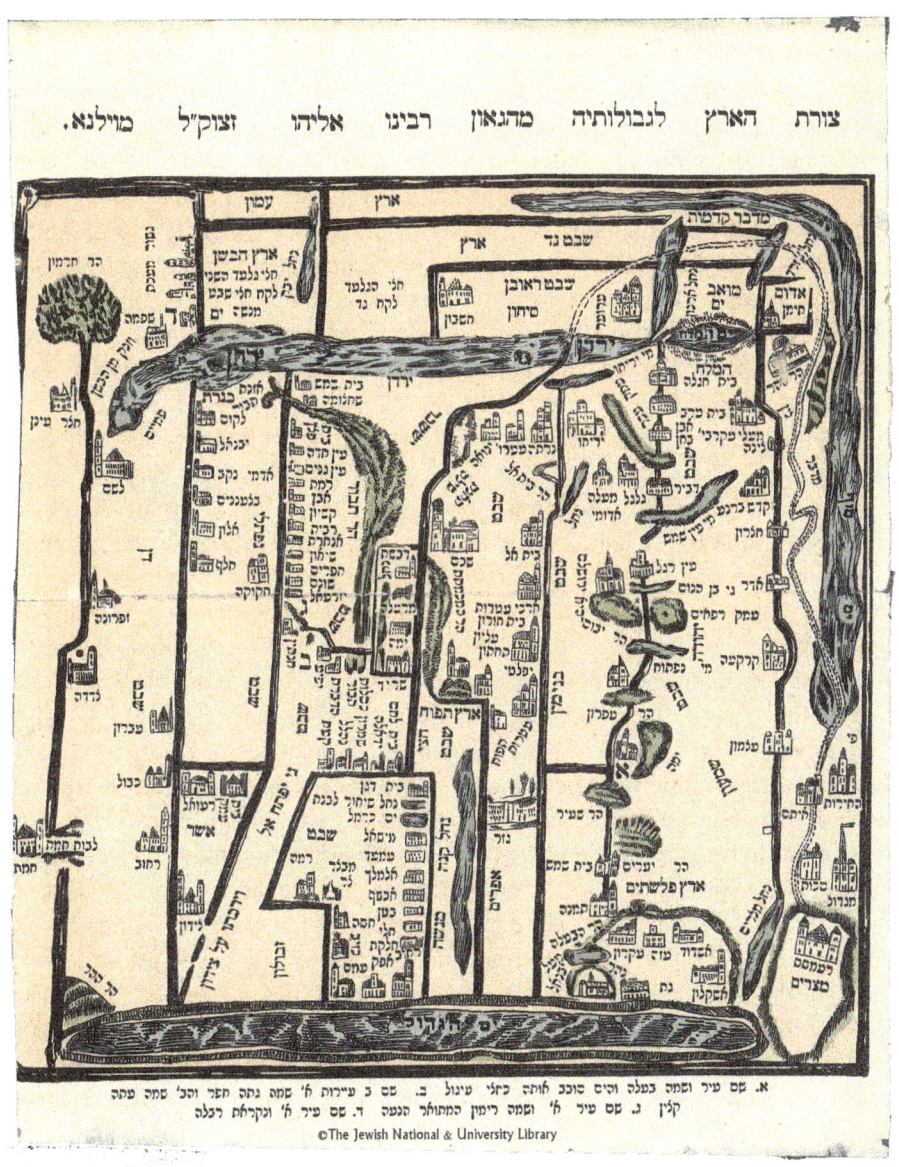

towns, from the Ascent of Akrabbim in the east to Azmon in the west. The final westernmost point on this border is the River of Egypt.

While the map basically adheres to Rashi's traditional, schematic design template, the Land is divided into tribal areas. As in the earlier maps discussed above, the allotments are delineated by straight lines. However, there are a couple of significant changes in both form and content that, in all likelihood, were introduced by the *Vilna gaon* or whoever brought this map to print. To begin with, the following sites were included in the vicinity of Jerusalem: Nephtoah, the Valley of Rephaim, Hinnom Valley, and En-rogel. Jerusalem itself takes the shape of a mountain capped by a spired urban complex. Adjacent to this illustration is a caption reading "The Mountain of Jebusite that is Jerusalem." The map also offers an in-depth account of the Land of the Philistines. That said, the arrangement of its cities is unrealistic from a geographical standpoint, as Gaza is inland and Gat on the coast. Unlike its predecessors, this map juxtaposes the areas of Simeon and Judah. Furthermore, the allotment of Dan is situated in Leshem, which the tribe conquered after heading north. To the west of Benjamin are, among other sites, Beit Shemesh and Timnah. The latter borders the Land of the Philistines and certainly Dan's original territory. Asher is boxed in by a strip of Zebulun, which reaches the Mediterranean. Positioned at the confluence of Naphtali, Asher, and Zebulun, Mount Tabor constitutes a prominent topographical feature.

The map under review includes many more cities on the allotments' borders than its predecessors. In some of the tribes, there are also inland cities. Whatever the case, all these sites are accompanied by an illustration of a city, house, or urban expanse. A few of these clusters of towns are reminiscent of those that inform Asher on the Teomim and Ziskind maps, namely the settlements that are amassed along the borders. Regardless of whether the map was compiled by the *Vilna gaon* himself, the author correctly assumed that Sidon is a port town, for along the length of the corridor linking Zebulun to the sea are the words "Its border will extend to Sidon" (Genesis 49:13). In addition, the city itself is on the western tip of Asher. From a design standpoint, the map is laden with pictograms of cities, both inside the tribal areas and outside Eretz Yisrael's borders. Moreover, bodies of water and mountain ranges are rendered figuratively, albeit in what remains an overly simplistic manner. On the bottom of the page is a short key with four items. The letters *beth*, *gimel*, and *daleth* refer to cities that were left out due to space constraints. Alternatively, the letter *aleph* signifies "the name of a city whose name is Baalah and the sea surrounds it in a semi-circle" (this description is commensurate with the illustration itself). Baalah was indeed a town on the border of Judah (Joshua 15:9), but the source for its proximity to "the sea" is unclear. What is more, the city is depicted inland. Despite the extensive detail on this map and even though it was printed in the early nineteenth century, it is utterly devoid of modern or geographically realistic content. On the contrary, this map is an unequivocal link in the internal development of traditional Hebrew cartography. Be that as it may, the use of artistic elements, like spired cities and bodies of water, perhaps attests to an outside influence. Put differently, the author was likely exposed to artistic maps, be they Christian or Jewish.

Another version of this map turns up in a thick and large-paged manuscript (508 x 424 mm), which is housed at the Israel National Library's Laor Collection. Titled "Eretz Yisrael's Apportionment for All its Borders," this colored map is rather large. While this edition has merited some scholarly attention,[30] it nevertheless war-

---

30 Tishby, (ed.), *Holy Land in Maps*, pp. 124–125; Vilnay, *The Hebrew Maps of the Holy Land*,

rants a few words. From a structural standpoint, this map basically resembles the version discussed above. This similarity is indeed commensurate with the map's sub-heading: "Copied from [...] the famed Hasid[,] our teacher and rabbi[,] our teacher the rabbi [–] Elijah [...] from the holy community of Vilnius the capital." That said, there are a couple of interesting discrepancies:

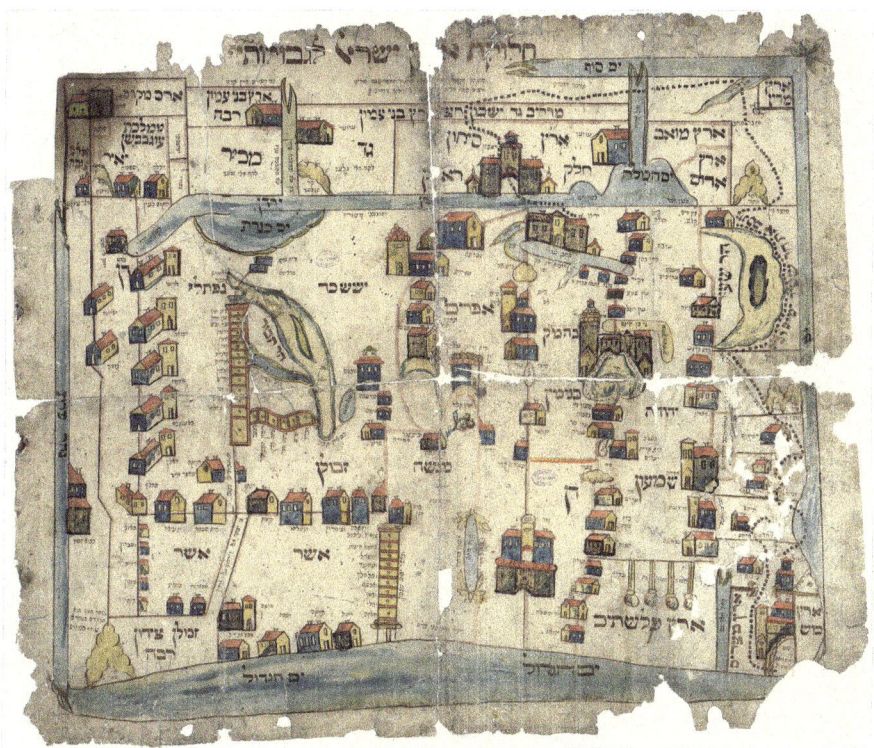

Fig. 99: Eretz Yisrael's Apportionment for All its Borders," 470 x 380 mm.

To begin with, lands that neighbored Biblical Eretz Yisrael appear along the map's outer circumference. Among the places are "the Land of Cush" to the south of Egypt; "the Land of Midian" in the southeast corner; and towards the northeastern corner are "the sons of Keturah [–] the Orient," "Geshuri and Maakathi," "Aram from the east, Paddan Aram," and "the Kingdom of Og in Bashan." Like the Rashi map, it naturally includes Moab, Ammon, and Edom. In addition, the Euphrates runs parallel to Eretz Yisrael's northern border (Teomim offered a similar account of this river).

---

pp. 22–23 [Hebrew]; Laor, *Maps of the Holy Land, a Cartobibliography of Printed Maps 1475–1800*, p. 120, no. 887. For a similar drawing in a manuscript, see Ya'ari, "Miscellaneous Bibliographical Notes 41," p. 215 [Hebrew].

Secondly, the manuscript version depicts a partial, rather than full, crossing of the Red Sea (see the relevant discussion in the previous chapter). Thirdly, the route of the Israelites' journey is presented in considerable detail. Most notably, figurative illustrations of Mount Sinai, Mount Horeb, and Mount Seir are arrayed along the length of the route. The latter also merits the caption "And for many days we made our way around the hill country of Seir" (Deuteronomy 2:1).

Jerusalem is portrayed as a large city atop a mountain, which is labelled "Mount Moriah." Etched on a quasi-ribbon alongside the mountain are the words "Zion that is Jebus that is Jerusalem." A high tower is identified in abbreviated form as the Jewish Temple. With respect to Dan, the author signified both of its territories. Adjacent to Jericho are the Valley of Achor and "the waters of Jericho." There are two illustrations that pertain to Bethel: a boulder that is perhaps a rendering of the rock that Jacob slept on; and an expansive city, which is labelled "Bethel that is Luz." The Philistine towns are oddly shaped like four hanging scepters or pillars. Of the five, Gaza was inexplicably omitted. All told, the map is rich from an artistic standpoint. Many sites are deftly portrayed as spacious houses. Especially intriguing are the list of settlements in Naphtali and Asher. Such lists already informed earlier maps; however, instead of drawing cities, the author integrated stacks that are reminiscent of wooden index boxes. Neither the source nor meaning of these singular illustrations is clear.

To the best of our knowledge, the maps that are attributed to the *Vilna gaon* had an impact on two contradictory developments. On the one hand, the fact that R. Elijah's sons and students published schematic maps in his name, rather than maps with realistic graphical characteristics, bolstered the standing of the traditional Hebrew map. In so doing, they distanced themselves from modern and Maskilic trends in cartographic design. On the other hand, as noted in the previous chapter, there were authors of realistic maps who sought to ride on the coattails of the savant. More specifically, for the purpose of boosting their readership, they claimed in their preambles that the maps were crafted in accordance with R. Elijah's commentaries. In summation, the maps that are ascribed to the *Vilna gaon* by his progeny and students are entirely lacking in realistic cartographic representations. As such, they remained true to a schematic, traditional framework that incorporates erudite and Halakhic knowledge exclusively on a theoretical plane.

The Map of Aryeh Leib ben Isaac of Sejny in the Second Edition of Kişvai Areţs

In 1813, the second edition of Teomim's *Kişvai areţs* came out in Grodno.[31] This version includes a map that was prepared and updated by one Aryeh Leib ben Isaac of

---

31 Joshua Feivel ben Israel (Teomim) from Tarnogród, *Sefer kişvai areţs*: [...] an Explanation of the Land of Canaan for All its Surrounding Borders in the Portion of These are the Voyages [*Mas'ei*]

Sejny. As evidenced by the tribute—"a righteous person of blessed memory"—that was affixed to his name, the map was published posthumously.[32] The title is "An Explanation of LC [the Land of Canaan] for All its Borders that are in the Bible[,] Drew and Corrected [by] the R.[,] the Exile's Source of Light in Torah and Piety[,] the Honor of the Sanctity of Our Teacher the Rabbi[,] Ary[eh] Leib, a Righteous Person of Blessed Memory from the Holy Community of Sejny as Written in *Sefer Kişvai Areţs*." The preamble emphasizes that this map is not a copy, but a new adaptation of the original: "[In[particular we added commentary and a correction of the drawings of the borders as per the wisdom of this book." Adhering to the traditional approach, the design of ben Isaac's map is schematic and plain, but features many innovations.

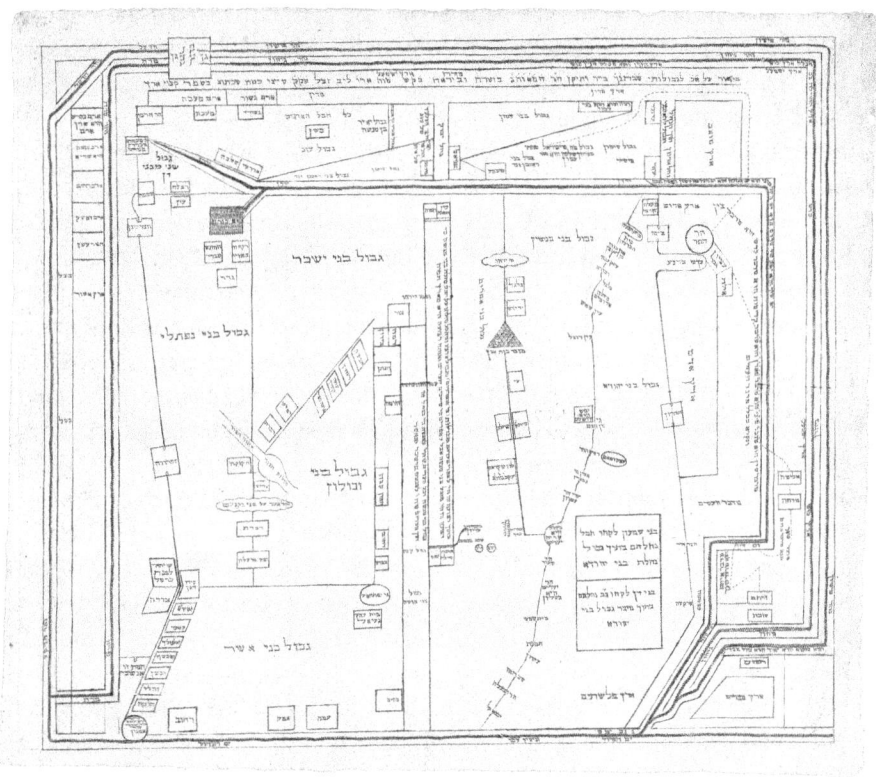

Fig. 100: Aryeh ben Isaac's map in the second edition of *Kişvai areţs*, 378 × 326 mm.

---

and the Arrangement of the Land's Division [...], Grodno: Simcha Zimel ben Menachem Nachum Press, 1813 [Hebrew].
32 Sejny is a town in northeastern Poland.

As in the first edition of *Kiṣvai areṣ*, the map is oriented eastwards. In the northeast corner is a plain square representing the Garden of Eden, from which emanate the four rivers of Paradise. The Tigris and Euphrates demarcate the Land of Israel's northern border (left). Conversely, the Pishon and Gihon flow to the east and the south before linking up with the Nile on the bottom-right corner. All four rivers seemingly empty into "The Great Sea and the islands therein," which naturally form Eretz Yisrael's western boundary.

The outer parameter contains neighboring lands. On the northern side is Babylon, and to its right is a strip divided into the following compartments: "Aram-Naharaim that is Paddan Aram;" "Aram-Zobah that is Syria;" Aram-Rehob; Aram-Damascus; and the Land of Assyria. Running alongside the Gihon and Pishon are two clauses: "The Land of India and it encompasses the Land of Cush" to the east; and "The land of Havilah and it encompasses the Land of Cush" to the southeast. The Land of Ishmael turns up both in the top-right corner and opposite the caption reading "Gihon" on the southern border. Beneath the map's title are Aram-Maakah, the Land of Geshur, and Midian.

Transjordan is divided into several sub-units, all of which are labelled. The northernmost unit, Bashan, includes the Hermon on its top-left corner; south of the mountain are Maakath and Geshur; and opposite these settlements are Edrei and Salekah. This area also includes "The border of Og" and "The border of Jair ben Manasseh." Southwest of Bashan is a triangular compartment representing the western half of the Gilead. Ben Isaac also identifies this area with Sihon and "the border of the sons of Gad and Reuben." In the eastern corner of Transjordan is "Og's half of the Gilead" and the "border of Machir ben Manasseh." The Jabbok River is juxtaposed to both halves of the Gilead. However, the author left out the Yarmouk River, which in the real world separates the Gilead from the Bashan and the Golan.[33] The expanse between the Jabbok and Arnon River is divided in two: the eastern half borders the Ammonites and contains "Rabba namely Rabbat of the Ammonites." Among the elements in the western half are Peniel and Succoth, Aroer and Jaazer, "The border of Sihon, and "The border of what Israel witnessed[34] in Sihon [,] which he took from the Ammonites." The Land of Moab turns up to the south of the Arnon River.

As per the account in *Mas'ei*, the northern border of Eretz Yisrael runs from Shepham and Zedad, on through "Hamath that is Antioch," to "Mount Hor in the language of the Bible and in the language of the Mishna Tor Amnon." Along the eastern border is the Jordan River. It begins at a diagonal (perhaps under the influence of Elijah Mizrachi) near "Leshem that is the Pamyas Cave." The Jordan does not empty into the Sea of Galilee, but is connected to the latter via another fictitious and unlabeled tributary. In the south, the river links up with "The Red Sea that is

---

33 Perhaps the reason for the Yarmouk's omission is that the river is not mentioned in the Bible.
34 The text reads *sahadu*, which is Aramaic for *he'īdu* (witnessed) in Hebrew.

the Lake of Sodom, that is the Primordial Sea, that is the Dead Sea[,] that is the Sea of Arabah[,] namely to the Negev." This lengthy and complex caption alludes to Joshua (3:16; 12:3), Zechariah (14:8), and Joel (2:20). As discussed in chapter 4, the list of the Dead Sea's names in the Zaddik map, which was influenced by Adrichom, is even longer. Against this backdrop, scholars would be hard-pressed to determine whether ben Isaac personally coined this lengthy caption or drew heavily on some external source. In any event, this body of water encloses the Land from both the east and south, as the author considered the Red Sea and the Dead Sea a single body of water.

Ben Isaac's southern border of Eretz Yisrael also conforms with the account in *Mas'ei*, as it extends from the Ascent of Akrabbim to Azmon via Kadesh Barnea. On the other side of the border is the Sinai Desert. Beginning at Etham, the route of the Israelites' voyage passes through Marah and Elim. At this juncture, the author inserted a lengthy caption: "The wilderness of Sin that is the Wilderness of Sinai that is the Paran Desert that is the wilderness of Kedemoth that is the wilderness of Kadesh that is the Wilderness of Zin and it is called in general the Wilderness of the Peoples." As discussed earlier, the name "Wilderness of the Peoples" obviously alludes to Rashi's exegesis and might also betray a familiarity with the Mizrachi map. Thereafter, the route continues on to Eilat, Ezion-Geber, and Mount Hor before entering Transjordan. Between the Sinai and the realm of Judah is the Land of Edom.

The tribal allotments in Eretz Yisrael proper are rife with captions and toponyms, the majority of which naturally derive from the Book of Joshua. Two contiguous square frames in Judah read thus: "The Simeonites took their area of inheritance within the borders of the inheritance of the sons of Judah;" and "the Danites also took the bor[ders] of their inheritance within an isthmus in the boundary of the sons of Judah." Alternatively, a caption in the north reads "The second border of the sons of Dan." In other words, Ben Isaac refers to both the Danites' original and later allotments. "Jebus that is Jerusalem" is situated on the border between Judah and Benjamin. Two unique elements grace the Ephraim-Benjamin border: a lake representing the waters of Jericho; and the wilderness of Beth Aven in the form of a sizable, diamond-patterned triangle.

Two small circles depicting Mount Gerizim and Ebal surface in the heart of Ephraim. On the tribe's northern border is a compartment with the following text: "In this strip Ephraim had differentiated cities. . ." Needless to say, this is a reference to the tribe's above-mentioned cities within Manasseh's borders. The territory of Manasseh constitutes an elongated rectangle; and to its left is another strip denoting the tribe's cities in Issachar and Asher. The majority of Asher's territory is on the coast, but the tribe also possesses a narrow corridor that darts between Manasseh and Zebulun. Here too, Tyre and Sidon are located on the eastern edge of Asher, quite a distance from the Mediterranean shore. Spanning the Zebulun-Naphtali border is a relatively large, mandolin-shaped object hosting two captions: "Aznoth Tabor" and "Kisloth Tabor." This emphasis on Mount Tabor and its surroundings bears witness to the sway of the *Vilna gaon*'s maps. The Sea of Galilee is the most prominent site in the allotment of Naphtali. Adjacent to the sea is the aforementioned Gerar, "Rakkath that is Sippori," and "Hammat [i.e., Hammath] that is Tiberias." The identification of Rakkath with Sippori is only corroborated by a single view in tractate Megillah of the Babylonian Talmud. In contrast, arguments whereby it is synonymous with Tiberias can be found in that very same discussion in Megillah, other Jewish sources, and the modern literature as well.[35]

The Map of Isaac ben Phinehas Berman

In 1851, one Isaac ben Phinehas Berman published his *Sefer ḥidushai yiṣhak* (the Book of Isaac's Innovations).[36] Among the book's components is a traditional, Rashi-inspired Hebrew map. Berman's version is rectangular and its design is rather plain. The vast majority of its lines are straight, and all the tribal allotments are shaped as either rectangles or triangles. Atop the map is a lengthy preamble:

> Here before you are the borders of EY as explained in the p[ortion] of *Mishpatim* (Laws) from the Red Sea to the Sea of the Philistines and this is a length [sic] from east to west of 400 parasangs [as per] the British or German [unit of] measurement which is a total of 1200 versts [as per] the Russian measurement; and from the wilderness that the Israelites went into and until the river that is the Euphrates, [is] also a length of 400 British parasangs, which is 1200 versts, this is from south to north; and so it is said in Psalms 72[,] May he rule from sea to sea, and from the river to the ends of the earth; and from Jerusalem to the Euphrates is a walking distance of 15 days, which is 450 versts; and this is clear.[37]

Berman's map faces east. The four winds are arrayed on the margins in large letters. On the right side are Egypt, the Red Sea, and the route of the Israelites' peregrinations, which is marked with a fine dotted line from Raamses to Jericho. The stations along the way are Succoth, Pi Hahiroth, the wilderness of Sin, the Etham wilderness, Mount Horeb, Mount Hor, and Mount Seir. From hereon in, there are no toponyms until the final stop (Jericho). Along Eretz Yisrael's southern border are Moab, Edom, Amalek, and the River of Egypt. Midian appears far to the south, adjacent to Mount Seir. The lower strip is comprised of the Great Sea, which he also calls the "Sea of the Philistines." Slightly protruding from the left of the map's framework are the Euphrates River and Aram-Zobah. Rabbat of the Ammonites and Gilead turn up on the northeastern corner. The Red Sea abuts this compartment to the right, thereby circuiting the Land from the east as well. Though unpartitioned, Transjordan consists of Gad, Reuben, half of Manasseh, and Mount Abarim.

---

35 Babylonian Talmud, tractate Megillah, 6a. There is also an argument therein over whether Rakkath is synonymous with Tiberias or Sippori.
36 Isaac ben Phinehas Berman, *Sefer ḥidushai yiṣhak*, Königsberg: A. Samter [Press], 1851.
37 Dating back to the Middle Ages, the verst or versta is a measure of distance that has survived into the modern era. Over time, its length has expanded from about 2000 to 3500 feet. Berman's estimates notwithstanding, there is no real significance to the distances on the map itself.

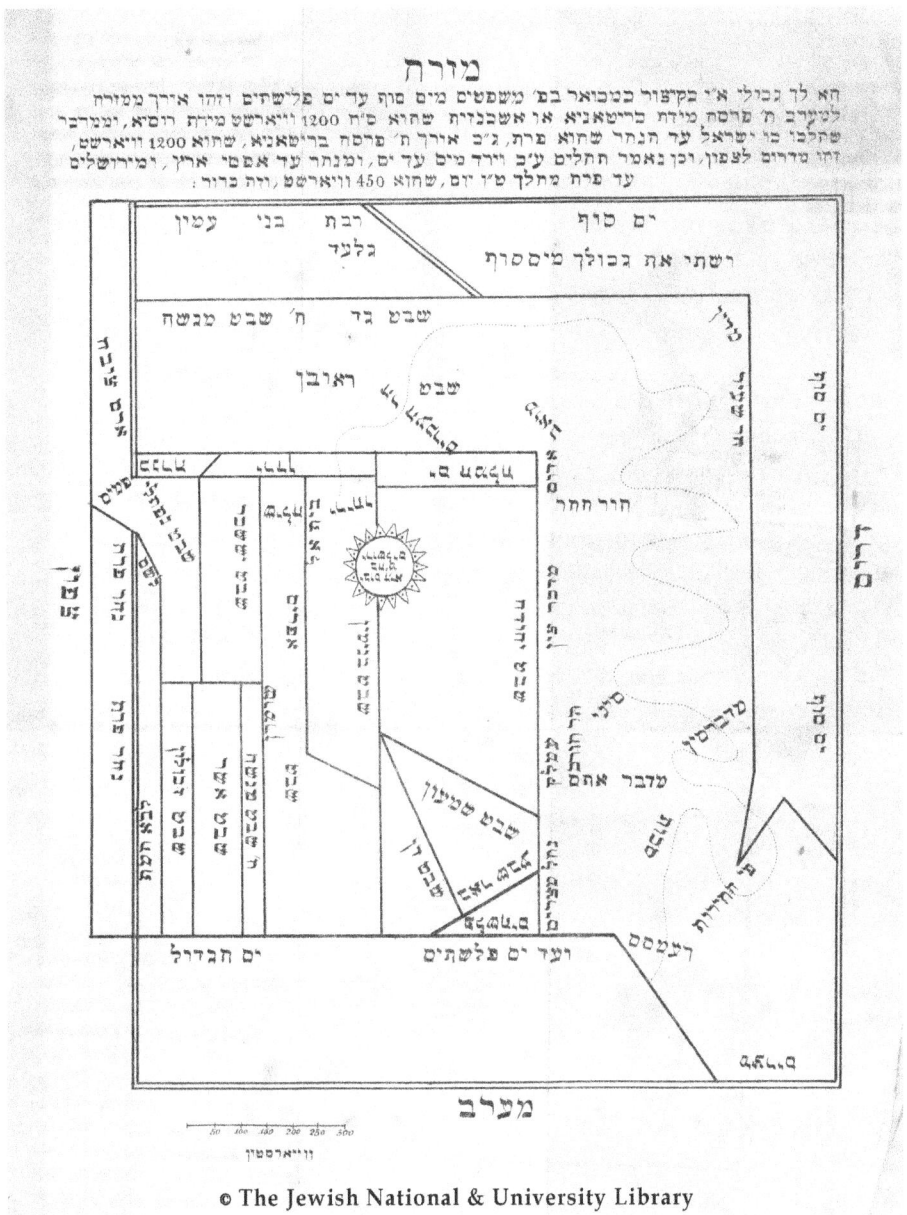

Fig. 101: Isaac ben Phinehas Berman, the "Here Before You are the Borders of EY" map, 180 x 205 mm.

Eretz Yisrael's eastern boundary consists of the Sea of Galilee, the Jordan, and the Dead Sea, which are drawn as a succession of three rectangles; each of their names is printed upside-down. The Land itself is divided into the tribal allotments. With the exception of the triangular areas of Dan, Simeon, and the Philistines to the west of Judah, the territories are basically rectangular. The only item that stands out on this map is Jerusalem, which is rendered as the sun and bears the following, inverted caption: "Jebus that is TJT [the Jewish Temple] Jerusalem." Besides Jerusalem, the author provided captions for Leshem, Pamyas, Samaria, Shechem (Nablus), Beit El, and Jericho. However, these toponyms lack any point of reference or other mark.

Although the map is exceedingly plain and schematic, Berman's pretension of estimating the size of the territory that he covered using modern units of distance ("a British and German parasang" and a Russian verst) and the linear scale on the bottom-left corner attest to a certain spirit of modernization.

Jacob Emden and His Maps

Jacob Emden (1697–1776) was considered one of the leading Jewish lights of his generation. Known by the moniker YaAVeTz (a Hebrew acronym for Yaakov ben Zvi), he was a rabbi, halakhic authority, kabbalist, and staunch opponent of Sabbateanism. Following in his father's path, Emden took an interest in both Jewish religious studies and the sciences. Moreover, he knew multiple languages, such as German, Dutch, and Latin. Throughout his lifetime, the YaAVeTz barely held a rabbinic post. This lack of official duties enabled him to unwaveringly maintain his nonconformist views. For the sake of disseminating his ideas, Emden owned a private printing house where many of his works were indeed brought to press.[38] In his collection *Mor ve'kṣiah* (Myrrh and Dried Figs), the German scholar added an unpretentious map of Eretz Yisrael,[39] which is not directly related to the book's content. This cartographic sketch is unlike any other documented map in the history of cartography, be it Jewish or Christian. Hence, there is little doubt that the map is a completely original work by the said author. Chronologically speaking, the map antecedes the others in this chapter; and from a design standpoint, it is unaffiliated with this sub-genre. That said, there are similarities in content between this group and Emden's map. Given the importance and singularity of the latter, I have decided to present it as an archetype that stands apart from the rest.

While the map faces west, some of its captions are aligned in other directions. Therefore, it is less than certain that the map has a defined orientation. At first glance, Emden's map appears to have realistic characteristics. A systematic analysis, though, reveals an utterly schematic construction that entails a few significant geographical errors. The map's purview extends from Egypt in the south to Mesopotamia ("Aram-Naharaim") and Syria in the north, and from Cyprus ("Kiprus Island") in the

---

[38] Samet, "Emden, Jacob;" Schacter, *Rabbi Jacob Emden: Life and Major Works*; Cohen, *Jacob Emden, a Man of Controversy*; Bick (Shauli), *Rabbi Jacob Emden: The Man and His Thought* [Hebrew]; Katz, *Rabbanut, Hasidut and Haskalah,*, vol. 1, pp. 149–166 [Hebrew].
[39] Jacob Emden, *Mor ve'kṣiah*, part two, Order Zmanim from *Tor Orekh Ḥayyim* [...], Altona Press in the author's home, 1761–1768, volume two, between pages 96 and 98.

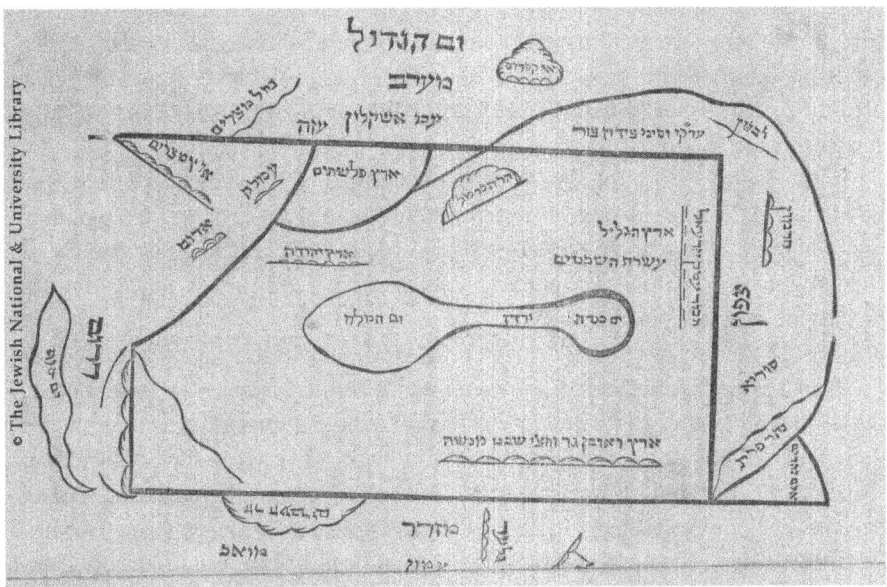

Fig. 102: Jacob Emden's map of Eretz Yisrael, 260 x 180 mm.

west to Moab, Ammon, and Bashan in the east. The Land of Israel—the two and a half tribes included—is drawn within a rectangular framework. Captions reading "the Ten Tribes" and "Land of Reuben and Gad and half the tribe of Manasseh" can be found on Eretz Yisrael's eastern and western sides, respectively. The partial inclusion of Transjordan occasioned two egregious errors: first, Gilead, Ammon, and Moab are completely detached from the allotments of Reuben, Gad, and half Manasseh; second, Mount Abarim is situated far east of the Jordan, so that it is inconceivable that Moses could have gazed upon the Promised Land from such a distance.

Similar to the other maps discussed in this chapter, the Euphrates River is positioned in the northeastern corner. The river's source is in Aram-Naharaim, but the direction of its flow is vague. For instance, it is unclear whether it spills into the Mediterranean. Alternatively, there is no sign of the Tigris River. Lebanon appears in the northwest corner; to its immediate west are Tyre and Sidon as well as "the Arkites and the Sinites" (Genesis 10, 17). The Red Sea takes the form of a landlocked and isolated lake in the south. Rather than flowing into the sea from the interior, the River of Egypt is drawn within the area of the Mediterranean. The two banks of the Jordan are unmistakably divided by a unified body of water consisting of the Sea of Galilee, the Jordan, and the Dead Sea. There is also a logical distance between Judah in the south and "the land of the Galilee" up north (adjacent to "the Ten Tribes"). Surprisingly, the caption "Tabor[,] the Jezreel Valley" is north of the Galilee.

In referring to Judah, Reuben, Gad, half of Manasseh, and the ten tribes, Emden's map constitutes a nascent stage in the Jewish map's signification of the tribal allotments. To the best of my knowledge, though, this map was not replicated, nor did it leave a trace on subsequent maps. Therefore, it would be inaccurate to view it as the source that heralded the emergence of the tribal areas in Hebrew cartography. At any rate, the YaAVeTz's renowned independence and unique worldview come to expression in this one-of-a-kind map.

In the fourth section of *Leḥem shamayim* (Bread of Heaven), Emden's work on the Mishna,[40] the famed rabbi inserted another map. This cartographic diagram is part of his commentary on tractate Shevi'it (Seventh Year), which focuses on the Jewish agricultural sabbatical.[41] According to its preamble, the map is "for the benefit of the close reader [for whom] I will draw the borders of EY as per what is understood by us in this Mishna of ours." Located on the other side of the page, the map itself is a schematic, northern-oriented representation of Eretz Yisrael. The four winds are displayed on the outer edges of this practically square template. Each corner features a geographical object: the Dead Sea in the southeast, the River of Egypt in the southwest, Amanus in the northwest, and the Euphrates River in the northeast. Aside for the Great Sea, the western end includes Kziv and Acre; and between these two sites is the word "Strip." The eastern boundary is comprised of the Jordan and the Sea of Galilee, which meets the Euphrates River. A slack line between the northeast corner of the map and Kziv separates the "First Conquest" of the Land from the "Second Conquest."

The content of this map and its inclusion in a commentary on tractate Shevi'it tie it to the topic of *shalosh arṣot le'shevi'it* (three lands or zones for the sabbatical year), which was also explored in Shlomo Adeni's *Melekhet Shlomo* (see the discussion in chapter 2). This issue first comes up in the beginning of the tractate's sixth chapter. In essence, nearly all the toponyms in Emden's diagram appear in the Mishna: "All that the returnees from Babylon possessed from Eretz Yisrael to Kziv—was neither eaten nor worked; and all that the returnees from Egypt possessed from Kziv to the [Euphrates] River and until Amanah [i.e., Amanus]—was eaten but not worked." The above-mentioned "strip" between Kziv and Acre hints to another matter originating in tractate Gittin, which was discussed in the second chapter.[42] In the map under review, Acre and Kziv are clearly on the Mediterranean coast. That said, Acre is located by the River of Egypt in the south (instead of its true location up north), perhaps in order to emphasize the magnitude of the "Strip."

---

40 This particular part of the book is called *Mishneh leḥem* (Double Bread).
41 Jacob Emden, *Sefer Leḥem shamayim*, [...], vol. 4], with the book *Mishneh leḥem*, a commentary on tractate Shevi'it, p. 13, Altona [Press], 1768 [Hebrew].
42 See chapter two.

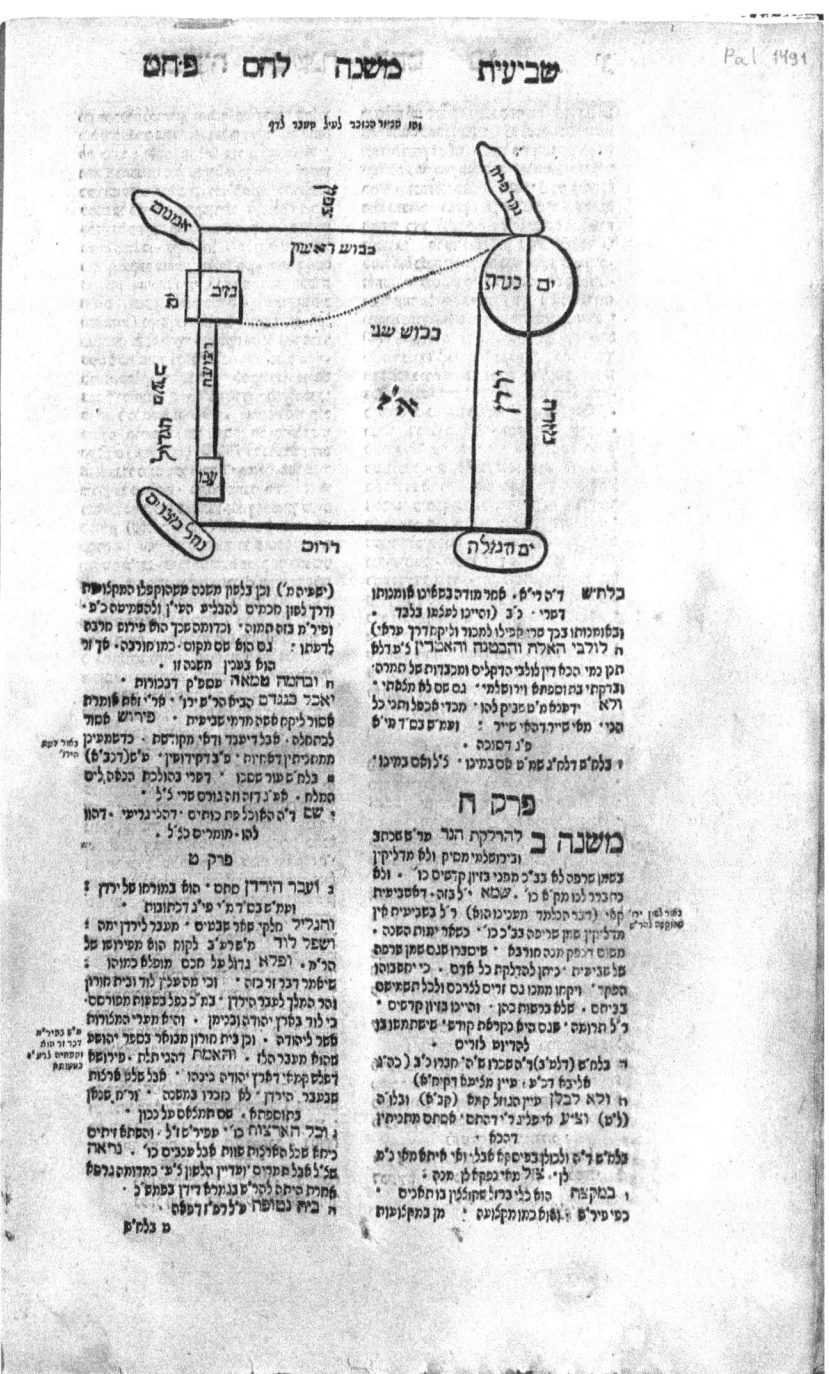

Fig. 103: Jacob Emden, the map in *Sefer leḥem shamayim*, ca. 115 x 115 mm.

Conclusion

Throughout this chapter, I have scrutinized the development of maps that present the tribal allotments within the schematic framework of the traditional Hebrew map. This sub-genre's chronological purview ranges from the map of Joshua Feivel ben Israel (Teomim) in the late eighteenth century to the early twentieth-century versions of maps that are attributed to the *Vilna gaon*. All these maps depict the tribal areas through the prism of the Book of Joshua, the second half of which deals entirely with the allotments. This subject matter first appeared, practically from naught, on traditional Jewish maps in a more or less standard format towards the end of the 1700s in Eastern Europe. Moreover, this phenomenon was not limited to the description of the tribal areas, for it also involved two transitions: from a minimalistic map offering a small handful of toponyms to one replete with names; and from a map based on a few verses in the portion of *Mas'ei* to one that gleans information from various strata of the Bible, the literature of the Sages, and their commentaries. These developments may be connected to the Jewish mapmakers' exposure to Christian cartography in which the tribal allotments were already a major theme for several centuries (see the relevant discussion in chapter 4). Alternatively, they might have been influenced by Jewish cartographers, like bar Yaacov, who fashioned maps on the basis of Christian sources (see the preceding chapter). Whatever the case, the traditional map underwent significant changes, as it adopted new themes from Christian-authored Holy Land maps.

At least some of the maps in question devoted a great deal of attention to Jerusalem and the building of the Jewish Temple. Examples include the cartographic representations of the Book of Ezekiel in which the Temple Mount merits top billing; Altschuler's Jerusalem-centric map; and the rendering of Jerusalem as the sun in Berman's map. All these maps hint to the fact that their Jewish authors drew a correlation between the tribal allotments and the perception of Jerusalem and the Temple as the heart of the Land of Israel.

Eretz Yisrael's apportionment into tribes debuted on these maps toward the end of the 1700s, concomitant to the publication of Shlomo of Chelm's Hebrew adaptation of Latin works and Yehonatan ben Yaakov's replication of a German map. This timing hints to the possibility that this topic reached our sub-genre from Christian maps. Just as the latter two figures predicated their own works, cartographic or otherwise, on Christian sources, it also stands to reason that Teomim and Ziskind had access to and contended with similar materials. Furthermore, the gradual inroads of the Haskalah might have also left its mark on Jewish scholars, even those who remained loyal to the traditional worldview.[43]

---

[43] A more detailed discussion, as well as bibliographical references, on the Haskalah can be found in the beginning of chapter 7.

The doubts concerning the *Vilna gaon*'s outlook notwithstanding, one can surmise that his openness to general studies, or at least his willingness to view them as ancillary disciplines for the main goal of Torah knowledge, facilitated the penetration of secular works into his circle of colleagues.[44] In any event, there was a substantial difference between those Jewish authors who copied the said Christian maps and the main protagonists of this chapter. The former emulated an outside graphical approach and introduced it to Hebrew cartography, whereas the latter remained true to the schematic framework of the Rashi maps. In the process, the "old-school" authors engendered a hybrid tradition. From the standpoint of form, these mapmakers leaned on the established template and completely ignored the geographic reality in the Land of Israel. At one and the same time, though, they adopted the new Jewish theme of the tribal allotments and occasionally incorporated other "foreign" elements into their maps. This trend was bolstered sevenfold upon its embracement by the *Vilna gaon*'s disciples, who attributed these innovations to their revered teacher.

A few surprises that run counter to the literal interpretation of the Bible or the period's common geographical knowledge turn up on these maps. The most prominent example is the depiction of the Tigris and Euphrates spilling into the Mediterranean from northern Eretz Yisrael. Other elements of this sort include the Red Sea as a continuation of the Dead Sea, or as a body of water that boxes in the Land of Israel from the east; the placement of Tyre and Sidon far from the Mediterranean coast; and the absence of the Yarmouk River, even though it is mentioned in the Mishna. In my estimation, the recurrence of these phenomena in traditional maps is indicative of the emergence of replication as a key tool for transferring knowledge, both in general and in all that concerns the sub-genre under review. On many occasions, the mapmakers' efforts to copy and recycle information appear to have dwarfed their interest in coping with geographic knowledge or independently analyzing the Biblical texts.

---

[44] The question of the extent to which the *Vilna gaon* was indeed open to the Jewish Enlightenment and the wisdom of the nations is examined at length by Etkes, *The Gaon of Vilna—The Man and His Image*, pp. 44–83 [Hebrew].

6 Cartographic Tableaux of the Holy Places

The nineteenth century bore witness to a series of Jewish-authored cartographic tableaux that were far-removed from modern cartographic trends. More specifically, this genre consisted of old-fashioned, partially schematic maps of the Land of Israel in which representations of holy places were arrayed in table form. Correspondingly, these tableaux contend with Eretz Yisrael's scenery and geography in a picturesque manner. While depicting geographical elements like Mount Carmel, the Jordan, the Sea of Galilee, the Mediterranean coast, and settlements, their main emphasis is on holy sites, including *qivre ṣaddiqim* (tombs of the righteous).[1]

The classification of these works as maps stems from the fact that they graphically represent a space with a defined easterly orientation and a somewhat realistic, if highly imprecise, alignment between its various components. The Mediterranean Sea fills the bottom (west) portion of these tableaux, and the Jordan and the Dead Sea are located on top; to the left (north) are Lebanon and the Galilee; and on the right (southern) side are Jerusalem, Hebron, and Sodom. In all likelihood, this series was an advanced phase of the traditional Holy Land tables painting genre. We will get back to the link between these groups at the end of this chapter. At any rate, the sites in this particular corpus are arrayed in four or five rows, with occasional smaller lines squeezed in between them. In the series' first work, the artist clearly arranged the sites into quasi-rows. The lines steadily blurred as the genre evolved, but they continued to influence the division of the tableau's space and the alignment of their sites. This rigid framework stands behind the inaccurate spatial relation between many of the sites, including their location and the distances between them. For example, Tiberias is located above Jerusalem instead of to its left; and Acre is beneath Jerusalem instead of being beneath Safad.

All the tableaux in this series bear a strong resemblance to each other, as they comprise a defined group with relations of source and replication between the constituent parts. Nevertheless, it is difficult to establish the order in which the tableaux came out, for not all of them have been authoritatively dated. In addition, the earlier ones are in manuscripts and the later ones were printed; and there is an appreciable gap between the periods in which these two sub-groups were rendered. It also bears noting that there may very well be other tableaux of this sort that have not come to our attention such as one that was recently on auction. At any rate, this genre can be said to have evolved over time, as the painters gradually integrated new content. In the pages to come, I will survey these tableaux, take stock of their cartographic elements, attempt to establish their chronological order, and analyze their sources.

---

[1] These tableaux were touched on by Peled, "Seven Artists," pp. 146–150 [Hebrew]; Ilan, *Tombs of the Righteous in the Land of Israel*, p. 56 [Hebrew]; and most recently by Salomons, "Tracking the Vanished Mapmaker" [Hebrew]. I am indebted to Mr. Reuven Salomons and Dr. Doron Bar for sharing valuable information with me.

The Tableaux of Haim Shlomo Luria

The earliest tableau in this series is by one Haim Shlomo Luria of Tiberias. According to Luria, he created this painting after the earthquake that pummeled Tiberias and Safad in 1837. Unfortunately, its current location is unknown. All that we have is a hazy photograph that was published by Haviva Peled.[2] On the bottom of the map is a lengthy text describing the deadly earthquake and the tribulations in its wake. The final line offers several pieces of information: the author's name; the fact that he is a progeny of Isaac Luria (ha'ARI); and the painting's dedicatee—one R. Shalom Alflalo. Wrapping around the other sides is a title and preamble, which is drawn in large font: "The Layout of the Land for All its Borders; with all its distinctions and its boundaries[,] its seas, its waters[,] and its rivers; its cities and its villages[,] with all its hidden treasures;[3] its prophets and its pious; its Tannaim and its Amoraim; its righteous and its holy; its geniuses and its rabbis; its sages and *rabanan saborai*;[4] take note pleasant reader how I presented before you and signified its characteristics in a lucid and precise manner[;] and observe its ways and learn." A detail that sets this tableau apart from its successors is the representation of Gaza near the bottom-right corner in the form of a built-up city with a wall and towers.

To the best of our knowledge, Luria drew another tableau. However, only a segment of it (depicting Safad) was published, of which there is but a single extant photograph. According to the writing on the side of this photo, the tableau was made for a "Lord Yaacov ben Aziz" and is housed in Oxford.[5] That said, I did not find any more information regarding these maps or their author. Luria is not mentioned in the list of Tiberias' rabbinical emissaries, nor does R. Shalom Alflalo's name come up on the list of the city's distinguished rabbis.[6] For this reason, and due to the poor quality of the photographs, I was unable to expound upon these paintings.

The Tableau of Moshe Ganbash

The second tableau was crafted by Moshe Ganbash, another obscure figure. The manuscript is held by the Jewish Museum in New York. For this reason, it is acces-

---

[2] Peled, "Seven Artists," photograph 122, p. 150. According to Peled, the director of the National Maritime Museum in Haifa saw the tableau in London and gave her the photograph.
[3] This is an allusion to Deuteronomy 33:19: "the treasures hidden in the sand."
[4] *Rabanan saborai* was an honorific for Talmudic sages in Babylon between the fifth and seventh centuries.
[5] Benayahu, "Devotion Practices of the Kabbalists of Safad in Meron," *Sefunot* 6 (1962), plate V (the plate is lodged between pp. 32 and 33).
[6] Avissar, *The Book of Tiberias: City, Region and its Settlement over the Generations*, esp. pp. 189–192, 278–312 [Hebrew].

sible and more amenable to meticulous examination.[7] It is clearly related to Luria's tableau, and it stands to reason that both were drawn at around the same time. Even if we cannot definitively establish the dates, it stands to reason that Luria's is older. As noted, Luria was a resident of Eretz Yisrael. Moreover, he laments over the earthquake of 1837, which he personally lived through, in his text.

Ganbash's tableau is painted in gouache and ink on a large paper folio (1041 x 864 mm) and includes several paper collages. On the bottom-right corner is a decorative inscription providing some basic information: "The young writer Moshe Ganbash may God protect and preserve him[,] written in Constantinople[,] may the Lord protect it." On the opposite side, the year of publication is encrypted in a chronogram, which is based on Psalms 27:14: "In the year [']Wait for the LORD; be strong and take heart and wait for the LORD.[']" The numerical value of the Hebrew letters is 599, namely the Hebrew year 5599 or 1839. As evidenced by the large letters spelling out the Tetragrammaton atop the map, Ganbash adopted the "Shiviti" style—an ornamental template that was quite prevalent in Jewish homes and synagogues.[8] Integrated within the Tetragrammaton headpiece are two smaller words: *adonai* (my Master) and beneath it *tamid* (eternally). Moreover, the tableau is surrounded by a decorative frame that includes various elements. On both of the upper corners are incantations for banishing "any evil eye." Beneath these texts are identical strips that feature a running vine, which climbs out of a grand portal. Further down are rectangular mazes reading "Jericho closes and is enclosed."

Between the outer frame and the map itself is the preamble: "This very drawing encompasses all of Eretz Yisrael and all its hidden treasures[:] those residing in the dust of its foundations[,] those sleeping in its ashes[,] its prophets and its holy[,] its righteous and its pious[,] its Tannaim and its Amoraim[,] its sages and its upright[,] its priests and its rulers[,] its kings and its judges who are ensconced in the ashes of its foundations[,] as well as the layout of the Land for all its borders [with] all its distinctions and its boundaries and a selection of its seas and its rivers[,] its mountains and its hills[,] and all its treasures." The preamble ends along the bottom margin: "Take note dear reader how I noted for your eyes everything with a fine explanation in coherent writing regarding its [i.e., Eretz Yisrael's] distinctions and with [...] its characteristics in a lucid and precise manner[;] and give to the wise and he will become wiser [...] the wisdom."[9]

The body of this painting is divided into four main strips. The lower one depicts the Mediterranean Sea. In the foreground is a wheel-powered steamship flying the Turkish flag—a hint to the author's identity.[10] This compartment also includes the coast of Eretz Yisrael, beginning with Beirut, on through Tyre, Sidon, Acre, and

---

[7] Mann and Bilski, *The Jewish Museum*, p. 58, il.
[8] Juhasz, *The 'Shiviti-Menorah': A Representation of the Sacred—between Spirit and Matter* [Hebrew].
[9] The brackets denote those places where the inscription was frayed and the text is illegible.
[10] It is worth noting that steam boats with wheels on both sides only began to be produced some twenty years before the painting of this tableau.

Fig. 104: Moshe Ganbash, "This Very Drawing Encompasses All of Eretz Yisrael" map, Constantinople, 1839, 1041 x 864 mm.

Mount Carmel, and ending with Jaffa. Overlooking the shoreline are the Lebanon Mountains, Baalbek (Heliopolis), and Deir el Qamar.[11] The second strip covers the inland region. To the left are sites in the Galilee: Peq'in, Simeon bar Yochai's cave, and the misplaced Beit She'an. Jenin and Shechem (Nablus) can be found in the center, while Jerusalem and its neighboring pilgrimage sites take up the right wing. The principal sites in Jerusalem are "Solomon's Midrash" (i.e., al-Aqsa Mosque) and "The Building of the Holy of Holies, place of the Foundation Stone" (the Dome of the Rock). Between these monumental buildings are the Western Wall and four large cypress trees. The third frame accentuates places in the Galilee. To the left is Safad "before the quake" and myriad *qivre ṣaddiqim* in the vicinity. Tiberias is located in the center of this row and Hebron on the far right side. The depiction of Safad—a mountain with a citadel on its peak and buildings stacked along its slopes—is quite realistic. To the left of the citadel is an enormous tree, which recurs on a later tableau as well. Above the city is a caption reading Mount Hermon.

---

11 Deir el Qamar is a small town in the center of the Chouf Mountains. It long served as the place of residence of Lebanon's rulers.

The upper compartment is arranged into a few, detailed sub-strips. To the left are Upper Galilean villages, such as Ammuqa and Gush Halav, and a bevy of revered tombs, including those of Deborah the Prophetess and her army commander Barak, along with the location of Kadesh (Barak's hometown in Naphtali). The middle section consists of Biblical toponyms that are tied to Eretz Yisrael's borders and Transjordan, like Ziphron, Heshbon, Azmon, and Hazar Addar. Between these sites is an illustration of a city bearing two labels: "City of Haran" and "Terah". Put differently, the settlement is tied to the Patriarchs' origins in Mesopotamia. The Dead Sea, Lot's wife, and "Sodom which was overturned" (Lamentations 4:10) appear to the right. This strip includes many toponyms that were clearly unfamiliar to Ganbash as tangible sites, for their relative locations are off target. For instance, Zanoah and Adullam, which were in the Judean lowlands, are drawn to the east of Jericho; "the village of Janoah" is mis-identified with "that is the Ascent of Akrabbim;" and Socoh mis-identified with "that is Jarmuth." From a graphical standpoint, the profusion of sites whose locations were unknown to the author is responsible for the fragmented appearance of the upper strip, especially its right side.

I will not expand on all the pilgrimage sites in this painting, for the evolution of the customary pilgrimage route has already commanded a great deal of research attention.[12] That said, I will mention those traditions whose representation and placement attest to the link between Ganbash's tableau and those that ensued, such as the following sites: the tombs of Phinehas ben Eleazar and Joshua ben Nun to the left of Shechem (Nablus); the burial places of Joseph the Righteous and Shechem son of Hamor to the city's right; Safad, its citadel and large tree; the pillar of Lot's wife; and the sites on Mount Lebanon. Content-wise, this map encompasses toponyms that belong to several different categories: Biblical sites like Haran or the Ascent of Akrabbim that were completely unfamiliar to Ganbash as actual places; unfamiliar sites that the author nonetheless had a geographical inkling of their location, such as Sodom and the pillar of Lot's wife;[13] integral stops on the pilgrimage route, like the tombs of the righteous and *arbah arṣot ha'kodesh* (four lands of holiness, namely Jerusalem, Hebron, Tiberias, and Safad[14]); sites that were relevant to everyday life in the region during Ganbash's lifetime, such as Tyre, Sidon, Beirut, and Deir el Qamar. The map's compiler treats all of the above as actual sites, even though the majority of

---

12 Reiner, "'Oral versus Written:' The Shaping of Traditions of Holy Places in the Middle Ages" [Hebrew]; idem, *Pilgrims and Pilgrimage to Eretz Yisrael, 1099–1517*, [Hebrew]; Ilan, *Tombs of the Righteous in the Land of Israel*; Goren, "An Eighteenth Century Geography: 'Sefer Yedei Moshe' by Rabbi Moshe Yerushalmi" [Hebrew].

13 In all likelihood, the pilgrims never laid eyes on "Lot's wife" or even made it as far as the Dead Sea. However, they were quite familiar with the literary tradition that pertains to this region. A case in point is the testimony given in Haim ben Dov Dvrish Horovitz, *Sefer ḥibat yerushalayim* (The Adoration of Jerusalem Book), p. 4 [Hebrew].

14 These were the cities in which most of the Jews resided in Palestine in the long period between the 15th and 19th centuries.

them were neither Jewish pilgrimage destinations nor personally familiar to him or to his sources.

As we have seen, Luria and Ganbash employed the traditional template for holy places—a table in which the sites are arranged into rows.[15] Nevertheless, their tableaux were highly innovative from a design standpoint, and this novelty endowed the two works with their strong cartographic feel. Both depict components of Eretz Yisrael's landscape, such as its mountains and valleys, the Mediterranean Sea, the Sea of Galilee, the Jordan, and the Dead Sea. Moreover, their geographic elements are basically arranged in a realistic fashion. Notwithstanding the lack of evidence for this hypothesis, it stands to reason that the earlier east-oriented, painted maps of the Holy Land—be they Christian-authored or Jewish adaptations by Abraham bar Yaacov and his ilk—influenced the design of Luria and Ganbash's tableaux.

Tableau of Joshua Alter and Mosheh ben Pinhas Fainkind

A later tableau is housed at Israel's National Maritime Museum in Haifa.[16] Printed on paper, a block of text on the map's lower-right corner states that it "was brought to print by our teacher the rabbi[,] Joshua Alter Beri[,] and by the l[ad] Mosheh ben R. Pinhas Fainkind of Turek."[17] Along the bottom margin is a pair of Russian sentences. The one to the left constitutes the printing permit, which was issued in Turek by a clerk named Parshin. The other text notes the lithography studio that printed the map, but its name is cutoff in the middle and cannot be fully reconstructed.[18]

The year of publication, which was not included on the tableau itself, is the object of debate. The date cited in the literature is 1875, but there are serious doubts as to its veracity.[19] A memorial book on the Jewish community of Turek briefly mentions the two compilers. Mosheh Fainkind (1935–1865) is described as a Zionist activist and Maskil. However, Abraham Benkel, the author of the relevant chapter, apparently confuses R. Joshua Alter with his son, the artist Chanoch Glicenstein (1870–1942),

---

15 Sarfati, The Illustrations of Yiḥus ha-Avot: Folk Art from the Holy Land.
16 Ben-Eli and Stieglitz, *The National Maritime Museum, Haifa—Spring 1972*, p. 38.
17 Turek is a city in Poland, north of Kalisz and west of Lodz.
18 The text to the left reads thus: "доз.печ. Нач. Зем. Стр. Тур. Уез. Паршин." I suggest that the following amendments be made: "доз[волено] [к] печ[ати] Нач[альник] Зем.[ского] Стр[...?] Тур[-екского] Уез[да] Паршин. This amended sentence can be rendered as "Approved for printing by Parshin, department head [...] in the administration of the County of Turek." The other Russian text is "Литогр. С. Ли...анович в гор. Турек" (S. Li[...]manovitz Lithography [Studio], in the city of Turek). I would like to thank Dr. Mitia Frumin for helping me translate this material.
19 This date (1875) is cited in the museum's catalog without any supporting evidence; Ben-Eli and Stieglitz, *The National Maritime Museum*, p. 38. Haviva Peled wrote that this year is noted on the folio's margins, but this claim is less than tenable. It appears as though the source for this date is Vilnay, *The Hebrew Maps of the Holy Land*, p. 32. However, neither Vilnay nor any other researcher backed up this hypothesis with firm evidence.

as the latter is mistakenly credited with contributing to the map's production.[20] The Yiddish lexicographer, Zalman Reisen, notes that Fainkind printed the tableau when he was seventeen years old—information that is consistent with his above-cited moniker "the lad."[21] If this is indeed the case, Fainkind's year of birth, 1865, rules out the possibility that the map came out in 1875, and the actual date should be no earlier than 1882.

The preamble, which is quite detailed and close in both style and content to those of the earlier manuscripts, wraps around the painting. "The Layout of the Holy Land for All its Borders[:] with all its distinctions[,] its cities and villages[,] its mountains and its hills[,] and all the treasures of its hidden mysteries[,] its prophets and its pious[,] its righteous and its holy[,] its Tannaim and its Amoraim[,] its geniuses and its rabbis; take note pleasant and coveted reader how I noted and displayed before you its characteristics in a lucid and precise manner and observe its ways and gain knowledge. . . ." (the end of this long preamble incorporates verses from Proverbs 25:11 and 6:6). On the bottom frame are parts of two more Biblical passages: "If I forget thee, O Jerusalem[,] if I do not remember you, if I do not exalt Jerusalem above my chief joy [Psalms 137:5–6][.] Zion shall be redeemed with justice[,] and the returnees with righteousness [Isaiah 1:27]." This combination also turns up in the later print versions. Over all, the wording of the preamble and the representation of the Land tie this tableau directly to the ones discussed below.

An in-depth look at this map's content and a comparison with that of Ganbash reveals strong links between the two. For example, both authors incorporated a large tree in Safad, offered a similar depiction of the Jordan River (it emanates from "the Pamyas cave" and does not traverse the Sea of Galilee in either tableau), and included Tyre, Sidon, Beirut, Deir el Qamar, and "Ramla that is Gat." On the other hand, there are many differences between the two. The map from Turek omits certain sites, while adding many others. In Transjordan, places that also surface on Ganbash's tableaux merit new details. "The Tabernacle of Shiloh" was inserted between Haifa and Jaffe; the Tombs of Zephaniah, Bezalel, and Aholiab were added near Beirut; Issachar was added next to Zebulun; and Joseph's tomb makes its debut in this genre, not far from Dinah's burial site.[22] As in the maps below, crescents conspicuously adorn the domes of Simeon bar Yochai's mausoleum. For the most part, though, the authors of these tableaux eschewed these sort of Muslim symbols even in places where they almost certainly existed, like minarets of mosques. Against this backdrop, we can conclude that the handwritten variations are tied to the subsequent printed

---

20 Seiffe, "Mosheh Fainkind" [Hebrew]; Benkel, "R. Yehoszua Alter (Glicenstein), Chanoch (Henrik, Enrico) Glicenstein" [Hebrew]. I owe a debt of gratitude to Reuven Salomons for bringing this information to my attention. He also proposed dating this map to 1882.
21 Reisen, *Leksikon fun der yidisher literatur, prese, un filologye*.
22 Placing Dinah's grave near Shechem (Nablus) contravenes the documented tradition of the time whereby the site is near Mount Arbel, in the Lower Galilee. See Ilan, *Tombs of the Righteous in the Land of Israel*, pp. 179–180 and the references therein to the pertinent sources.

versions. Be that as it may, the discrepancies between them indicate that the connection was somewhat loose. Consequently, Alter and Fainkind's map may have been based on another manuscript whose whereabouts are currently unknown.

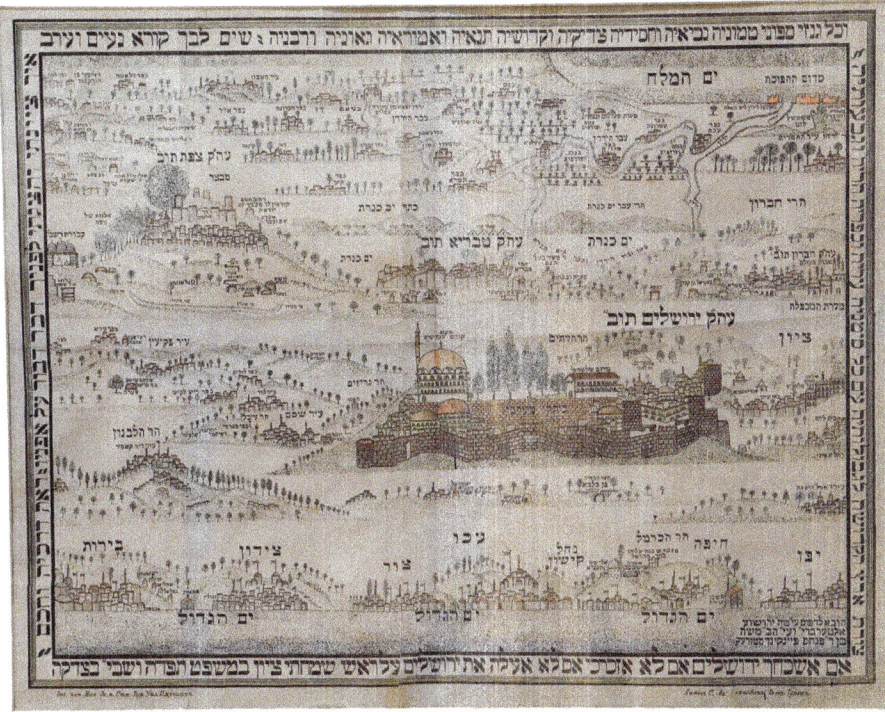

Fig. 105: Joshua Alter and Mosheh ben Pinhas Fainkind, "Layout of the Holy Land for All its Borders" map, 520 x 400 mm.

The Munk Tableau

A map that bears a close resemblance to Alter and Fainkind's apparently came out soon after in Vienna. The latter makes no reference to its author(s) or to those of its predecessor from Turek.[23] As noted on the lower righthand corner, it was printed at the Munk Lithography Studio.[24] Shadowing the Hebrew preamble and captions are

---

23 The Gross family collection in Tel Aviv holds a copy of this tableau, which was published by Tishby, *Holy Land in Maps*, pp. 132–133 [Hebrew]. I would like to thank Mr. William Gross for placing a digital version of this map at my disposal. Another copy, hitherto unknown to scholars, was recently purchased at a public auction by Mr. Simon Cohen. I also owe a debt of gratitude to Mr. Cohen for allowing me to examine and photograph this copy.

24 Lith. Ans. M. Munk in Wien. The studio was well known, especially in a somewhat later period.

German ones. Save for a couple of minor changes, the Hebrew preamble is identical to that of the Turek map.[25] The German text, captions excluded, consists of three parts. At the head of the map is a large title reading "Panorama of the Holy Lands."[26] The preamble runs along the bottom and left edges: "This tableau includes a topographic geographical account of Palestine with a greater emphasis on sites that are familiar from history, the Bible, and the Talmud, of the tombs of known figures and of other monumental buildings."[27] On the righthand side, there is another German text stating that the tableau was printed along the lines of the "original version by Rabbi Chaim Salomon Pinia of Safad."[28] This detail points to the fact that the Viennese author availed himself of an earlier tableau of this sort, be it a manuscript version (e.g. the tableaux of Luria and Ganbash) or one of the prints. The existing sources lack so much as a clue as to the biography of Chaim Salomon Pinia or the "original" map. Nevertheless, the literature has termed this genre "the Pinia maps."

A meticulous comparison between this tableau and its predecessors points to a strong resemblance between the former and Alter and Fainkind's tableau. However, there are also changes and additions to the map at hand. Most striking are the new wrinkles in the German captions. These developments are not merely lingual, for "panorama" and "topographic geographical account" are modern terms that betray an interest in the era's scientific advances. Moreover, new content was added to the map itself. For instance, there is a caption by the bottom-left corner reading "How Gaza once was."[29] Additionally, Deir el Qamar is identified as "the residence of the Emir Bashir." This is a reference to none other than Bashir Shihab II—the ruler of Mount Lebanon during the first half of the nineteenth century. The Hebrew caption next to the Dome of the Rock reads "Place of the Holy of Holies," while the corresponding German text offers the following explanation: "The Sanctuary, the place where the Holy of Holies of the Temple was in the past, a Turkish mosque."[30] Perhaps the reference to Deir el Qamar and Emir Bashir alludes to a contemporaneous event that sent shockwaves throughout the Jewish world: the blood libel against the

---

[25] "[I]ts distinctions and its boundaries" was added to the preamble. This sequence turns up in the Ganbash version, but not in the map from Turek. In addition, the Munk tableau refrains from describing the reader as "coveted."
[26] "Panorama des heiligen Landes."
[27] "Dieses Tableau einhält eine geographisch-topographische Darstellung Palästina's mit näherer Bezeichnung der aus der Geschichte, Bibel und Talmud bekannten Orte, der Gräber berühmter Männer und sonstiger monumentaler Bauten."
[28] "Nach originalzeichnungen von Rabbi Chaim Salomon Pinia aus Zefath."
[29] As in the adjacent Hebrew caption, the Jewish sources indeed identified Ramla with Gat. In consequence, there is a possibility that the author confused Gat with Gaza.
[30] On the image of the Dome of the Rock as the temple, see: Pamela Berger, *The Crescent on the Temple*: pp. 271–288.

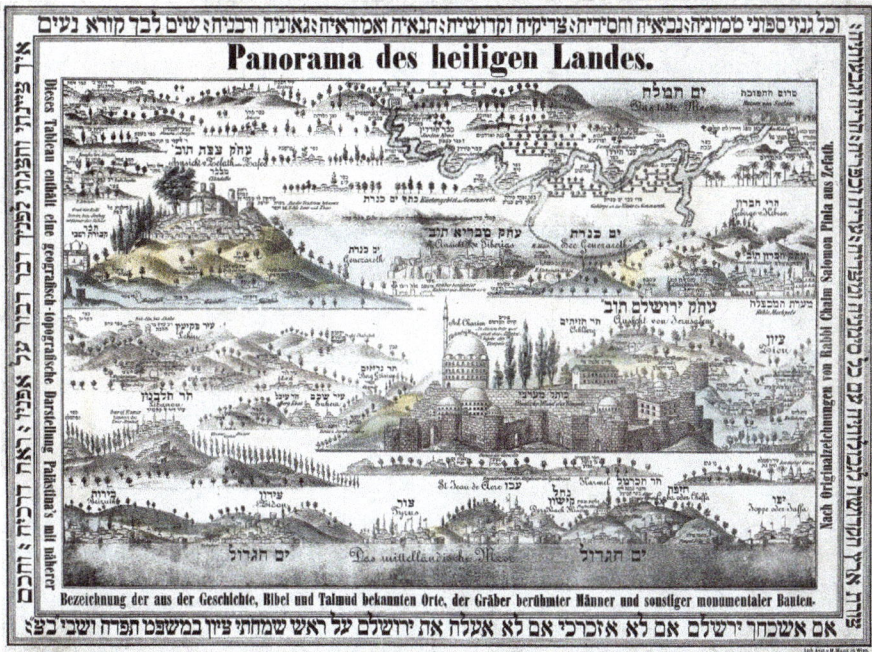

Fig. 106: The Layout of the Holy Land for All its Borders" map, Vienna, 600 x 480 mm.

village's Jewish community in 1847, which was resolved thanks to the intervention of Moses Montefiore (a British-Jewish philanthropist).[31]

Besides these German additions, there is a surprising Hebrew caption to the left of Ramla: "Bu Ghosh." This is clearly a reference to Abu Ghosh—a family that controlled the village of Kuryet el' Enab and thus the road to Jerusalem. In all likelihood, the common denominator between all these changes is that the authors cast their gaze beyond the confines of the Jewish world to the broader reality in Ottoman Palestine.

The Cloth Map

A map that closely resembles the Munk tableau was printed on a yellowish cloth sheet. The details that were revised in this edition are of secondary importance, and the template itself was most likely the fruit of a different block. In any event, the similarities suggest that they were both printed in the same Vienna press.

---

31 Braslavi, "The Jewish Community in Deir el Qamar and the Second Blood Libel (in the 19th Century)" [Hebrew].

There are three known copies of this tableau, at the Israel National Library,[32] the Israel Museum,[33] and the Danish Jewish Museum in Copenhagen. Since the map was printed on a cloth, we may surmise that it was intended to be used as, say, a small tablecloth or challah cover.[34]

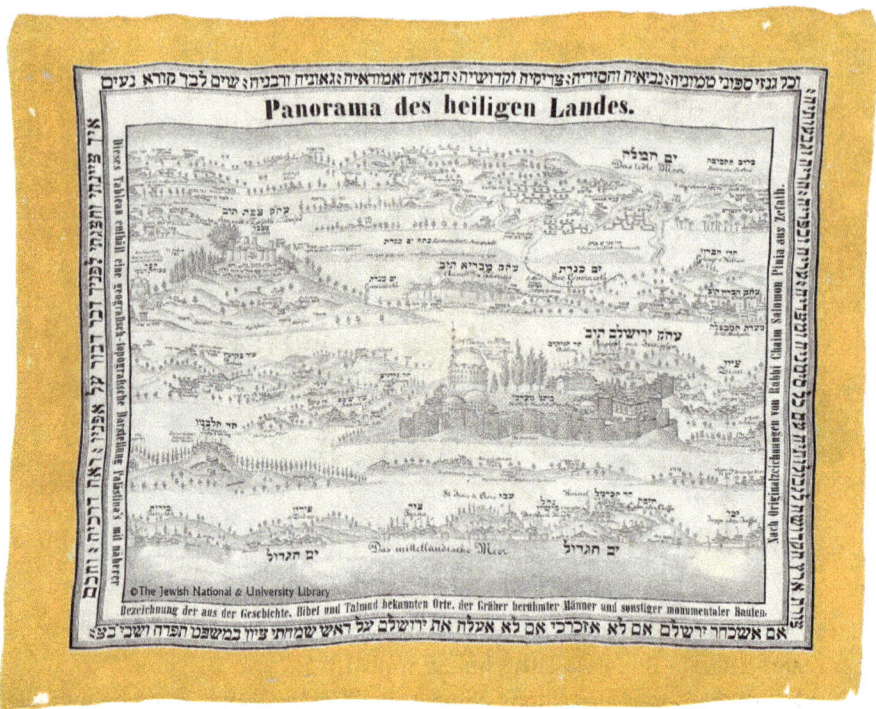

Fig. 107: The Layout of the Holy Land for All its Borders" cloth map, 620 x 474 mm.

The Map from the Schottlaender Lithography Studio in Breslau

Another lithographic tableau was subsequently printed in Breslau on a wide folio (542 x 370 mm). Like other works in this series, the author did not include the date of publication. At the top of this color map is a large German title: "*Panorama des heiligen Landes*" (Panorama of the Holy Lands).[35] While the name of the lithog-

---

32 Laor, *Maps of the Holy Land*, p. 81, no. 564.
33 Shachar, *Jewish Tradition in Art: The Feuchtwanger Collection of Judaica*.
34 It is worth noting that there are embroidered maps among the tables of holy places. Fischer, *Art and Artisanship in Nineteenth-Century Eretz Yisrael* [Hebrew].
35 A copy of this map is held by the Laor Collection at the Israel National Library. According to the testimony of Reuven Salomons, two other copies have been located in private collections.

rapher, Schottlaender, appears on the lower margin, there are no hints as to the author or the motivation behind this undertaking.[36] Across the bottom frame is a long German text in smaller letters. Though close in wording to the cloth map's preamble, there are a few substantial differences between the two, most notably the reference to the newly founded Jewish colonies.[37] The Hebrew preamble takes up the right and left frames: "The Layout of the Holy Land for All its Borders[:] its mountains and its valleys[,] its cities and its villages[,] with all the estates and the colonies in which the Jews returning from all the lands of their exiles have settled in [order] to work the land of their forefathers as in the days of yore. Zion shall be redeemed with justice, and its return[ees] with righteousness" (Isaiah 1:27). The final portion of this text obviously sets this tableau apart from the rest of the genre. More specifically, the author praises the incipient modernization of Eretz Yisrael, particularly the colonies of the First Aliyah (the initial wave of Zionist immigration to Palestine). Interestingly, the renewal and hope are expressed by means of a passage from the Bible.

While this tableau belongs to the genre under review, it is not merely an improved adaptation of the earlier works. Instead, it constitutes a new creation with many novel details—a sort of paraphrase that was assembled with a rather free hand. The map's purview ranges from Damascus, Mount Lebanon, and Beirut in the north to Hebron and Zoara in the south. Although the compiler was often unable to provide their exact geographic locations, it is evident that he was well-versed in the traditions about the Holy Land sites—both ancient and new alike. Despite the availability of modern maps of Palestine, the author clearly eschewed these sources. As in the earlier tableaux, the distribution of the places is less than realistic. This alignment is probably a remnant of the rows that undergirded the earlier maps, even if the lines are now all but unnoticeable. Insofar as the Land's general depiction is concerned, the mapmaker added Damascus and Haran in the northeast, positioned Beit She'an in the outer northeastern corner, replaced the toponym of Tyre with Zarephath (as per 1 Kings 17), and added Achzib. Moreover, he made note of Sodom and Gomorrah, but removed "Sodom which is overturned" and Lot's wife, who was artistically portrayed as a pillar of salt by his forerunners.

In contrast to the earlier tableaux, quite a few destinations on the customary Jewish pilgrimage route, especially many of its burial sites, were left off this map. Of all the *qivrei ṣadiqqim* in the Galilee, the author only incorporated those of Simeon bar Yochai and ha'ARI in Meiron (even though the latter is actually interred in Safad!); Rabbi Moshe Alshich, Hosea the Prophet, and Rabbi Joseph Karo in and around Safad, and Maimonides and Meir "ba'al haNes" (the miracle worker) in the vicinity of Tiberias. Furthermore, not a single tomb is denoted in Samaria, the Nablus region included. With respect to the greater Jerusalem area, the mapmaker sufficed

---

36 Breslau: Lith. u. Druck v. S. Schottlaender.
37 "Geographisch-Topographische Darstellung Palästinas mit genauer Bezeichnung der aus der Geschichte, Bibel und Talmud bekannten Orte, der Gräber berühmter Männer wie auch aller jüdischen Colonien Nach den neuesten Quellen bearbeitet."

Fig. 108: The Layout of the Holy Land for All its Borders" map, ca. 1900. 370 x 542 mm.

with the tombs of Samuel the Prophet, Simeon the Just, and Zechariah as well as the cave of Haggai and Malachi. Among the sites within the city's walls are Hurvat Rabbi Yehuda he'Hasid (a synagogue named after Rabbi Yehuda the "pious"), the Synagogue of R. Johanan ben Zakai, *Batei mahse* (the aegis houses),[38] and the School of Torah and Work. Established in 1882, the latter was a modern-style Hebrew school under the direction of Nissim Behar.[39]

The most significant innovation in this painting, though, is the "colonies of our Jewish brethren." In addition to the genre's usual repertoire of sacred places, this tableau incorporates Hebrew and German captions denoting over twenty of the colonies that were established during the First Aliyah.[40] Like all the places on this map, the colonies are not arranged in a geographically realistic manner, but one can nevertheless discern five primary areas of settlement:[41]

a. The outskirts of Jerusalem: Motza (Colonie Moza).

---

38 *Batei mahse* was an apartment complex that was built for destitute Jewish families in the Jewish Quarter of Jerusalem's Old City between 1860 and 1890.
39 Ben-Arieh, *A City Reflected in its Times: New Jerusalem—the Beginnings*, pp. 357–361 [Hebrew].
40 A similar, if unidentical, list of the colonies appeared in a booklet by Goldzweig, *Sefer igeret le'-kaiṣ ha'yamin*. This work, along with its map, will be discussed in the next chapter.
41 I would like to thank Dr. Ran Aaronsohn for helping me identify some of the colonies and establish the years in which they were founded.

b. Judea: Petah Tikva, Yehudiah (Colonia Jehudia),[42] Kfar Saba (Colonia Sabba), Nahalat Reuben (Colonie Wadi el Chanin—modern-day Ness Ziona), Mikveh Israel, Gedera, Rishon LeZion, Mazkeret Batya, and Be'er Tuvia (Colonie Beer-Tobiah Kastine).
c. Samaria: Zikhron Ya'akov, al-Hadera Colony (Colonie el-Cudera), Em al-Gamal Colony,[43] Em al-Tut Colony (Colonia Em el Tut),[44] Shfeya (Colonia [Meir] Schewja), and Tantura (Colonia Tantura).[45]
d. The Galilee: Yesud haMa'ala, Mishmar haYarden, Ein Zetim, Mahanayim, Rosh Pinna, and Metula.[46] As discussed below, none of the Lower Galilee's colonies were included on this map.
e. The Golan: Bnei Yehuda (Colonie Bene Jehuda) and Sachem Golan.[47]
f. Another Jewish settlement, Ezra (Colonia Esra), was displayed in the north, but I was unable to come up with any information regarding this site.

The addition of the colonies provides us with a tool for dating the information in the tableau and perhaps its year of publication. By 1900, all of the colonies on this map had already been established (and some were even abandoned by then). However, the settlements of the Lower Galilee, which were not included therein, had yet to be built. What is more, the inclusion and above-cited praise of the colonies ("in which the Jews returning from all the lands of their exiles have settled down in [order] to work the land of their forefathers as in the days of yore") point to a nascent change in the way Eretz Yisrael was grasped by certain Jews. From an expanse that was valued for its sacred tombs and distant historical past, it was becoming a locus of national renewal. Likewise, the inclusion of the School for Torah and Work in Jerusalem and Sha'arei Zion Hospital next to Jaffa, *inter alia*, also reflect a Zionist and modern conception of the Land. Needless to say, this shift is hardly limited to these tableaux or to the contemporaneous maps that are surveyed in the next chapter. For example, the era's literature and scholarly works reflect on trips to Eretz Yisrael and the daily life of its inhabitants.[48]

---

42 Yehudiah was a satellite of Petah Tikva that was founded in 1882 and abandoned eleven years later.
43 As of 1889, Em al-Gamal has gone by the name of Bat Shlomo.
44 Em al-Tut (or Um el Tut) was a temporary site to the east of Zikhron Ya'akov.
45 Founded in 1892, this small colony near the Arab village of Tantura was abandoned by 1896. All that is left of this settlement are the remains of a glass factory. Set up at the initiative of Baron Edmond James de Rothschild, the factory manufactured bottles for some time. Today it is within the boundaries of Kibbutz Nahsholim.
46 The Hebrew caption for this site was misspelled.
47 For a disquisition on Jewish attempts to settle the Golan Heights, see Ilan, *Attempts at Jewish Settlement in Trans-Jordan, 1871–1947* [Hebrew].
48 Bartal, "On Precedence: Time and Place in the First Aliya" [Hebrew]. For more on this topic, see the next chapter.

Conclusion: the Origins and Significance of the "Layout of the Land" Tableaux

Over the course of this chapter, I surveyed the following nineteenth-century tableaux of Eretz Yisrael: the manuscript versions of Luria and Ganbash; the printed maps from Turek and Vienna; the cloth variation of the latter; and the color adaptation that was printed in Breslau. The appreciable similarities between these six maps in all that concerns the Land's portrayal and the wording of their preambles is indicative of a well-defined genre whose constituent parts shared source-emulation relations. Conversely, their differences point to the evolution of this series over time. It stands to reason that there were other tableaux, still unknown. Given the fact that none of the extant print versions have been definitively dated, the missing links could have helped us decipher the order of replication. In any event, the information at our disposal has enabled me to suggest a chronological order. With respect to the manuscripts, Luria's map apparently preceded that of Ganbash. Among the printed tableaux, the one from Turek was followed by its bilingual (Hebrew-German) counterpart from Vienna, which came out in both paper and cloth editions. The tableau from Breslau is the final print version. This map stands out from the rest due to its portrayal of the fledgling Zionist settlement enterprise between 1882 and 1900. As per my findings, all the print versions came out in the final two decades of the nineteenth century.

Due to the recurrence of this basic template, it is incumbent upon scholars to try and reconstruct the origins of this genre and discern its early nineteenth-century sources. In my estimation, these maps are tied to the tradition of Jewish pilgrimage accounts. The development and crystallization of this corpus have been thoroughly researched. Elchanan Reiner has analyzed the canonization of these pilgrimage works and their evolution from a predominately textual to an illustrated corpus. As Reiner put it, "*yiḥus avot* [genealogy of the Patriarchs] underwent a transformation from a text to a picture, from a book to a scroll."[49] The drawings in that corpus' illustrated scrolls did not form an over-arching map of the pilgrimage route, but a chain linking graphic images of each particular site. Every venue in these scrolls merited an appropriate graphic symbol—a structure, cave, or place of settlement—along with a schematic description and a caption. These elements identify the site and explain its importance. This chain of illustrations began with an array of *yiḥus avot* scrolls in the latter stages of the sixteenth century. A prime example of these works was published in 1929 by Cecil Roth.[50] Moreover, Rachel Sarfati elaborated on seven manuscripts that project onto the entire corpus. Some of these texts are rendered in the artistic style that is ascribed to the Land of Israel, while the others are in

---

49 Reiner, "Traditions of Holy Places in Medieval Palestine—Oral versus Written," pp. 9–17 (the citation is from p. 15).
50 Roth (ed.), *The Casale Pilgrim: A Sixteenth-Century Illustrated Guide to the Holy Places*.

the so-called Italian style.[51] At a later stage, the holy places were presented on tables in a site-by-site manner. Put differently, they were arranged into individual compartments and collective rows. However, there was still no spatial connection between the sites or an over-arching cartographic template. These tables were drawn in manuscripts and appeared in prints from as early as the 1700s. At times, they took the form of Mizrah[52] and Shiviti tableaux and even embroidered maps.[53]

Tableaux depicting revered Christian sites also turn up in the popular Greek-Orthodox iconography. For the most part, these paintings were rendered on cloth and sold in the Holy Land to pilgrims as a memento of their devotional journey. In the literature, these paintings are referred to as *Proskynetaria*. The connection between the Jewish tableaux and their Greek-Orthodox counterparts has yet to be fully parsed and indeed warrants further research. However, this topic is beyond the scope of this book.[54] Yet another interesting parallel is the wall and ceiling paintings that were commonplace in the homes of well-to-do Arab families towards the end of the Ottoman period. The resemblance between these works and the Jewish tableaux is striking in all that concerns the portrayal of houses, boats at sea, and city-capped mountains, *inter alia*.[55] This same style also informed contemporaneous paintings of sacred places or areas, like Jerusalem, Safad and the Galilee, that graced the walls of synagogues and the homes of wealthy Jews in various communities throughout the world.

In all likelihood, these works paved the way for Luria and Ganbash's tableaux. A case in point is the Shiviti elements on the Ganbash map. Not only does the latter "set [*shiviti*] the LORD always before me" (the passage from Psalms 16:8 that underpins the Shiviti tradition), but the Holy Land and its sites as well. This sort of synthesis between the Shiviti template and accounts of sacred places indeed characterizes Ganbash's unique hybrid composition—a painting that depicts Eretz Yisrael, while also featuring an amulet against the evil eye. In essence, this same amalgamation stands at the heart of this one-of-a-kind genre in Hebrew cartography, which its authors headlined "the Layout of the Holy Land" series. This terminology links these nineteenth-century tableaux to much earlier maps, quite a few of which were discussed in the previous chapters. Likewise, the graphical roots of this archetype are the popular art that accompanied Holy Land pilgrimage from the fourteenth century onwards. It also bears noting that the high point of this genre is the colored litho-

---

51 Sarfati, "The Illustrations of Yiḥus ha-Avot," pp. 19–25.
52 Mizrah is a Jewish-European ornament that marks the direction of prayer (east).
53 Many examples are displayed in catalogs of the Israel Museum: Fischer, *Art and Artisanship in Nineteenth-Century Eretz Yisrael*; Sarfati, *Offerings from Jerusalem*.
54 Immerzeel, Deluga, and Łaptaś, "Proskynetaria-Inventory;" Skalova, "A Holy Map to Christian Tradition, Preliminary Notes on Painted Proskynetaria of Jerusalem in the Ottoman Era."
55 Sharif, *Hidden Places: Wall and Ceiling Paintings in Ottoman Palestine (1856–1917)*, especially the cover illustration and figs. nos. 15 and 16.

graphic tableau from Breslau with the patently cartographic traits. This map depicts Eretz Yisrael from its Biblical past until the dawn of the Zionist movement.

The purpose of these tableaux is less than clear, but the Ganbash map's Shiviti template hints to the possibility that they served as wall decorations for residences, sukkahs, or prayer houses. In other words, their use was similar to that of the widespread Mizrah tables (see note 52). On the assumption that the above-mentioned Chaim Salomon Pinia of Safad was a rabbinical emissary to Europe who raised funds for the Jewish community of Eretz Yisrael, it stands to reason that his paintings[56] served as a gift item that he and other messengers from the Land of Israel gave or sold to donors.

In the introduction, I noted that the Holy Land maps were a catalyst for imagined pilgrimage. To some extent, all the maps that we have thus far encountered in this book filled this role. As aforementioned, the tableaux under review were successors to the Holy Land tables and Shiviti plaques. Moreover, they depicted contemporary pilgrimage sites and other sacred places, such as the tombs of the righteous and Jerusalem, Hebron, Safad, and Tiberias (i.e., *arbah arşot ha'kodesh*). Against this backdrop, the function of triggering imagined pilgrimage is all the more pronounced. Consequently, the decision by the final tableau's author to use this particular format to portray the newly established colonies of the "Jews returning from all the lands of their exiles" at the dawn of the Zionist movement is remarkable.

---

56 As discussed above, none of Pinia's original tableaux have survived.

7 From Tradition to Modernity—the Hebrew Map between the Nineteenth Century and Early Twentieth Century

The Beginnings of a Modern Geographical Outlook among the Proponents of the Jewish Enlightenment

By the latter stages of the eighteenth century, the Jewish enlightenment, or *Haskalah* had made significant inroads among the Jews of Europe. This movement, which influenced every aspect of Jewish life, has been explored from many different angles, and an in-depth discussion on this vast phenomenon is certainly beyond the scope of the present book.[1] Instead, the chapter's emphasis is on the transition of the geographical research from a notional Eretz Yisrael—an imaginary place found between the pages of the Jewish Holy Scriptures and *siddur* (prayer book)—to the realistic Land of Israel. This shift left an indelible mark on the Haskalah movement's own literature.[2] For example, Zunz's extensive bibliographical study on Hebrew travel writing on Eretz Yisrael must be viewed in this context.[3] Even when the interest in the Land fell within the parameters of the debate between those Jews who sought emancipation and integration within the countries of Europe and those who leaned towards a solution to the "Jewish problem" within the Land of Israel, both sides were referring to a modern entity that was steadily being revealed in the works of contemporaneous Christian scholars.[4]

The question of when the Maskils (Jewish learned people) began to occupy themselves with geography is beyond the purview of this work;[5] and the same can be said for the connection between the Jewish Enlightenment's occupation with Eretz Yisrael and broader processes in European society, such as Biblical research among Protestants, Napoleon's campaign in the Middle East, and the rise of Europe-

---

[1] Etkes (ed.), *The East European Jewish Enlightenment*; Feiner and Israel Bartal (eds.), *The Varieties of Haskalah* [Hebrew]; Feiner, *Haskalah and History: the Emergence of a Modern Jewish Awareness of the Past* [Hebrew]; idem, *The Jewish Enlightenment in the Eighteenth Century*; idem, *The Jewish Enlightenment in the 19th Century* [Hebrew]; Zalkin, *A New Dawn: The Jewish Enlightenment in the Russian Empire—Social Aspects* [Hebrew]; Feiner, *Haskalah and History: The Emergence of a Modern Jewish Historical Consciousness*; idem, *The Jewish Enlightenment*; Feiner and Sorkin (eds.) *New perspectives on the Haskalah*.
[2] Barnai, *Historiography and Nationalism: Trends in the Research of Palestine and its Jewish Yishuv (634–1881)* [Hebrew]; Cohen, *From Dream to Reality: Descriptions of Eretz Israel in Haskalah Literature* [Hebrew].
[3] Zunz, "On the Geographical Literature of the Jews from the Remotest Times to the Year 1841."
[4] Bartal, "To Forget and Remember: The Land of Israel in the Eastern European *Haskalah* Movement."
[5] For a brief discussion on the place of geography, or what was then known as "cosmography," among Italian Renaissance-era Jews, see chapter 3 herein.

an Orientalism.⁶ Of course, the relation between Jewish and Christian scholarship was asymmetrical: the Christian literature by and large developed independently of the Jewish world, whereas the Maskilic research absorbed a great deal of information from the general Western discourse.⁷

Before evaluating the period's maps, let us touch on some of the earliest Jewish works of geography, especially those that pertain to the Land of Israel.⁸ *Tla'ot moshe* (Moses' Travails)—a Yiddish work by Moses ben Abraham (a convert to Judaism) that came out in 1711—is widely regarded as the first "Jewish book" of geography. However, as Bartal and Shemerok point out, large portions of *Tla'ot moshe* were copied from the following books: the aforementioned *Igeret orkhot olam*, by Abraham Farissol (the Renaissance-era Italian-Jewish writer); and a German translation of *Tabularum Geographicarum Contracarum* by the Dutch geographer Petrus Bertius.⁹ At any rate, *Tla'ot moshe* bears witness to the Maskilic circles' interest in geography from as far back as the early eighteenth century.

Penned about a hundred years after *Tla'ot moshe*, Samson Bloch's *Shvilei olam* (Paths of the World) is considered the first Hebrew book to present the geography of the world with modern tools. Bloch (1784–1845) leaned on general European books in this field. Although some of these sources were already outdated, their very use betrays the author's Maskilic bent. That said, he frequently tied geographic topics to questions of Halakha and custom. More than a few of these connections were superfluous, and Bloch was duly criticized by his contemporaries for this flaw. Nevertheless, Klausner summarized Bloch's enterprise thus: "He was nearly the first ... who attempted to write in Hebrew not books about the Haskalah but actual Haskalah books."¹⁰ In the sub-title of *Shvilei olam*, it is defined as "A book that encompasses the characteristics of all the lands for all their divisions, their climates, and their progeny[,] their seas and their rivers, their inhabitants' beliefs and religions, their government's practices, their wisdom and their knowledge, their language and their feats." The first volume, an account of Asia, and the second, on Africa, came out in 1822 and 1828, respectively. The third volume, which covers Portugal and Spain, was only published in 1856, eleven years after Bloch's passing.¹¹ This

---

6 This topic is raised by Barnai, *Historiography and Nationalism*, p. 13–16 [Hebrew].
7 Edwin Aiken's book is indicative of the extent to which one can delve into the construction of the Holy Land's Biblical image during the Victorian age without so much as referring to the Jewish canon; Aiken, *Scriptural Geography: Portraying the Holy Land*.
8 For a close look at these works, see Ben-Arieh, "Developments in the Study of *Yediat Ha'aretz* in Modern Times, up to the Establishment of the State of Israel" [Hebrew].
9 Shemerok and Israel Bartal, "'*Tla'ot Moshe:*' The First Geographical Description of Eretz Israel in Yiddish" [Hebrew].
10 Klausner, *The History of the New Hebrew Literature*, vol. 2, pp. 350–368, esp. p. 368 [Hebrew].
11 Samson ben Isaac Bloch, *Shvilei olam*, vol. I, Zholkva: Meyerhoffer, 1822; ibid, II, Zholkva: Meyerhoffer, 1828; ibid, part III, Lwów: Grosman, 1855.

groundbreaking work, which was indeed the first of its kind in Hebrew, merited many editions and was even translated into Ladino.¹²

Another identically titled Hebrew work, *Shvilei olam*, was released in 1865 (the same year as the second edition of Bloch's work). Its author, one Abraham Menachem Mendel Mohr, set out to describe "the nature of all the lands of Europe" for a Jewish audience. In the introduction, Mohr boasts that "There has never been one like it in all the multitude of books that were written in the holy tongue about all the wisdom and science to this day."¹³

Levisohn, Kaplan and Breslauer

Solomon Levisohn (1789–1821), a member of Prague's Maskilic circle, was quite active in the fields of the Hebrew language, Biblical poetry, and prayer. Furthermore, he penned *Mekh'karei ereṣ* (Studies of the Land),¹⁴ which came out in 1819. In Zunz's estimation, this book was the first Biblical geography in the Hebrew language."¹⁵ From Levisohn's standpoint, his treatise was written "in accordance with what I sought beyond the travel books and in accordance with what I extracted from the books of the distinguished sages of the nations." Although he never stepped foot in Eretz Yisrael, the Maskil felt as though he had roamed the Land and took in its views with his "own eyes."¹⁶ On the cover page, Levisohn opined that geographical knowledge is essential to understanding history: "Every educated person that reads in the chronicles about the annals that God occasioned for each nation and nation" should also "endeavor to find out the character of the land . . . that the nation has inhabited." That said, *Mekh'karei ereṣ* is expressly intended for a Jewish audience and concentrates on the Land of Israel. Most of the book consists of an annotated list of places in Eretz Yisrael, which are arranged according to the Hebrew alphabet (from Avel to Tishba). For each entry, Levisohn provided references to the Hebrew Bible and the literature of the Sages, as well as classical and Christian works. The source of this alphabetic arrangement, which also turns up in later Maskilic works, is the writing of Christian humanists and Hebraists, like Adrichom, Fuller, and Reland. Following in the footsteps of Eusebius and Jerome, these Christian

---

12 Samson ben Isaac Bloch, *Shvilei olam*, Salonica: Saadi Halevi Ashkenazi Press, 1853–1854 [Ladino].
13 Abraham Menachem Mendel Mohr, *Shvilei olam*, Lemberg: Shernetzel, 1856 [Hebrew]. The second, expanded edition was printed in 1880.
14 Solomon Löwisohn, *Mekh'karei ereṣ*, Vienna, 1819 [Hebrew].
15 Zunz, "On the Geographical Literature of the Jews," no. 139, p. 294.
16 These passages are taken from the author's introduction to the book. For an in-depth look at *Mekh'karei ereṣ*, see Klausner, *The History of the New Hebrew Literature*, pp. 261–274; Michael, *Solomon Löwisohn: A Selection of His Works* [Hebrew]; Sasson, "Solomon Levisohn and His Contribution to the Study of the Land of Israel in the Nineteenth Century [Hebrew].

scholars expanded on the two Church fathers output by availing themselves of sources from the Middle Ages and the beginning of the modern era. Likewise, *Mekh'karei ereṣ* clearly adheres to the worldview of the Jewish Enlightenment. More specifically, the work is predicated on exposure to science and expresses a desire to break out of the traditional framework and augment it with insights from non-Jewish scholars.

While the Hebrew edition of Levisohn's book is devoid of cartographic material, the German edition contains a map of the Bible lands. The boundaries of the latter are Egypt in the south, the Black Sea up north, Cyprus in the west, and from the Caspian Sea to the Persian Gulf in the east.[17] The map, which is referred to on the cover page, is an exact replica of a map that came out some nineteen years earlier in a German book titled *Biblische Archaeologie*.[18] It is possible that the map was incorporated into *Mekh'karei ereṣ* by Levisohn's German publisher. At any rate, it was neither designed by the author nor translated into Hebrew for the purposes of this book. In light of the above, it does not really fall under the category of a Jewish map and thus falls outside our stated purview.

A second edition of *Mekh'karei ereṣ* was published in 1839 by one Jacob Kaplan (1801–1841), who expanded the book's scope and source list. Kaplan dedicated many years to these activities and to editing the original.[19] Furthermore, he included a map of the tribal allotments therein.[20] Written in Latin, Kaplan's map apparently emulates an earlier one. However, it is evident that this version was especially adapted for the second edition of *Mekh'karei ereṣ*, for Kaplan's name appears in bold letters beneath the title.[21] The fact that the map was printed in Latin is indeed noteworthy. This decision hints to the possibility that the map was copied from a prior source. In my estimation, it also attests to Kaplan's ability to read Latin, at least to the extent necessary for reproducing this map.

A similar book, *Glilot ereṣ yisrael* by one Menachem Mendel Breslauer (Breza), came out the same year as the first edition of Levisohn's book.[22] The opening part

---

17 Salomon Löwisohn, *Biblische Geographie*, Wien 1821.
18 Johann Jahn, *Biblische Archaeologie*, Wien: C.F. Wappler, 1, 1797, Tab. I.
19 Ginsburg, "Individuals and Generations: A Hundred Years of the Annals of One Family," pp. 215–253, esp. pp. 220–235 [Hebrew].
20 J. Caplanio [Kaplan], Terra Sancta sive regio duodecim tribuum Israelis, quam illustrissimi descripta: secundum librum Erez Kedumim / ab ejusdem auctore J. Caplanio.
21 *Eretz Kedumin, Mekh'karei ereṣ (Land of Ancient Times: It is a Book of Studies of the [Holy] Land, which Talks about the Lands and Seas, Mountains, and Hills [...] Cities and Villages that were Mentioned in the Holy Scriptures* [...]) original by [...] R. Solomon Löwisohn of Prague; arranged as well as researched [and] brought to print for the second time [... by] Jacob ben Shlomo ha'Cohen Kaplan from Minsk, Vilnius: Menachem Mann and Simcha Zimel Press, 1839 [Hebrew].
22 Breslauer (Breza), Menachem Mendel ben Haim Yehuda Leib, *Sefer glilot ereṣ yisrael: Written about its borders, its Mountains, its Waters, and its Apportionment into Twelve Parts Like the Number of God's Tribes and for All its Properties [...] including a Map of the Land of Israel and a Map of the Israelites Voyages in the Wilderness*, by Menachem Mendel son of our teacher (and our rabbi) [...] the erudite[,] heaven-fearing Hasid the Breslauer, Broslev: Saltzbach Press, 1819 [Hebrew]. The

Fig. 109: Jacob Kaplan, "Map of the Land of Ancient Times," 240 x 345.

(pp. 1–49) is bilingual, that is Hebrew and Yiddish texts appear on alternate pages. It begins with a general introduction explaining that the planet is round as well as the following key concepts: the poles, the equator, and the degrees of longitude and latitude. Thereafter, Breslauer elaborates on Eretz Yisrael's location within the Middle East, its borders, and primary mountain ranges and rivers. Moreover, he discusses the Land's division into tribal allotments. Penned entirely in Hebrew, the second part (pp. 51–131) consists of a list of toponyms, from Avel to Tirzah, "all of which are mentioned in the holy books and also a few in the books of the *aḥronim*,[23] which are most worthy of knowing, according to the order of the a-b-c." Under each entry is a brief explanatory paragraph as well as Biblical and Talmudic references. On occasion, Yiddish transcriptions of Latin names have been added in parenthesis, such as "*Kafarna'um*" (Capernaum) and "*Selaptelis*" (a misspelling of Beth She'an or Schythopolis). There is also a Hebrew-Yiddish transcription of the German *Wüsten Arabien* (the Arabian Desert), which naturally derives from the Latin *Arabia Deserta*. These examples testify to the use of non-Hebrew sources. In fact, certain sites on the map indicate that Breslauer leaned on Reland's *Palaestina ex monumentis veteribus illustrate*. For instance, the passage "Hadad-rimmon, the name of a city in the plain of Megiddo near Jezreel [Zechariah 12:11], that is *Maxsimianapolis*" strongly resembles its analogue in the Dutch scholar's work.[24]

According to the cover page of *Glilot*, the book contains "A map of the Land of Israel and a map of the Israelites' journeys through the wilderness." The two maps are absent from most of the known copies of this work, but the first was preserved in a copy that was put on the public auction block in 2008.[25] This particular map helped me equate both of Breslauer's maps with two identical maps of unknown provenance that were bound into a Pentateuch, which is held by the Israel Museum.[26] My hypothesis is bolstered by a comparison between various elements of the maps and Breslauer's book. For example, he writes that "The length of the Land [is] from fifty-two and until fifty-five degrees [longitude] and its width [is] from thirty one to thirty

---

work is listed in the biographical study of Zunz, "On the Geographical Literature of the Jews," no. 140, p. 294.
23 Literally "the last ones," the term *aḥronim* refers to halakhic adjudicators from the mid-sixteenth century and on.
24 Reland, *Palaestina ex monumentis veteribus illustrate*, p. 891: "Ita legit Hieronymus & notitia Ecclesiastica, aliique, Non Maximianopolis, quam seriptionem alii praeferunt Urbs olim dicta fuit Hadadrimmon, sita juxta Jezraelem. Heiron. Adc. Xii Zacharias."
25 *Jerusalem Judaica* (catalog of a public auction held in April 2008), no. 386, p. 243.
26 The pair of maps are integrated into the following edition of the Pentateuch: *Ḥumash makor ḥayyim*, with the Targum onkelos [i.e., Aramaic translation], Rashi C[ommentary], Targum ashkenaz [Yiddish translation], and Nahmanides' Explanation, and Kitzur tikun sofrim by the RaShaD [the Hebrew acronym for Rabbi Shlomo Dubno] […], Berlin: [publishing house], 1831–1833 [Hebrew]. Since other extant copies of this same edition of the Pentateuch do not include the maps, it stands to reason that they were subsequently added to this particular copy by its owners. I am indebted to Ariel Tishby and the Israel Museum for showing me the maps and placing photos of them at my disposal.

three and a half"—the same coordinates that are listed on the map of Eretz Yisrael.[27] The book's descriptions of Mount Abarim and Mount Nebo also dovetail neatly with their portrayal on the map. In addition, both *Glilot* and the map use the appellation "Sea of the Philistines" for the part of the Mediterranean Sea opposite the Land's southern coast and the name "La'antas" (i.e., the Litani) for the river that runs its course adjacent to Tyre.[28]

Facing north, Breslauer's map of Eretz Yisrael covers both banks of the Jordan River and ranges from Damascus in the north to the area just beyond the Dead Sea in the south. Its general perspective is Biblical, as the Land is divided into the tribal allotments and the toponyms derive from the Hebrew Bible. Furthermore, the legend on the bottom-right corner includes the symbols "City of Levites" and "City of Refuge." It also bears noting that these sort of marks were commonplace in European Holy Land maps during the previous centuries. In fact, the influence of these Christian-authored sources on the maps under review is appreciable in other respects as well. To begin with, as opposed to its actual contours, the western shoreline is circuitous, rife with bays, and slants out into the Mediterranean from Jaffa on down. Eretz Yisrael's southern border is "Sihor or the River of Egypt," and a bit to its north is a caption reading Rhinocorurah (i.e., al-Arish). Especially compelling is the fact that the Dead Sea takes the form of an elongated object featuring a small, two-pronged dry tongue on its southeastern side. This element ties the map to the discoveries from Ulrich Jasper Seetzen's expedition, news of which was then reaching Europe. More specifically, the German explorer had fully circuited the Dead Sea and was the first European researcher to document its tongue. Predicating himself on this information, Franz Xaver von Zach (a Hungarian born astronomer who founded several well-respected scientific journals), portrayed the Dead Sea in this fashion as early as 1810.[29] While I was unable to identify any single primary map, we may surmise that Breslauer gleaned data from several maps. In any event, it is obvious that he was abreast of the latest developments in the field.

The second map, which turns up in that same Pentateuch, tracks the Israelites' peregrinations through the Sinai Desert. Though depicted in a completely inaccurate manner, the map's outer western boundary is the Nile and its eastern delta. The lion's share of the map consists of the Sinai Peninsula, beneath which lies "the Arabian Sea that is the Red Sea." Alternatively, the righthand portion of the map depicts southern Eretz Yisrael, from Moab until Jericho. Winding through these areas is the route of the Israelites' journey and the stations along the way. As we have seen, this path is consistent with the accepted European cartographic representation

---

27 Breslauer, *Glilot ereṣ yisrael*, p. 18 [Hebrew].
28 Ibid., pp. 27–30.
29 For more on Seetzen and his expeditions, see Goren, *'Go View the Land:' German Study of Palestine in the Nineteenth Century*, pp. 29–39 [Hebrew]. The book includes a photo of the Seetzen-Petermann map (p. 38). The original map is held at the archive of the Geographical Institute of Gotha, which Petermann founded in Thuringia, Germany.

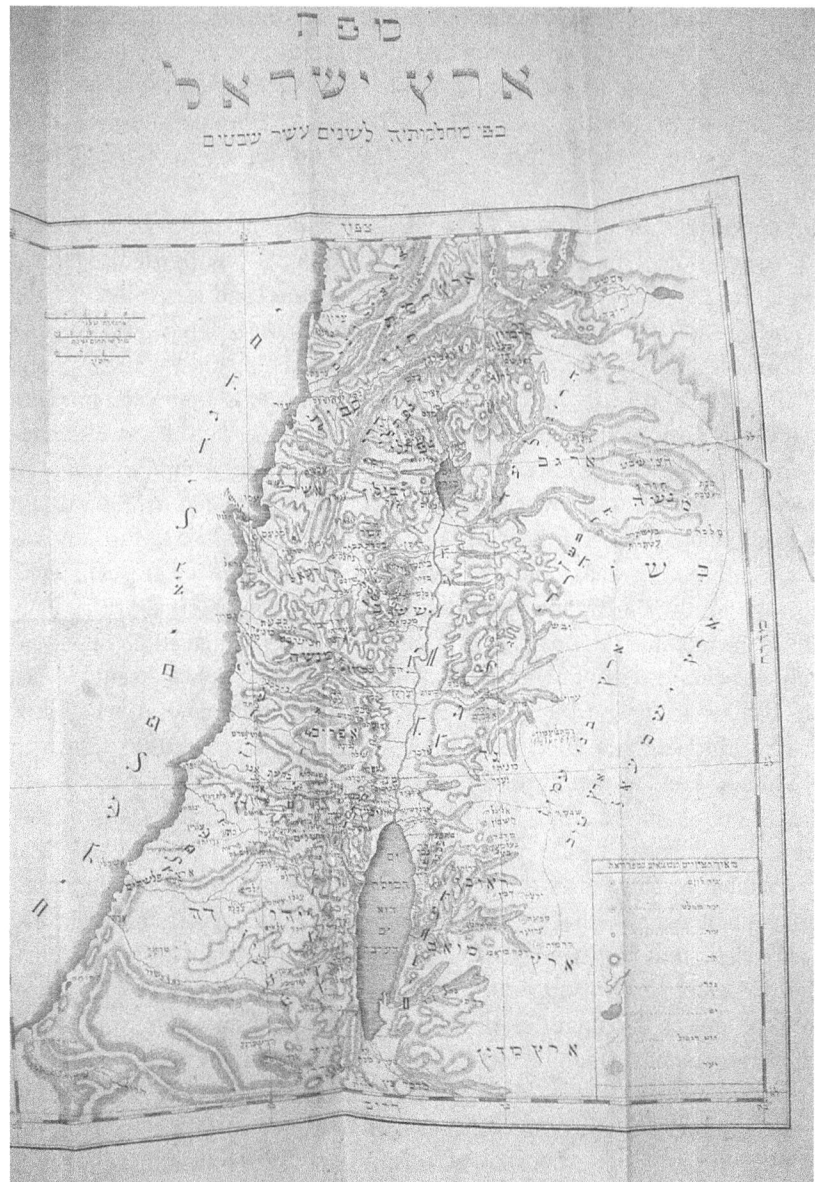

Fig. 110: Menachem Mendel Breslauer (Breza), "Map of Eretz Yisrael," 272 x 300 mm.

of this theme. The words "the Sea of the Philistines that is the Great Sea" trace the Mediterranean coast. Moreover, a caption reading "the Way to the Land of the Philistines" marks the route that was *not* taken by the Israelites. It is worth noting that the portrayal of the Dead Sea herein is less updated than on Breslauer's other map, for this version merely hints to the tongue, which is located south of the point where the Zered Brook spills into the sea. I did not manage to discern the source of this map either, but it draws heavily on the European genre of exodus maps.

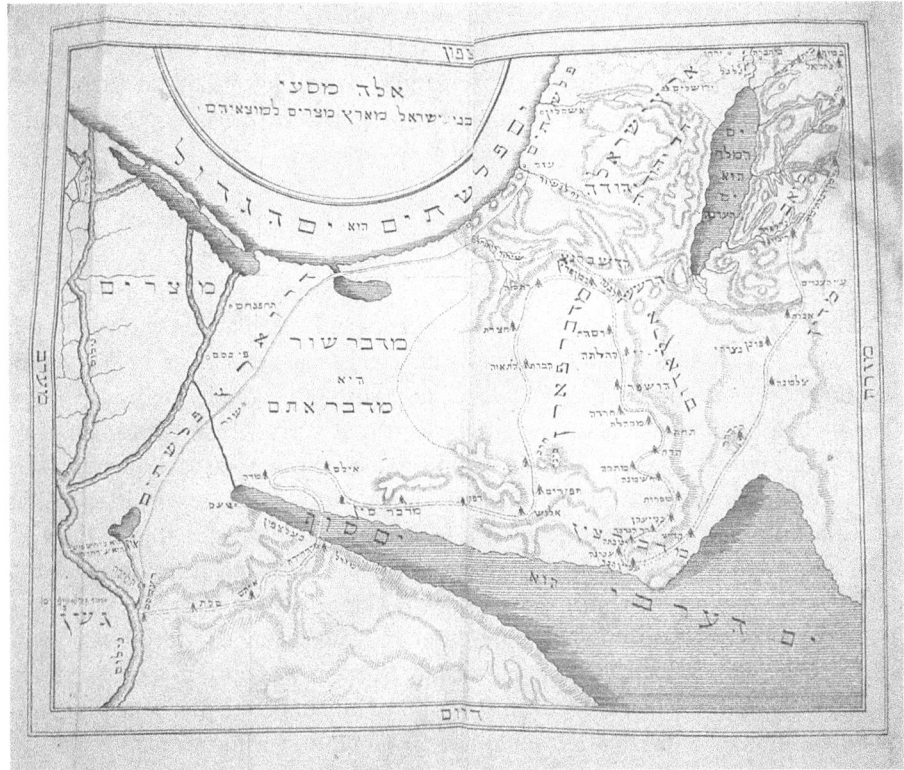

Fig. 111: Menachem Mendel Breslauer, map of the Israelites' peregrinations, 220 x 180 mm.

In sum, the research of Levisohn, Breslauer, Kaplan, and their Maskilic successors on the geography of Eretz Yisrael, particularly its ancient places of settlement,[30] is informed by a synthesis between the rabbinical-cum-traditional Haskalah and the authors' expertise in the Jewish Holy Scriptures, on the one hand, and a heavy dosage of Christian-authored studies, on the other. For instance, Levisohn made extensive use of Reland's book, Jerome's scholarship, and other Christian sources. While adopting a similar strategy, this did not stop Kaplan from rebuking Levisohn for copying from the books of the sages of the nations.[31] This sort of fusion between the inner-Jewish cultural world and the imperatives of the Age of Enlightenment, foremost among them respect for general scientific knowledge, indeed characterize

---

[30] Ben Arieh, "The Character of Hebrew Geographic Literature about Eretz-Israel during the Nineteenth Century and until World War I" [Hebrew].
[31] This barb was leveled at the Prague Maskil in the introduction to Kaplan's version of *Mekh'karei ereṣ*; Löwisohn, *Eretz Kedumin* (=*Land of Ancient Times*).

the Maskilic outlook. True to this ideology, proponents of the Haskalah did not limit themselves to Jewish sources in the fields of geography and the Land of Israel.[32]

The personification of this amalgam between traditional rabbinic erudition and modern scholarship in all that concerns the geography of Eretz Yisrael and its ancient settlements is, perhaps Yehoseph (Joseph) Schwarz, who many consider to be the very first Jewish researcher of the Land of Israel. Regardless of whether he merits this title, Schwarz was certainly the first Jew to personally explore the Land's mountains and valleys with his own two eyes.

Yehoseph Schwarz

Yehoseph (Joseph) Schwarz (1804–1865) was born in Floss, Bavaria and received a traditional Jewish education. While continuing his religious studies in Würzburg, he enrolled at the local university where he took up geography and astronomy, even though the former was not an accepted field of study within the Jewish community.[33] In those days, few Jews attended German universities, and Schwarz was the only Jewish student on his campus. Biographical sources emphasize that Schwarz continued to meticulously observe Halakha throughout his time in Würzburg.[34] Moreover, he studied Torah on a daily basis and insisted on eating only "Jewish fare." To this end, he occasionally skipped meals. This lifestyle points to an ambivalence between religious devotion and intellectual curiosity. While often fasting and subjecting himself to mortification, Schwarz was open to a modern, secular education. Contemporary scholars thus deem him to be the first Maskil in Jerusalem, but it is doubtful that he would have been pleased with this moniker.[35]

In 1833, Schwarz immigrated to the Land of Israel and indeed settled down in Jerusalem. He is best known for *Tvu'ot ha'areṣ* (Grains of the Land), a Hebrew book that came out in 1845.[36] Schwarz described this work and the rest of his research as the fruit of his expeditions across the length and width of Eretz Yisrael and his deep knowledge of the Jewish canon. *Tvu'ot ha'areṣ* subsequently merited a few more Hebrew editions[37] and was also translated into German and English.

---

32 Zalkin, "Scientific Literature and Cultural Transformation in Nineteenth-Century East European Jewish Society."
33 Brawer, "Schwarz, Yehosep."
34 Frumkin, *The Annals of the Sages of Jerusalem* [Hebrew]; Sasson, "The Life and Enterprise of Rabbi Yehoseph Schwarz[,] Author of '*Tvu'ot ha'areṣ*'" [Hebrew].
35 Jeff Halper, *Between Redemption and Revival, the Jewish Yishuv of Jerusalem in the Nineteenth Century*, p. 269.
36 Yehoseph Schwarz, *Sefer tvu'ot ha'areṣ, the Book of Joseph's Words*, part II, Jerusalem: Y. Beck Press, 1845.
37 Ibid, Lemberg: Y. M. Shtand Press, 1865; ibid, Jerusalem: Abraham Moses Luncz [Press], 1900; ibid, Jerusalem: the Committee to Publish the Writing of R[abbi] J[ospeh] Schwarz, 1978; ibid, Jerusa-

Schwarz's Map of Eretz Yisrael

In 1829, while still in Germany, Schwarz compiled and published a map of the Land of Israel, which netted two more editions within three years.[39] Penned in Hebrew, the map's title, its author's name, and the year of publication can be found in an oval frame on the lefthand side of the folio. The map stretches from Upper Egypt in the west to the heights of Transjordan in the east, and from the Red Sea's two northernmost bays in the south to Mount Hermon and Sidon up north. Eretz Yisrael is divided into the tribal allotments, and Schwarz charts the Israelite's travels through the wilderness. The map, especially the area denoting the Land of Israel itself, is teeming with toponyms from the Hebrew Bible, the majority of which are located in accordance with the Book of Joshua. In the upper-left corner is a German inset map, which offers a contemporaneous account of the Land. It ranges from Damascus up north to the area just below the Dead Sea in the south. On the righthand side of the page is an illustration of Moses and Aaron. At their feet is the opening verse from the portion of *Mas'ei:* "These are the journeys of the Israelites, which went forth out of the Land of Egypt by their companies under the leadership of Moses and Aaron" (Numbers 33:1). Further down is a linear scale, which measures distances in parsangs, along with a legend comprised of the following categories: Royal City, City of Refuge, City of Levites, Small City, and Israelite Camp in the Wilderness.

On closer review, the Mediterranean coast is replete with bays and shaped in a completely unrealistic manner. A case in point is the cupped inlet near Jaffa. The shoreline is not the only deviation from reality. The Dead Sea is elongated, and its pointy southern edge juts out to the west. The Sea of Galilee is elongated as well, and the sources of the Jordan River appear to flow from the foothills of the Lebanon Mountains into the Waters of Merom (i.e., Lake Hula). In contrast to the main map, the Mediterranean coast on the inset bulges into the sea at more than a few locations, and the Dead Sea's tail protrudes to the southeast. Schwarz also introduced several interesting components: perched atop Mount Sinai are the two tablets of stone; the Israelites' peregrinations differ from the familiar winding routes in Euro-

---

lem: Ariel, 1979; ibid, Jerusalem: the M. Schwarz Family, 1998. Various corrections and additions were integrated into these later editions

38 Frumkin, *The Annals of the Sages of Jerusalem*, p. 232; Schwarz, *In the Gates of Jerusalem: Documents on the Annals of Jerusalem and its Inhabitants* [Hebrew].

39 Avi Sasson notes that the second edition of this map was printed in 1831 in Vienna and the third in 1832 in Trieste; idem, *Hebrew Authors and Scholars in the Nineteenth Century and Their Contribution to the Historical-Geographical Study of the Land of Israel in the Nineteenth Century*, p. 110 [Hebrew].

pean maps; unlike many of the maps from the previous era, this one is not apportioned into the regions of Judea, Samaria, the Galilee, and Perea.[40]

These innovations notwithstanding, the map contains numerous elements that attest to the fact that it is essentially a compilation of European maps from the 1700s. For example, the long and tapering Dead Sea adheres to the much emulated tradition of the Adrichom map; the Calmet map, which also merited many editions and imitations, is most likely the inspiration behind the shape of the Gulf of Suez and the Bay of Eilat; and the Dead Sea's above-mentioned contours on the inset map derive from Blaeu's map of the Holy Land.[41] Furthermore, Schwarz's most prominent ornamentation—the figures of Moses and Aaron—is nearly identical to the one on the Homann map.[42]

In Sasson's estimation, the above-mentioned symbols on the map's legend were no less than "an innovation in the history of cartography, both Jewish and Christian alike."[43] However, as we have seen, three of the categories—Royal City, City of Refuge, and Levite City (*Urbs Regia*, *Urbs Levitica*, and *Urbs Refugii*)—already graced Breslauer's Hebrew map. What is more, they turn up even earlier on countless non-Jewish maps. In fact, these marks appear in the same exact format on Homann's map, which is indeed Schwarz's primary source. It also bears noting that the Hebrew title of the map under review, "Holy Land," is in all likelihood a direct translation of the popular Latin name for European maps of this sort—*Terra Sancta*.

Sasson avers that Schwarz borrowed from various sources, among them the French engineers' map of Egypt and Palestine that was surveyed and drafted in the midst of Napoleon's Middle Eastern campaign. The latter is also known as the Jacotin map, in recognition of Pierre Jacotin—the director of the surveyors and geographers that were attached to the Napoleonic forces in Egypt. According to Sasson, the crossed swords near Acre on Schwarz's inset are taken from the French map, where the mark signifies battle sites or military encampments.[44] Although this symbol may have originally been used for these purposes, Schwarz's map has little in common with the engineers' cartographic survey. Moreover, there are serious doubts that the Hebrew compiler even availed himself of this source. To begin with, his elongated Sea of Galilee, the lack of a bay between Haifa and Acre, and the sharp inden-

---

[40] In many of the Christian-authored Holy Land maps, there is a dual apportionment: the division into the tribes represents the early Biblical era and the First Temple period; and the breakdown into the regions of Judea, Samaria, the Galilee, and Perea represents the Second Temple period—the time of Jesus and the Gospels.
[41] Willem Janszoon Blaeu, "Terra Sancta quae in sacris Terra Promissionis olim Palestina Amsteldami," 1629.
[42] Johann Baptist Homann, "Iudaea seu Palaestina [...] hodie dicta Terra Sancta prout olim in Duodecim Tribus divisa separatis ab invicem Regnis Iuda et Israel," collecta ex Tabulis Guil. Sansonij [...] a Joh. Baptista Homanno, Norimbergae, 1707.
[43] Sasson, *Hebrew Authors and Scholars in the Nineteenth Century*, p. 112 [Hebrew].
[44] Ibid., p. 111.

Fig. 112: Yehoseph Schwarz, "The Holy Land and its Borders" map, 500 x 645 mm.

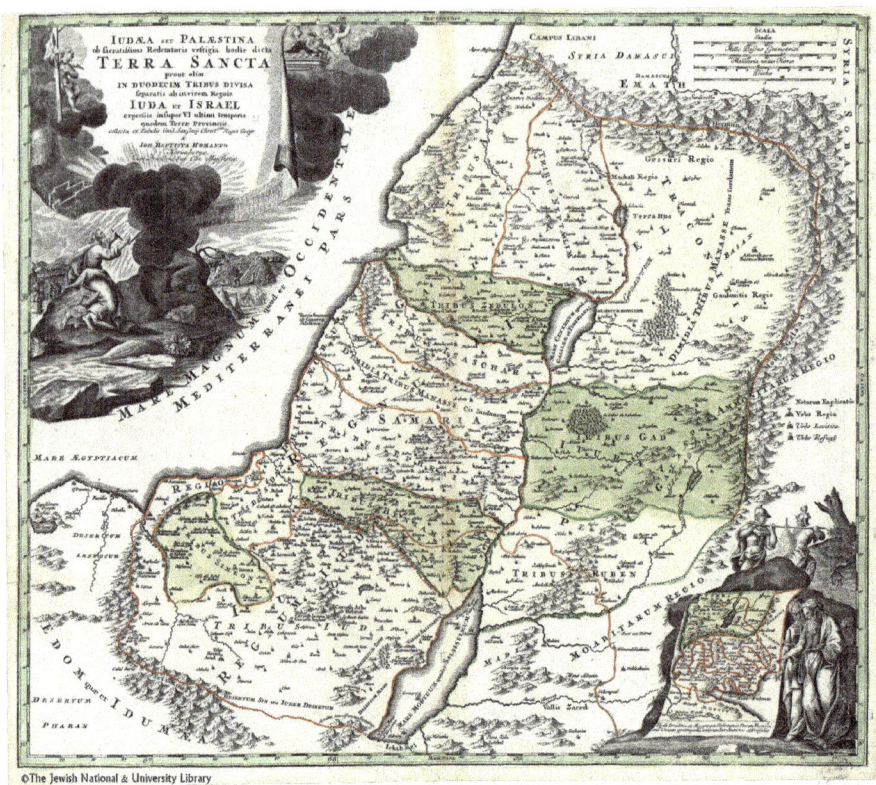

Fig. 113: Johann Baptist Homann, "Judea that is Palestine" map, 560 x 484 mm.

tation next to Jaffa run counter to the analogous elements on the Jacotin map.⁴⁵ Furthermore, the absence of the French engineers' novel and accurate components from the Schwarz map must be viewed as a surprising fault considering that its author studied geography at a German university soon after the publication of the Napoleonic survey. Beyond this, the form of the Dead Sea in the ancillary map was copied, as already mentioned, from the map of Blaeu.

Yehoseph Schwarz's reputation as both an open Maskil and devout Jew aside, my findings show that his map was neither original nor very Jewish in style or content. More specifically, he copied and recycled numerous elements from Christian Holy Land maps that were rendered within the framework of the Humanism and Hebraism movements in seventeenth- and eighteenth-century Europe. In fact, components that could ostensibly be viewed as Jewish and unique, such as the illustration

---

45 Pierre Jacotin, "Carte topographique de l'Egypte et de plusieurs parties des pays limitrophes," *Description de l'Egypte*, XX, Paris 1818, Sheet 43–47. For a discussion on the cartographic enterprise of the French engineers, see Anne Godlewska, "The Napoleonic Survey of Egypt, a Masterpiece of Cartographic Compilation and Early Nineteenth-Century Fieldwork."

Fig. 114: The figures of Moses and Aaron on the Schwarz and Homann maps.

of Moses and Aaron or the "City of Refuge" symbol, were all taken directly from the maps of Homann and his circles. Although Schwarz could very well have been a Maskil who was receptive to new ideas and opinions, his first map was by and large a recycling of old ideas from Christian cartographers. Given its reliance on the European cartographic tradition of the 1700s, the map under review is in no way, shape, or form a pioneering or distinctly Jewish creation.

Grains of the Land

As already noted, Yehoseph Schwarz's crowing scientific achievement was his book *Tvu'ot ha'areṣ* (Grains of the Land), which has been touted "the first geographical book on the Land of Israel to be written in Hebrew during the modern era."[46] *Tvu'ot ha'areṣ* explicates the borders of Eretz Yisrael and all its major geographical elements. In addition, it divides the Land into regions in accordance with various chapters of the Hebrew Bible and texts from the literature of the Sages. Likewise, the book's description of the greater ancient world is commensurate with the annals of the nations as per Genesis 10. It also includes a chapter each on Jerusalem and the Temple. Schwarz's scientific approach is quite evident in his discussions on the Land of Israel's fauna and flora, its climate, geology, and history. All these geographic top-

---

46 Ben Arieh, "The Character of Hebrew Geographic Literature about Eretz-Israel during the Nineteenth Century and until World War I," p. 38 [Hebrew].

ics, however, are presented exclusively with words, for the book does not contain any maps.

In the introduction, Schwarz emphasizes that *Tvu'ot ha'areṣ* is predicated on three types of sources: field research and expeditions that he personally conducted throughout the Land ("I shall write what I saw with my eyes [–] and I went up the mountain and descended into the vale"); his expertise on the Bible and the rabbinic literature; and "on the words of the [non-Jewish] sages [– their] research on the Land from the time of Josippon to this age." In a letter to his brother, Schwarz writes that he met Gotthilf Heinrich von Schubert while the renowned German naturalist was in Jerusalem, thereby confirming his amenability to non-Jewish scholarship.[47] Moreover, Schwarz availed himself of the writings of "Isahbius" (i.e., Eusebius) and "Hiranimus" (*Hieronymus*, namely Jerome).[48] The author criticizes the exegetes—Jewish and otherwise—who were unfamiliar with the Land "and for this reason sometimes interpret the opposite of the truth and reality." Moreover, he reproaches "the sages of the Jewish people" for not knowing "the land of their forefathers' estate."[49] While criticizing his co-religionists, Schwarz completely ignored his Jewish predecessors' work on the Book of Joshua; this includes the commentaries of the aforementioned Altschuler, Teomim, and the *Vilna gaon*, which were illuminated with maps (schematic as they may be). Thereafter, Schwarz scoffs at "the wise of the nations who came here [i.e., Eretz Yisrael] from afar to explore and excavate the entire Land." Giving the non-Jewish explorers "chapter and verse," he showed them that their efforts to comprehend the Land were in vain—akin to "a blind man" groping in the dark—unless they "familiarize themselves with the teachings of our [i.e., the Jewish] sages of blessed memory."

These words notwithstanding, it appears that Schwarz unsparingly borrowed from the findings of Christian scholars who explored the Land before him. Save for the above-cited passage from the book's introduction, though, he refrains from crediting these sources. Ben-Arieh, who has taken stock of *Tvu'ot ha'areṣ*, notes that its author relied heavily on Reland's book. For instance, like Breslauer before him, Schwarz followed Reland's lead in identifying Hadad-Rimmon (Zechariah 12:1) with Maximianopolis.[50] It also bears noting that the first to identify these toponyms with each other was Jerome, in his commentary on Zechariah. At any rate, Schwarz cited neither Reland nor Jerome.

In my estimation, the source that had the greatest impact on *Tvu'ot ha'areṣ* was Edward Robinson and Eli Smith's *Biblical Researches in Palestine, Mount Sinai and Arabia Petrea in 1838*. The latter came out in English in 1841 and in German the fol-

---

47 This letter can be found in Schwarz, *In the Gates of Jerusalem*.
48 Schwarz, *Tvu'ot ha'areṣ*, the Luncz edition, p. i (Author's Foreword).
49 These words seal another of Schwarz's works—*Tvu'ot ha'shemesh* (Grains of the Sun). Moreover, he cites from this passage in the Author's Foreword to *Tvu'ot ha'areṣ*. See idem, *Tvu'ot ha'areṣ*, the Luncz edition, p. ii.
50 Ibid, p. 195–196.

lowing year, namely at the same time Schwarz was working on his own book.[51] *Biblical Researches* was indeed the most cited work by Holy Land explorers during the mid-nineteenth century; and many of the region's historical geographers consider Robinson to be one of the key figures in the field to this day. There are indeed a few similarities between the descriptions in *Biblical Researches* and *Tvu'ot ha'areṣ*. For example, both works identify Tayibe (a village northeast of Ramallah) with biblical Ophrah, on the grounds that Eusebius located the settlement five miles east of Bethel.[52] Leaning on Robinson, Schwarz identified Taanach with Ta'annuk;[53] and wrote that local inhabitants referred to Megiddo as "Lagian" (his transcription for the Arabic *al-Lajjun*)—an identification that was the object of debate among researchers in the mid-nineteenth century.[54]

The Maps of Eretz Yisrael in the English and German Editions of *Tvu'ot ha'areṣ*

In 1850, an English version of *Tvu'ot ha'areṣ* came out in Philadelphia.[55] The fact that it was translated and edited by Isaac Leeser, the city's chief rabbi and among the leaders of American Jewry, speaks to the importance of this edition. Leeser indeed devoted a great deal of time and energy on translations and editing. He had previously rendered the Jewish Bible and *siddur* into English and was the founder of *The Occident*—the very first journal put out by the American-Jewish community. Therefore, it was only natural for the American rabbi to take this assignment upon himself.[56]

In the introduction to the English edition, Schwarz continued to reproach both Jewish exegetes and Christian researchers. Unlike the earlier versions, though, he expressly mentioned that Reland was one of his sources.[57] Two maps were incorporated into the English edition of *Tvu'ot ha'areṣ*: a relatively large map of the entire Land of

---

51 Robinson and Smith, *Biblical Researches in Palestine, Mount Sinai and Arabia Petrea in 1838*; Robinson, *Palästina und die südlich angrenzenden Länder: Tagebuch einer Reise im Jahre 1838 in Bezug auf die biblische Geographie unternommen von E. Robinson und E. Smith*; mit neuen Karten und Plänen in fünf Blättern, Halle 1841–1842.
52 Robinson and Smith, *Biblical Researches in Palestine*, vol. 2, pp. 124–125; Eusebius, *The Onomastikon of Eusebius*, the Schwarz edition, p. 28, line 4; Schwarz, *Tvu'ot ha'areṣ*, p. 151.
53 Ibid, p. 184; Robinson and Smith, *Biblical Researches in Palestine*, vol. 3, pp. 156–157.
54 Schwarz, *Tvu'ot ha'areṣ*, p. 184; Robinson and Smith, *Biblical Researches in Palestine*, vol. 3, pp. 177–180.
55 Schwarz, *A Descriptive Geography and Brief Historical Sketch of Palestine*.
56 Riemer and Michael Berenbaum, "Leeser, Isaac;" Meyer, *Response to Modernity: A History of the Reform Movement in Judaism*, pp. 234–235 [Hebrew].
57 Schwarz, *A Descriptive Geography and Brief Historical Sketch of Palestine*, p. xii-xiii.

Israel; and another map that depicts the Land's borders according to Numbers 34.[58] A German edition of the book came out in 1852.[59] This version combines the two maps from Leeser's edition into one, as the map of the promised borders is displayed in an inset within the large one.[60]

The larger map in the English translation bears the title "Map of Palestine, by J. Schwarz of Jerusalem." Beneath the title is a legend of site categories along with an explanation of the Arabic Geographic terms that the author used. The map's purview stretches from Beirut to the southern extremities of the Dead Sea and "Rafia," and from the Mediterranean Sea to Transjordan. The major mountain ranges are signified with hatching, and the rivers and brooks with a thin line. Both the Biblical and Arabic names are provided for most of the toponyms. Despite the rash of inaccuracies, the map seems to have been constructed with modern tools. Moreover, its design was in vogue at the time.

The other, smaller map depicts Eretz Yisrael's borders as per Numbers 34, thereby adhering to the established cartographic tradition that dates back to Rashi's diagrams. In line with Schwartz's identification of the sites in that chapter of the Bible, this map's geographical scope is a bit more expansive than the first. The northern limits are Tripoli ("Trablos") and, further right, Zedad, while the southern boundary is al-Arish River or the River of Egypt. As is only natural, there are fewer toponyms on this map. Some are transcribed from the Hebrew, others from the Arabic, and a few derive from the Greek (e.g., Sebasté and Antipatris).

A comparative analysis of the principal map in the English and German editions shows that they were replicated virtually without modification from the Holy Land maps of Heinrich Kiepert, the era's leading cartographer. Kiepert's maps showcase the findings of Robinson and Smith and were prepared for the above-cited editions of their book.[61] The relation between the Schwarz and Kiepert maps is evident from, *inter alia*, the shape of the coastline, the Sea of Galilee, and the Dead Sea, especially its tongue. Both maps also contain a roundish mountain signifying the Gilead, to the south of the Jabbok River's confluence with the Jordan. In addition, the mountain ranges are designed in a similar fashion. These shared elements bolster my argument that, notwithstanding his silence on this matter, Schwarz leaned on Robinson and

---

58 Idem, "Map of Palestine, by J. Schwarz of Jerusalem 5607," 1847 T. Sinclair's lith. 1850; idem, "Boundary of Palestine according to Numbers XXXIV 1–15, by Rabbi Joseph Schwarz, of Jerusalem," 1847, T. Sinclair's lith. 1850.
59 Idem, *Das Heilige Land* [...], Frankfurt am Main 1852.
60 Idem, "Palaestina von J. Schwarz in Jerusalem," 1847, Lith. Anst. v. A. Lippenberger [...] Gest. v. Ch. Schein. Frankfrt $^a/_m$., [1852]. The title of the ancillary map is "Die Gränzen Palästina's nach 4B, Mos, XXXIV. 1–15."
61 Heinrich Kiepert, "Map of Palestine, Chiefly from the Itineraries and Measurements of E. Robinson and E. Smith [...];" Robinson and Smith, *Biblical Researches in Palestine*. For more on Kiepert's Holy Land map, see Goren, "The Present as an Expression of the Past: The Holy Land in the Work of the Historical Cartographer Heinrich Kiepert" [Hebrew].

The Maps of Eretz Yisrael in the English and German Editions of *Tvu'ot ha'areṣ* — **239**

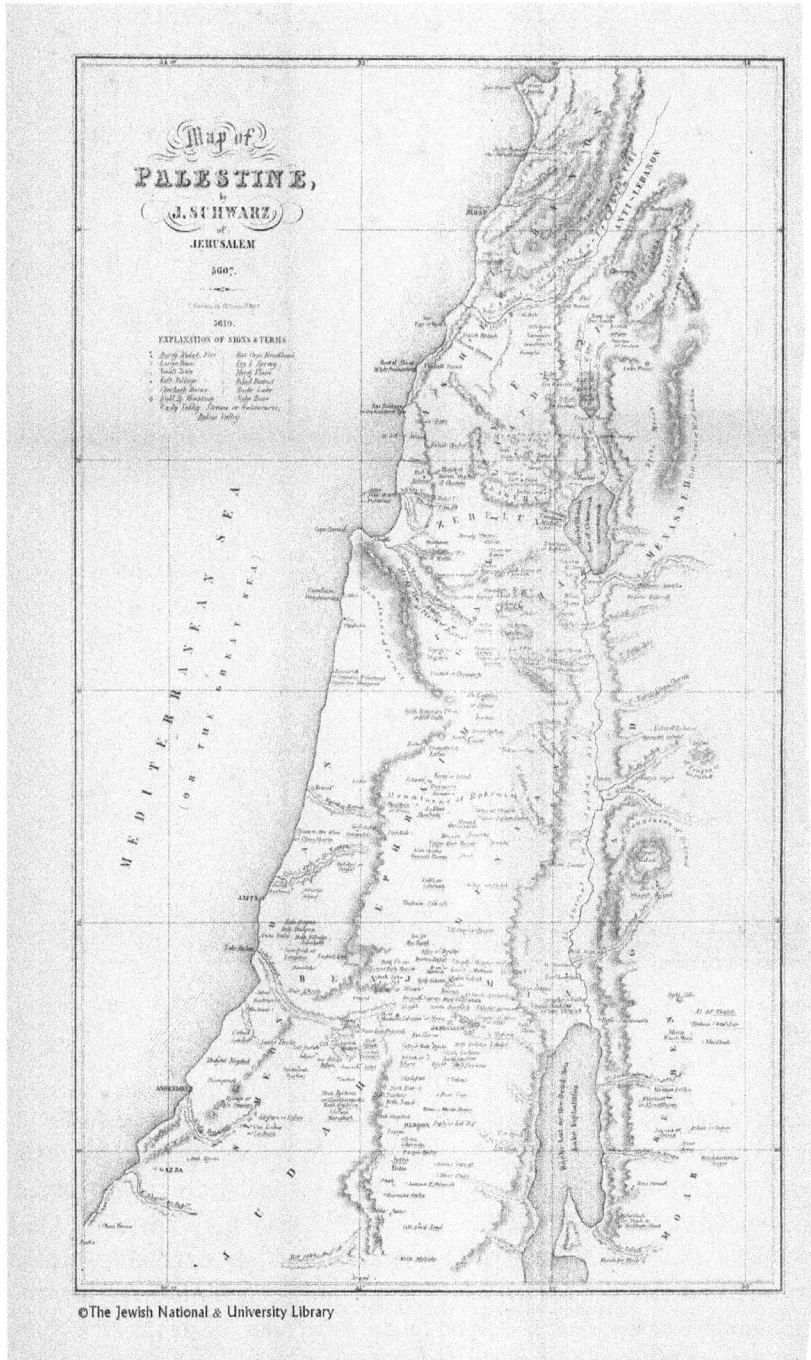

Fig. 115: Yehoseph Schwarz, map of the Land of Israel in the English edition of *Tvu'ot ha'areṣ*, 290 x 490 mm.

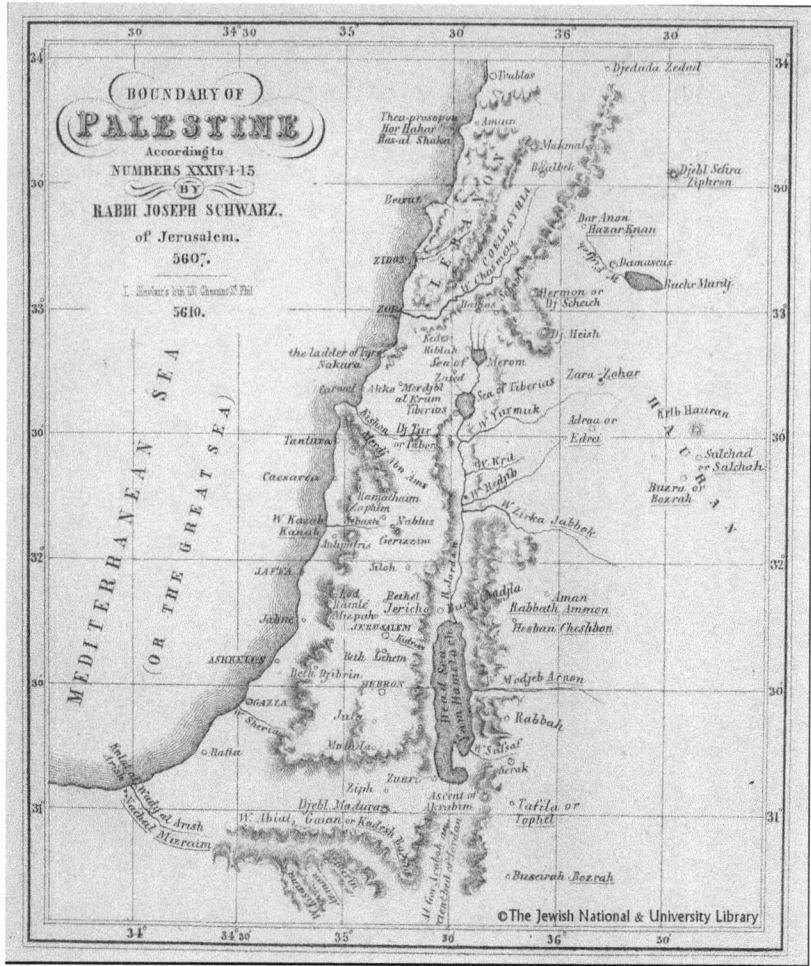

Fig. 116: Yehoseph Schwarz, map of Eretz Yisrael's borders in the English edition of *Tvu'ot ha'areṣ*, 140 x 170 mm.

Smith's book. Moreover, this omission epitomizes the Jewish author's persistent, if unwarranted, denial that "the sages of the nations" had any impact on his work.

Against this backdrop, Yehoseph Schwarz's works, particularly his maps, contain two distinct undercurrents: scientific and rabbinical. Despite his academic training and despite being classified as a Maskil by certain scholars, the German Jew's principal focus was on the Bible and the rabbinic literature. In other words, he perpetuated the traditional framework of Jewish erudition. Soon after the publication of the English and German editions of *Tvu'ot ha'areṣ*, Heinrich Graetz claimed that its author lacked the good sense to take advantage of the classical texts and the pertinent research on these sources. Moreover, the accomplished historian bemoaned the lack of a distinction in *Tvu'ot ha'areṣ* between toponyms from Hebrew sources that denote

The Maps of Eretz Yisrael in the English and German Editions of *Tvu'ot ha'areş* — 241

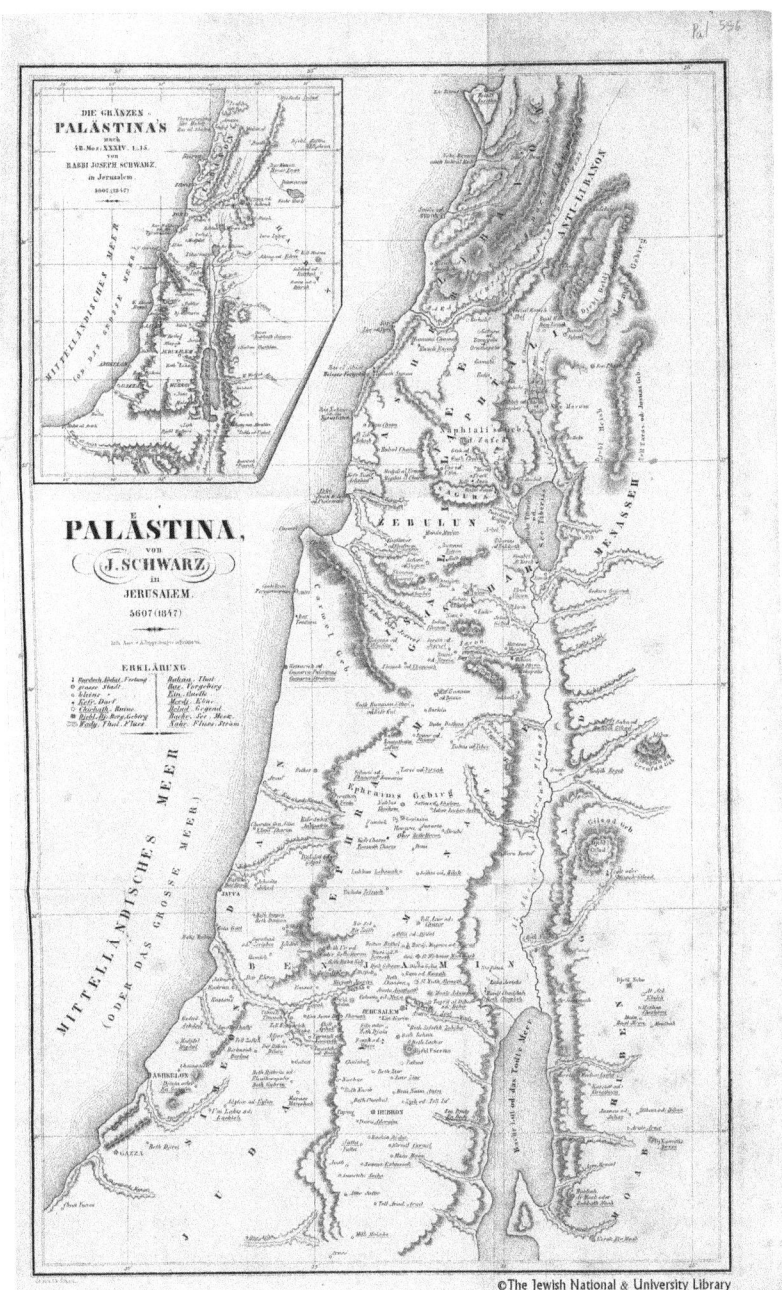

Fig. 117: Yehoseph Schwarz, map of the Land of Israel in the German edition of *Tvu'ot ha'areş*, 290 x 484 mm.

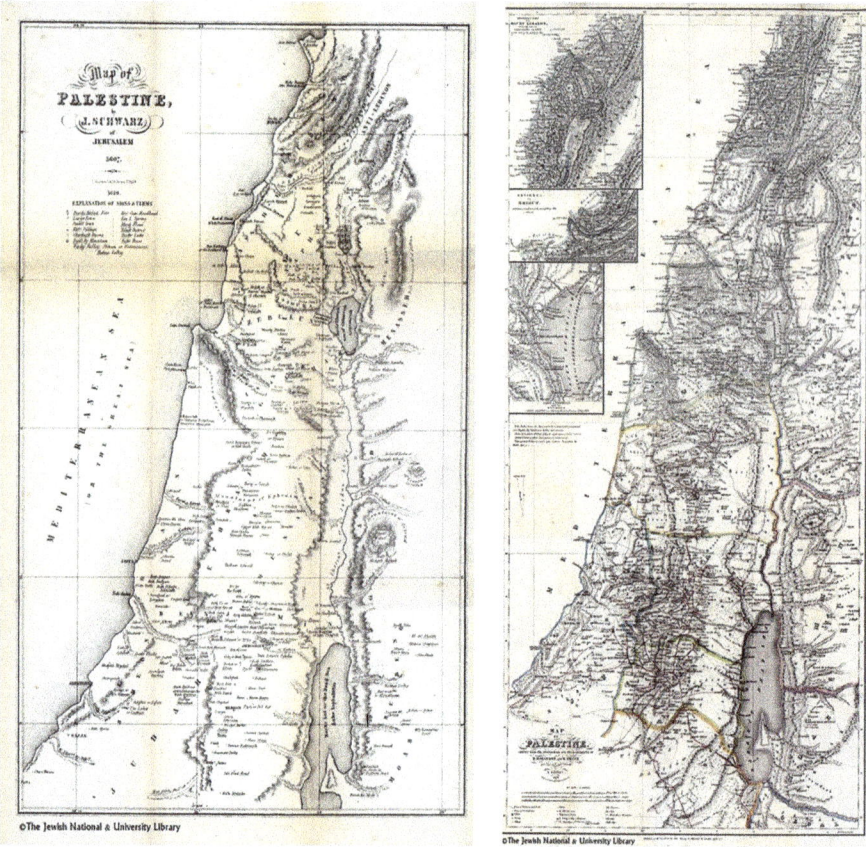

Fig. 118: The *Tvu'ot ha'areṣ* map (English version), 380 x 426 mm; and Kiepert-Robinson "Map of Palestine, Chiefly from the Itineraries and Measurements of E. Robinson and E. Smith [...]," 376 x 426 mm,

actual places and those of a mythical character. Accordingly, he felt that Schwartz's attempts to pinpoint the latter sites on a realistic cartographic representation of Eretz Yisrael were misguided.[62] Even when the mapmaker availed himself of the general literature and the fruits of the modern research, he tended to hide or blur this fact —an approach that contravened the Haskalah's scientific ethos.

These shortcomings aside, Schwarz was the first Jewish author to insinuate a modern map into the Hebrew and Jewish literature. Put differently, his maps set a precedent by taking into account the geographical knowledge that was steadily accumulating during the 1840s. In so doing, he embodied the transition from schematic and artistic maps to scientific ones that are supported by accurate geographical

---

[62] Graetz, "Die talmudische topogrpahie."

data. That said, and despite his reputation as a pioneering figure, Schwarz's maps were devoid of original ideas or significant innovations, for he merely recycled the ideas and forms of his predecessors.

The Maps of Salomon Munk

Salomon Munk (1803–1867) was a German-born Maskil and prolific researcher. After studying at universities in Bonn and Berlin, he immigrated to Paris, where he eventually became the director of the French National Library's Semitic manuscripts department. Moreover, he authored numerous studies on Judaeo-Arabic and Hebrew literature that was produced during the golden era of Jewish culture in Spain. Among his published works was a scientific edition and French translation of Maimonides' *Moreh nevukhim* (Guide to the Perplexed).[63] With respect to the topic at hand, Munk's *Palestine: Description géographique, historique, et archéologique*—a thick volume on the Land of Israel—came out in 1845 (the same year as *Tvu'ot ha'areș*).[64] It netted three more editions in French as well as a German translation (the latter, though, was not accompanied by any maps).[65] Although Munk's *Palestine* was part of a general historical series on the nations of the world, its principal topics are the history of the Land during the First and Second Temple periods, namely its quintessentially "Jewish" epochs. Out of the book's roughly seven hundred pages, around ten percent are devoted to the age before Joshua ben Nun's conquest of Canaan and some five hundred pages to the period from the Israelites' settlement of the Land to the Romans' destruction of Jerusalem. Within this context, Munk also explores the dawn of Christianity. The centuries between the destruction of Jerusalem and the mid-1800s are relegated to a twenty five-page appendix. While catering to both Jewish and non-Jewish readers alike, *Palestine* can be viewed as the work of a Maskil on the Land of Israel during the periods of Jewish autonomy. A biographical anecdote that strengthens the impression that Munk was committed to the Jewish world is his expedition to Egypt in the company of Adolphe Crémieux (a leading figure in the French Jewish community) and Moses Montefiore in 1840.[66] At the end of this book, there are 68 woodcuts of scenes from the Land of Israel, including unmistakably Christian sites (e. g., the Church of the Holy Sepulchre and the Church of the Nativity). Interspersed among the woodcuts are five maps on the following topics: the Israelites' peregrinations; Jerusalem at the time of its destruction at the hands of Titus; the tribal allotments; the Land under Roman rule; and contemporary Palestine.

---

[63] "Munk, Solomon," *Encyclopaedia Judaica*, 2nd ed., vol. 14, p. 615.
[64] Salomon Munk, *Palestine: Description géographique, historique, et archéologique*, Paris 1845.
[65] Salomon Munk, *Palästina: geographische, historische und archäologische Beschreibung dieses Landes, und kurze Geschichte seiner hebräischen und jüdischen Bewohner*, nach dem Französischen von S. Munk bearbeitet von M. A. Levy, Leipzig 1871–1872.
[66] Zunz, "On the Geographical Literature of the Jews," p. 302.

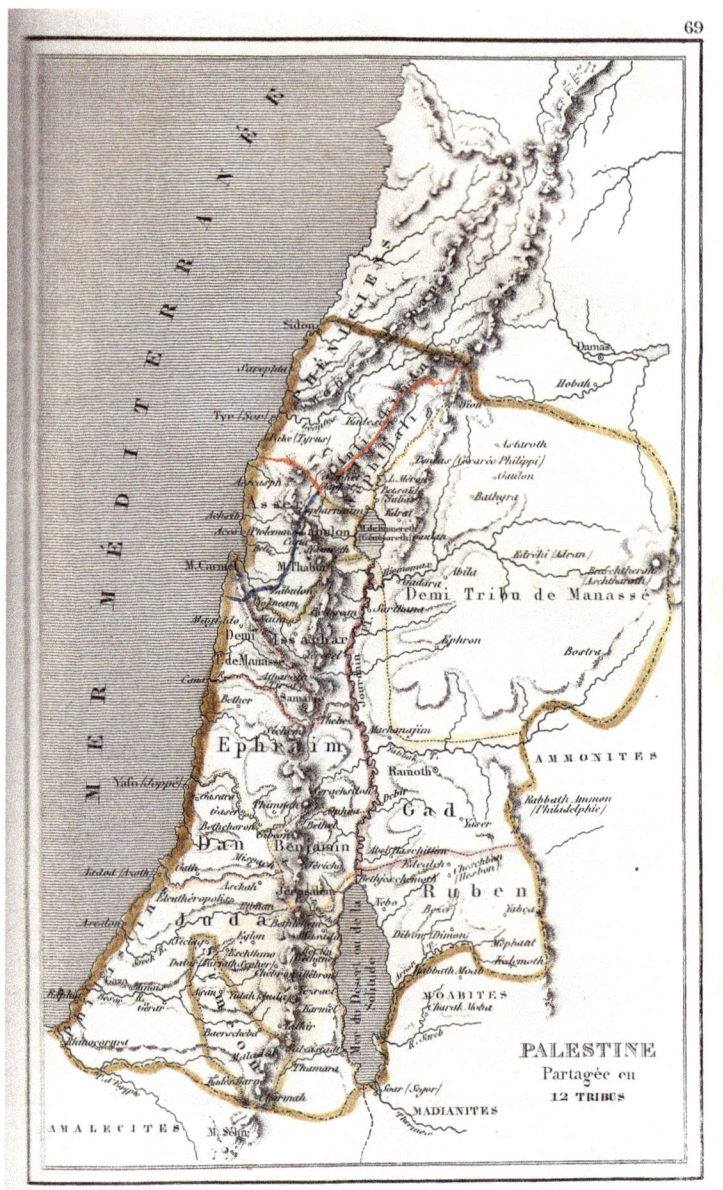

Fig. 119: Salomon Munk, "The Land of Israel and its Division between the Twelve Tribes," 105 x 172 mm, in idem, *Palestine: Description géographique, historique, et archéologique*, tab. 69, facing p. 224.

All of Munk's maps share the same cartographic basis, possess modern designs, and are placed within a system of global coordinates. His maps of the entire Land run from Lebanon in the north to al-Arish River in the south. The Mediterranean

Coast still has a surfeit of bays. While the Dead Sea continues to be depicted in an inaccurate manner, there is evidence that Munk was aware of its tongue; however, his rendering thereof, which was probably compiled after Seetzen's map, is exponentially smaller than its true size.[67] The mountain ranges are rendered as a succession of peaks, which are marked by hatch lines. Turning to the map of the Israelites' peregrinations, the route passes by Mount Sinai (the largest mountain on the peninsula). From there, it heads north until "Kadesh Barnea," which according to Munk is adjacent to the southern tip of the Dead Sea.[68] The route then dips back south and reaches Eilat ("Akabah"), via the Arabah. Thereafter, it loops back north through the mountains of Edom and Moab. The map of the tribal allotments places the area of Dan along the coast, namely its original location. With the exception of its toponyms and regional breakdown, the map of the Roman era is identical to the others. The contemporaneous map is bigger than the rest, so that its toponyms are easier to read. Some of the place names are transcribed from the Arabic, like Yafa (i.e., Jaffa) and Askalan (Ashkelon). Others are signified by French derivatives of the Greek and Latin, such as Batanee (Bashan) and Lac Asphaltite (the Sea of Asphalt). Alternatively, a handful of sites are denoted by their conventional European names: Jerusalem, Acre, and Hebron, *inter alia*.

The map of Jerusalem highly differs from the rest of Munk's maps. Following in the footsteps of maps from the earlier period, it is loosely connected, at best, to the city's topographical reality. While I could not authoritatively identify a primary source for this map, it contains elements that are apparently taken from the Pococke and d'Anville maps.[69] That said, the new maps of Jerusalem that were crafted in the early 1800s, like that of Sieber or Catherwood, had no appreciable influence on this one.[70]

In sum, Salomon Munk's book and maps present the geography and history of the Land of Israel through the eyes of a learned, mid-nineteenth century Maskil

---

[67] See Goren '*Go View the Land*,' pp. 29–39, including the reference to Seetzen and his map therein.
[68] In all likelihood, the placement of Kadesh Barnea in the northern Arabah is based on one of Edward Robinson's famous mistakes. The explorer identified the site with Ein Yahav, rather than Ain el-Qudeirat in the eastern Sinai. This erroneous location subsequently turned up in other works, including some of the maps discussed below. See Robinson and Smith, *Biblical Researches in Palestine*, vol. II, p. 582, 610.
[69] "Plan de la Ville de Jerusalem Ancienne et Moderne," Par le S.ʳ d'Anville, Ecrit par G. Delahaye, Jean Baptiste Bourguignon d'Anville, *Dissertation sur l'étendue de l'ancienne Jérusalem et de son temple, et sur les mesures hébraïques de longueur*, Paris [1747]; Richard Pococke, "A Plan of Jerusalem and the Adjacent Country," T. Jefferys Sculp, in Pococke, *A Description of the East and Some Other Countries*, 2 vols. in 3 London 1743–1745, vol. 2, part 1: "Observations on Palestine, or the Holy Land, Syria, Mesopotamia, Cyprus, and Candia," p. 7.
[70] For a discussion on Catherwood and his map, see Ben-Arieh, "The First Surveyed Maps of Jerusalem" [Hebrew]; Ben-Arieh, "The Catherwood Map of Jerusalem," *The Quarterly Journal of the Library of Congress* 31/3 (1974): 150–160. Sieber's map is expounded on in Karte von Jerusalem und seiner næchsten Umgebungen, geometrisch aufgenommen von F.W. Sieber im Jahre 1818, Aufgenommen und entworfen von F.W. Sieber, ins reine gezeichnet von J. Wach, gestochen von J. Stölzel, Franz Wilhelm Sieber, *Reise von Cairo nach Jerusalem und wieder zurück [...]*, Prag and Leipzig 1823.

Fig. 120: Salomon Munk, "The Route of the Hebrews' Journey upon Traversing the Desert," 105 x 163 mm, in idem, *Palestine: Description géographique, historique, et archéologique*, facing p. 122.

who worked outside the traditional Jewish framework and wrote for a broader French and European audience.

Land of Israel Maps from the Second Half of the Nineteenth Century

During the latter half of the 1800s, two trends left their imprint on the design of the Hebrew map of Eretz Yisrael: the advances in European Holy Land research; and the spread of the Haskalah among European Jewry. The former enriched all those interested in the Land with an abundance of information on its attributes and geography, much of which was cartographic. Among the fruits of this labor was the evolution of an accurate and trustworthy map. This process was ushered in by the above-mentioned maps of the French military survey. It then continued with the maps that Kiepert drafted on the basis of Robinson's findings and other maps by scholars like van de Velde.[71] This enterprise reached its peak in 1880 with the release of the Palestine Exploration Fund's Survey of Western Palestine (SWP).[72] Covering the Upper Galilee to the Be'er Sheva Valley, the surveyors employed systematic measurement techniques and conducted a full triangulation of the entire area. The end result was a comprehensive and readily accessible portrait of the subject at hand. Furthermore, the survey touched upon the majority of ancient sites and mounds. In so doing, it shed light on both the contemporaneous expanse and the realm of the Holy Scriptures.[73] On the one hand, then, the SWP served the political-cum-military interests of the European powers, whose aspirations in the Holy Land and throughout the region dovetailed neatly with the survey's spirit of modernization; and on the other hand, the work was imbued with Orientalist and religious feelings, not least the millennial ardor that had gained traction in many and manifold groups in European society during the second half of the nineteenth century. This dual perspective also suited many of the Hebrew cartographers who not only viewed the country as their sacred and ancient homeland, but as a realistic destination for pilgrimage and perhaps settlement as well. With the rise of the Haskalah and the idea of a Jewish national reawakening, the significance of this two-pronged approach to Eretz Yisrael steadily grew.

The spread of the Jewish Enlightenment in the latter part of the 1800s placed geographic and cartographic information at the disposal of many Jews throughout Europe and led to the creation of works with geographic content, not least maps in Hebrew, Yiddish, and other European languages. Some of these maps were still devoted to traditional Jewish topics, such as the borders of Eretz Yisrael and the trib-

---

[71] Carel Willem Meredith van de Velde, "Map of the Holy Land Constructed by C.W.M. van de Velde from His Own Surveys in 1851 & 1852, from those made in 1841 by Majors Robe and Rochfort Scott, Lieut. Symonds [...] and from the Results of the Researches made by Lynch, Robinson, Wilson, Burckhardt, Seetzen," Engraved by Eberhardt and by Stichardt, Gotha 1858.

[72] Conder and Kitchener, "Map of Western Palestine in 26 sheets 1:63, 360," *Survey of Western Palestine*. Headquartered in London, the Palestine Exploration Fund (PEF) is active to this day.

[73] Hopkins, "Nineteenth-Century Maps of Palestine: Dual-Purpose Historical Evidence;" Moscrop, *Measuring Jerusalem: the Palestine Exploration Fund and British interests in the Holy Land*.

al allotments, whereas others depicted the contemporaneous Land of Israel. Regardless of the topic, the geographical and cartographic infrastructure was steadily modernized and updated, as the Jewish mapmaker increasingly leaned on the findings of the European research. A few of these maps even presented modern developments, such as the embryonic stages of the Jewish settlement in Eretz Yisrael.

The Map of Benjamin II

In the mid-1800s, Israel Joseph Benjamin (1818–1864)—also known as Benjamin II, after the famed globetrotter from Tudela—spent nearly a decade traveling through Asia and Africa. Starting his odyssey in Eretz Yisrael, he subsequently visited Damascus, Kurdistan, Iraq, India, China, and Iran. From there he proceeded to North Africa, where he sojourned in Egypt, Tunisia, Algiers, and Morocco. In all these lands, Benjamin went to the most remote communities of, as he put it, "our brethren the Jewish people" and reported on their situation. At different stops, he presented himself as a sage from Jerusalem. Under this assumed identity, he taught local Jews Halakha and corrected a plethora of customs. Upon returning to Europe, he described his experiences abroad in a 130-page book, *Five Years in the Orient*, which first came out in French in 1856.[74] The German edition was published in 1858,[75] while the first English edition[76] and the Hebrew translation were released the following year.[77]

The book's focus is indeed on Jewish communities, but also reflects Benjamin's Maskilic attitude, especially his appreciation for geography. This outlook is exemplified by the inclusion of letters of praise from Alexander von Humboldt, Carl Ritter, and August Petermann—three German pillars of the international geographic community—in the opening pages of his book. In the Hebrew version, *Sefer masei yisrael* (the Book of Israel's Travels), these letters were displayed alongside *haskamot* (imprimaturs) of distinguished rabbis. The juxtaposition of letters from scholars and rabbis bears witness to the book's multifaceted character, as Benjamin divided his attention between the far-flung communities of his Jewish "brethren" and the latest breakthroughs in modern geography.

A fold-up map of Benjamin's journey was attached to the end of the German, English, and Hebrew editions of *Eight Years in the Orient*.[78] The large map (375 x

---

[74] Benjamin, *Cinq années de voyage en Orient, 1846–1851*.
[75] Idem, *Acht Jahre in Asien und Afrika: von 1846 bis 1855*, Hannover 1858.
[76] Idem, *Eight Years in Asia and Africa from 1846–1855* (preface by Dr. Berthold Seemann), 1859. The second English edition came out in 1863.
[77] Israel Joseph Benjamin, *Sefer masei yisrael*, rendered into the Hebrew language by David Gordon, Lyck: Tsvi Hirsch Pettsall Press, 1859 [Hebrew].
[78] Two woodcuts were added to the English edition: a view of Jerusalem from the Mount of Olives; and a sketch of the city's walls. I was unable to definitively identify the source of the former. However, the source of the latter is a lithographic tableau consisting of drawings of Jerusalem titled

230 mm) ranges from Constantinople in the northwest and the Nile and Red Sea in the southwest to the Indian Ocean in the east. It is accompanied by secondary maps of the Maghreb and of India and China. In all the editions, the map is written in German and consists of the same elements. However, there is also a Hebrew title in the English and Hebrew editions: "And Israel Wrote the Destinations of his Travels, and these [Constitute] the Itinerary of his Travels for All its Destinations." The cover page informs the reader of the map: "Also accompanying [the book] at its end [is] an itinerary of all the regions of the expanse that the traveler passed through." The map is devoid of cartographic innovations. More specifically, it is based on the geographic knowledge that existed in Germany during the mid-1800s, when the basic alignment of the lands of the earth was already common knowledge.

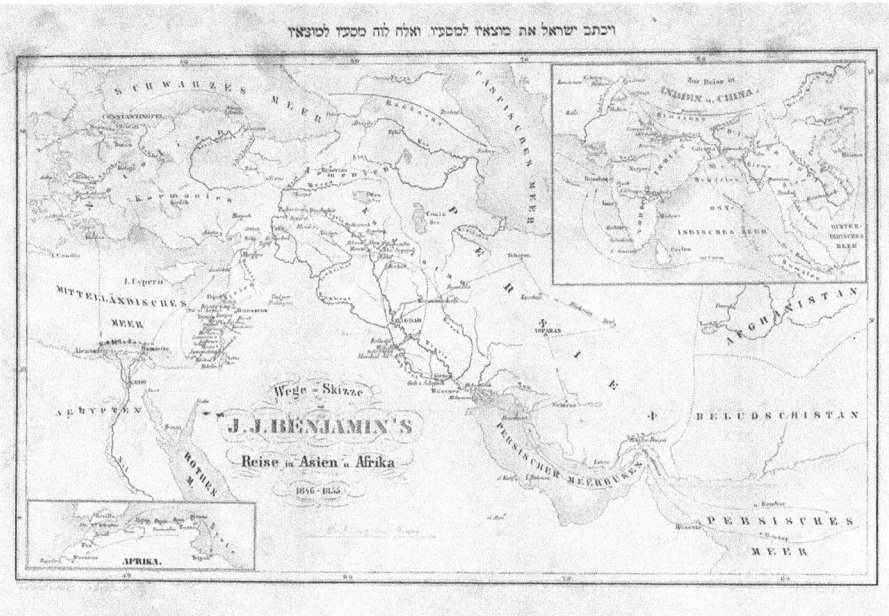

Fig. 121: Israel Joseph Benjamin, "And Israel Wrote the Destinations of his Travels, and These [Constitute] the Itinerary of his Travels for All its Destinations" map, in idem, *Sefer masei yisrael*. 378 x 226 mm.

"Drawings of the Holy Land and All the Places that are Sacred to Our Brethren, the Jewish People[,] the Well-Wishers of Zion and Jerusalem [,] and Everyone that Mourns its Destruction may They Witness its Construction [Anew]." Scholars attribute this particular illustration, which is housed at the Israel National Library's Laor Collection, to Yehoseph Schwarz.

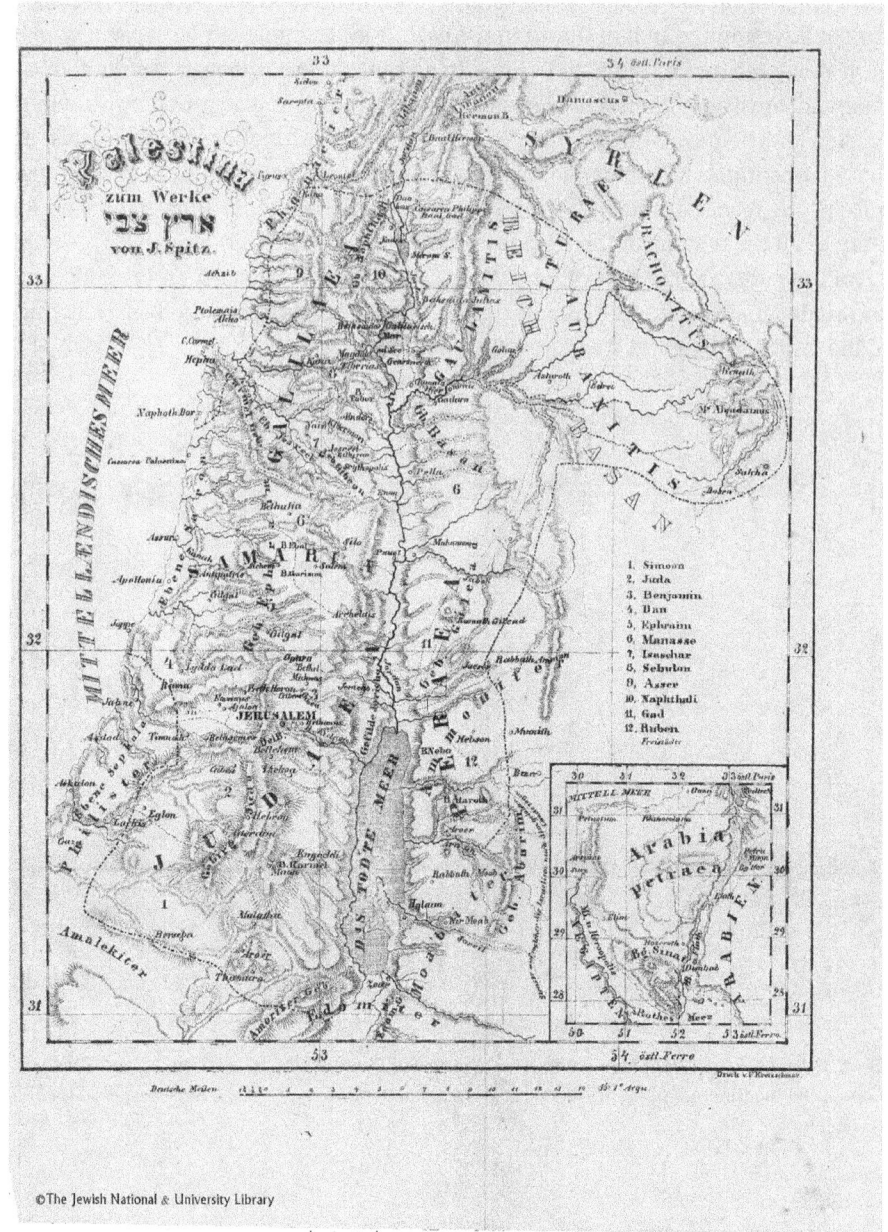

Fig. 122: Jonas Spitz, "Palestine in the Work *Eretz Ṣvi*" map, 175 x 215 mm.

The "Good Land" Map of Jonas Spitz

In 1853, Jonas Spitz published his German-language book, *Eretz ṣvi [Good Land]: Biblical Geography*,[79] on conception of the Land of Israel in the Jewish Holy Scriptures. This rather short book takes the form of a highly modern geographical survey and includes chapters on the following topics: the history of Eretz Yisrael; its borders, climate, mountains and hills, rivers and lakes; chapters on the Land's fauna and flora; and its Biblical sites. Some of these places are arranged by tribal allotment and others in alphabetical order. Many Hebrew toponyms and concepts are integrated into the German text. At the end of the work is a folded map, which is printed on a page that is a bit larger than the rest. The map depicts the Land in a modern cartographic format whose distinguishing features are a set of coordinates and hatched relief. From a geographic standpoint, the representation of the Dead Sea is realistic, for it is consistent with the era's research. However, drawing on the cartographic tradition of the eighteenth century, the Mediterranean Coast is still rife with bays. Spitz's map ranges from Sidon in the north to Zoar in the south, and from the Mediterranean coast to the deserts far to the east of Damascus and Basra. In the lower-right corner is an in-set map of the Sinai Peninsula, which is labelled *Arabia Petraea*. Eretz Yisrael is divided into regions on several levels: the tribal allotments are listed on a key to the right and signified with numbers at the relevant locations; the names of the ancient nations that bordered Eretz Yisrael, such as the Philistines and Amalek, are denoted with relatively small captions; and the provinces during the Hellenistic and Roman periods—Judea, Samaria, and the Galilee—are showcased in large font. Though not expressly stated, the boundaries approximate Solomon's kingdom at its height. The author transcribed place names that were in use during antiquity into German letters. More specifically, there are toponyms from the Biblical period, such as Gilgal, Jezreel, Timnah, and Ramah, as well as from the Second Temple era and the days of the Mishna and Talmud, like Yavneh, Lydda, Antipatris, Archelaus, Betulia, and Caesarea. The non-Biblical toponyms indeed evoke their Greek and Latin names. On the other hand, there are no references to the state of affairs in the Land at the time of the map's publication. In this respect, Spitz perpetuates the traditional approach to the Land's past, albeit in a modern guise.

Ephraim Israel Blücher and His Maps

A Maskilic rabbi and scholar, Ephraim Israel Blücher (1813–1882) taught at the University of Lemberg (Lwów) and was also active in Vienna and Budapest. As evi-

---

[79] Jonas Spitz, *Eretz ṣvi Das biblische Palästina: zum bessern Verstaendnisse der heiligen Schrift in geographischer Beziehung* [...] Prag 1853.

denced by his Aramaic book of grammar,[80] Blücher's main fields of research were Hebrew and Aramaic. Concomitant to his linguistic-scientific enterprise, he served as the rabbi of Wadowice Province and published works on Jewish life. For instance, I found the Maskil's German translation of the Scroll of Ruth and his book on the place of the synagogue in Jewish communal life. The cover page of the second work lists Blücher's other writings, along with his professional titles and posts.[81]

Housed in the Israel National Library's Laor Collection is an undersized, stand-alone map (90 x 130 mm) of the tribal allotments.[82] The size of this map perhaps suggests that it was an integral part of a book. However, on the bottom-left corner is a notice that the map can be purchased directly from the author via the post, thereby implying that it was printed on an independent basis. Written in German, Blücher's map has a rather simple design. Its boundaries are the greater Beirut and Damascus areas up north and the outer limits of the Negev down south. The representation of the Dead Sea and the Mediterranean appears to be highly realistic, but the latter still has an excess of bays. Nearly imperceptible, light brown hatching denotes the Land's mountain ranges. The area of the map is teeming with Biblical toponyms and also includes the names of the tribes (in a larger font). Ringing the margins is historical data. Three of the sides note the years of activity of Biblical figures before the Common Era, rather than Hebrew dates. On the right margin is a list of ancient weights and measurements.

In 1858, Blücher printed a detailed, Hebrew-German map of Eretz Yisrael on a large folio (420 x 515), which was apparently folded within a book. On the top-left corner is a lengthy Hebrew-Talmudic Aramaic title: "Map of the Land of Israel by Ephraim Israel Son of Rabbi Shlomo Blücher, Head of the Court of the Region of Wadowice; in Honor of Our Teacher (and Our Rabbi)[,] Rabbi Joseph Hirsch[,] May his Candle Illumine, the Precious Gem of the Holy Community of Prague, 1858." A German heading is tucked into the opposite corner: "Map of Palestine in Hebrew Script, edited according to the best sources, by the rabbi, Dr. E. J. Blücher ..."[83] These titles inform the reader of the author's full name, academic title, and rabbinic post as well as the map's dedicatee (Rabbi Joseph Hirsch). The upper-right corner hosts a small inset map of the Land of Israel and Sinai, which depicts the route of the Israelites' peregrinations. Unlike most other maps on this topic, the said route heads north

---

80 Ephraim Israel Blücher, *Grammatica aramaica in qua non solum de aramaismo biblico, sed etiam de illis idiomatibus, quae in variis Targumium et Talmud uti etiam in aliorum scriptorum vetero-rabbinorum [...]*, 1838.
81 Ephraim Israel Blücher, *The Scroll of Ruth: With an Ashkenazic [German] Translation and Hebrew Commmentary*, Vienna: Franz Adler von Schmid Press, 1849 (Hebrew, with a German cover page); Ephraim Israel Blücher, *Die Synagogenfrage für deutsche Israeliten*, Wien 1860.
82 Ephraim Israel Blücher, "Kannan & Nebenländer das heilige Land Palaestina in der ältern biblischen Zeit," Grätz bei Posen, (n.d. [ca. 1858]).
83 Karte von Palästina in hebräischer Schrift: nach den besten Quellen bearbeitet, von Rabbiner Dr. E. J. Blücher, Vormals öffentl. Universitätslehrer an dor K.K. philosoph. Facultät zu Lemberg, gegenwärtig, Wadowitzer Kreisrabbiner, 1858.

Fig. 123: Ephraim Israel Blücher, map of Canaan, 69 × 99 mm.

along the length of the Transjordan, perhaps hinting at the Israelites' war against King Og of Bashan. Positioned in the Great Sea is a legend with the symbols for Cities of Refuge, Royal Cities, and Jerusalem. "A Measurement for German Parasangs," namely a linear scale, can be found on the bottom-left corner.

The geographical format is modern and largely accurate. In consequence, it is evident that Blücher relied on the latest research findings. The Land of Israel itself extends from Baalbek, Lebanon in the north to a borderline running from Zoar to "the River of Egypt" in the south. Beyond this boundary are Hormah and Kadesh. The placement of the latter in the eastern Negev is based on Edward Robinson's erroneous identification. The majority of toponyms are Biblical, but there are also Mishnaic sites: Bnei Brak, Usha, and Sippori, *inter alia*. While all the place names are written in Hebrew, the prominent ones also merited German transcriptions or translations. The names of the tribal allotments are printed in large font, but their borders are not demarcated. The entire territory of Eretz Yisrael is delineated by a fragmented borderline on the north, east and south. In contrast to the traditional maps of this sort, the borderline encompasses the two and a half tribes in Transjordan.

Apropos to Blücher's Maskilic background, both of his maps combine age-old Jewish knowledge with the German language and the fruits of modern geography and cartography.

The Map of M. Löwi

Circa 1870, an enormous, multi-colored, and richly detailed Hebrew-German lithograph map titled "The Land of Israel Following the Land's Division into the Twelve Tribes by M. Löwi" was brought to press.[84] According to the map, it is the second and revised edition, but we do not know where and when the first came out. There are also doubts as to whether the map under review was part of a book or was published on its own. While the map's fold marks suggest that the former is the case, its size points to a stand-alone work.

On the upper-right corner is a small map of "Jerusalem at the Time of the First and Second Temple;" and on the lower right-corner is a somewhat larger map of the Sinai Peninsula that enumerates, among other things, the Israelites' stations in the wilderness. Like Blücher's map, the geographical descriptions are highly accurate.

The focal point of the main map is the tribal allotments, the borders of which are delineated with colored lines. Roman numerals signifying each of the tribes correspond with the key on the bottom right. Names of places and regions are labelled

---

[84] "Palæstina nach der Vertheilung unter die 12 Stämme [...] Gez. von M. Löwi]," n.d. The only possible information that I found on the map's compiler is an item on one Marcus Löwi. The latter fashioned a wall map of the Land of Israel, which came out in Pressberg (i.e., Bratislava) in 1878. However, there is no evidence that he is the author of the map under review. See Röhricht, *Bibliotheca Geographica Palaestinae*.

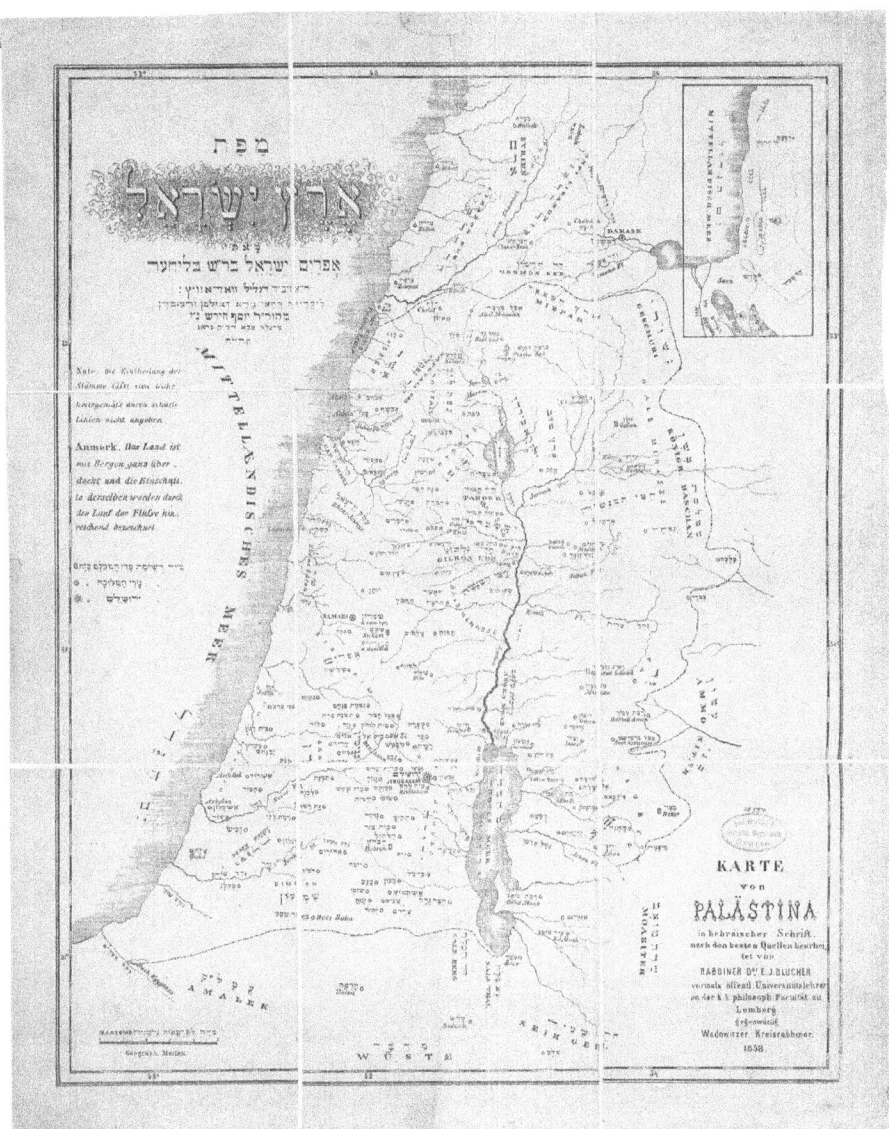

Fig. 124: Ephraim Israel Blücher, "Map of Eretz Yisrael," 420 × 515 mm.

in German (black letters) and Hebrew (red). Most of the toponyms are Biblical, but there are also later names, such as Caesarea, Pameas, and even the rare, Midrashic name of Tarnegola the Upper.[85] The left and right columns are filled with detailed, German information on the toponyms, which Löwi arranged by region: Judea, the

---

85 *Yalkut Shimoni [for the portion of] Ekev*, p. 1114 [Hebrew].

Fig. 125: The Land of Israel Following the Land's Division into the Twelve Tribes by M. Löwi," circa 1870, 755 x 938 mm.

Land of Samaria, the Land of Galilee, and Transjordan. The lower frame takes stock of mountains, sources of water, and places on either side of the Jordan River. Needless to say, the location and identification of many of the sites diverge from the modern research. Furthermore, they are grounded on neither a meticulous study nor firsthand knowledge of the Land's geography. That said, the author was clearly well-versed in the Biblical and Rabbinic literature as well as classical sources. Löwi's decision to print the map in German betrays his wavering between tradition and innovation.

Juda Funkenstein's "Land of Canaan" Map

In 1873 or 1874, one Juda Funkenstein came out with a multi-lingual—Hebrew, English, French, and Yiddish—map of Eretz Yisrael. I did not find any information on the author or the circumstances behind this undertaking, but the map's content speaks for itself. The voluminous map is comprised of eight folios, which are glued together on a cloth. Its title, "The Land of Canaan," takes the shape of a Hebrew-English guilloche. Below the main heading is a small French-English subtitle reading "The Promised Land." Studding the bottom corners are Yiddish and French cartouches.[86] Adjacent to the above-mentioned guilloche is a Yiddish-English legend of the map's symbols, followed by a list of revered Jewish tombs. The map ranges from Sidon and Damascus in the north to Egypt in the south, and from the Nile in the west to a bevy of lands and settlements to the east of Transjordan. Funkenstein's overall design is modern, even if many details, such as the bay-saturated Mediterranean coastline and the exaggerated twists and turns of the Jordan, are unfounded.

The map's modern appearance notwithstanding, its content revolves around the Biblical past. In the Sinai, the route of the exodus loops around the peninsula. At Kadesh Barnea, the route winds its way down to the Bay of Eilat, executes another U-turn, and then shoots back up north through Edom. As per Robinson's above-noted error, Kadesh Barnea is located in the northern Arabah. To the right of Ezion-Geber and Eilat, in an area Funkenstein calls "Madian," is an illustration of the Tabernacle surrounded by the flags of the twelve tribes. This rendering appears to be a modern version of the artistic template that was discussed in earlier chapters. The tribal allotments are signified by large, interlaced Hebrew-Latin captions. In fact, all of the sites are denoted using both languages. Most of the sites derive from the Bible, and many are furnished with references to apposite Biblical passages. In all likelihood, it was a desire to present the lands of the Bible that led Funkenstein to incorporate Haran and the Tigris and Euphrates Rivers due north of Damascus. The Biblical emphasis aside, a couple of hints to the modern world are sprinkled throughout the map, such as the site of Napoleon's battle near Gaza (for whatever reason, the author leaves out the better known Sieges of Jaffa and Acre). Another modern reference is the Yiddish text south of Mount Carmel: "Sand for glass." Soon after the map came out, Baron Rothschild established a bottle factory near Tantura (on the present-day site of Kibbutz Nahsholim) that took advantage of this sand. However, I am unable to point to any direct link between this caption and the factory.

An inset map of Jerusalem frames the upper-left corner. Like the main map, it integrates old with new. That said, the majority of its sites pertain to the First and Second Temple periods. A case in point is the drawing of the Jewish Temple and

---

[86] Le Pays de Promesse. Carte Hebraique de l'original [...] Dressé et Publie par Dr. Juda Funkenstein. Ce sont ici les voyages des Enfants d'Israel d'Egypte dans La Pays de Canaan. Idem, "Illuminated Hebrew Map," (Yiddish) London 5633 [=1873/74].

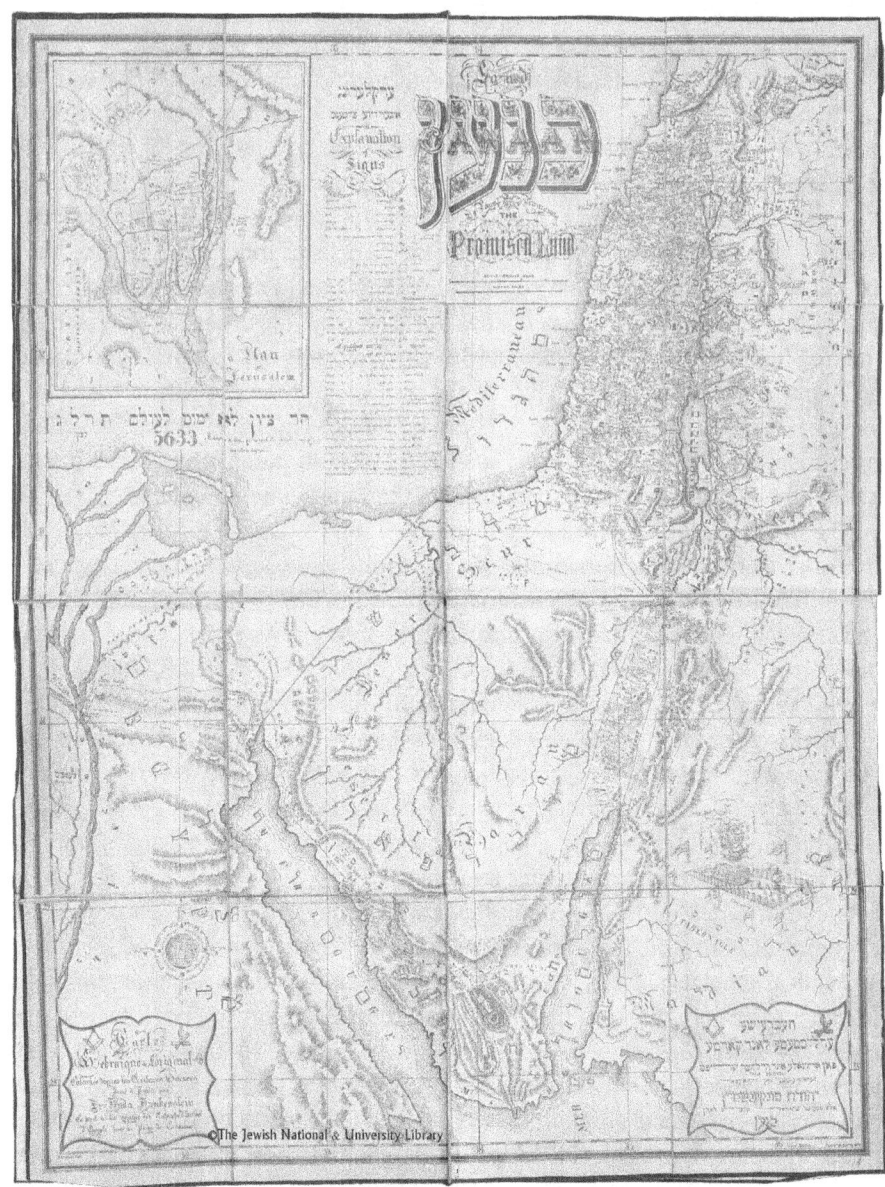

Fig. 126: Juda Funkenstein, "The Land of Canaan" map, 1003 x 1320 mm.

three walls from the days of the Second Temple, to the exclusion of the contemporaneous Islamic sites. Funkenstein also added other long defunct Jewish sites, like the Jackals' Spring and the Fountain Gate. Conversely, the map includes the Russian Compound, which only started to be built in 1860, and the Tombs of the Kings, which de Saulcy explored in 1863. The general geographic account is typical of the second half of the nineteenth century. Beneath the map of Jerusalem is the verse "Mount Zion—it will never fall" (Psalms 125:1) and the Hebrew date of publication (5633), in both Hebrew letters and Arabic numerals.

The copy of this map in the National Library of Israel is quite ornate and the map comes in a blue, hard-covered folder with gold-colored trimming and letters. On the back of the map itself is a handwritten dedication to Adolphe Thiers (1797–1877), the serving president of the French republic. However, the circumstances behind this gesture are unknown.[87] Circa 1876, Funkenstein apparently published another, Hebrew-English edition of this map, which is comprised of four folios.[88]

Linetski's Die Carta Eretz Yisrael

Yitskhok Yoyel Linetski (1839–1915) was a prolific Yiddish writer who split his time between Russia and Ukraine. Among the many genres that he experimented with was satire on the Hasidic Jews of his generation. In the early 1880s, Linetski relocated to Odessa where he moved among the circles of Hovevei Zion (Lovers of Zion)—an influential Zionist group—and began to espouse the idea of Jews settling the Land of Israel.[89] To this end, he penned a thirty-page booklet on the geography of Eretz Yisrael, which came out in 1882.[90] The work consists of a handful of short chapters on the Land's history; its borders, regions, mountains, and lakes; its soil, rock types, and minerals; agricultural produce; and wildlife and the like. In the opening and closing pages of the book, there are references to and purchasing information for a map of Eretz Yisrael that the author published. Even though both works were apparently tied to the same initiative, the map did not come with the booklet. In any event the two were most likely part of Linetski's Zionist and, perhaps, commercial endeavors.

The fairly large (459 x 653 mm) color lithograph map bears two headings. Tucked into its upper-left corner is a Russian title and scale (1:900,000). On the righthand side is the Yiddish title: "Map of Eretz Yisrael, Improved Third Edition, by Yitskhok Linetski." Aside for this note, there is no extant information about the two earlier editions. Under the Yiddish heading is a legend of symbols and an inset map titled "Old and New Jerusalem." The city's topography is depicted in an exceedingly up-to-date and accurate manner. As its title suggests, the map integrates sites from the Second Temple period with those from the late nineteenth century. For instance, Jerusalem's contemporaneous quarters (i.e., "the precinct[s] of the Armenians," "the Christians,"

---

[87] The dedication reads thus: "This book has been offered by the author as a gift to Mister Thiers, President of the French Republic on May 12th 1874."
[88] The catalog of the British Library suffices with these details and nothing else.
[89] Rejzen, *Leksikon fun der Yidisher Literatur, Prese un Filologye*, vol. 5, pp. 163–158 [Yiddish].
[90] Yitskhok Yoyel Linetski, *The Short Geography of Palestine* [die Kurze geographia fon Palestina], Odessa: P. A. Zeleni Press, 1882 [Yiddish].

"the Ishmaelites," and "the Jews") and thoroughfares (Jaffa Road and Hebron Road) share the landscape with "the Camp of Titus' Army," and "Bezetha[,] New City."[91] Linetski did not offer any illustrations of the buildings on the Temple Mount, but there is a compartment labelled *beit ha'mikdash* ("the Jewish Temple"). Alternatively, the map is devoid of references to Jerusalem's late nineteenth century Muslim sites, not least the Dome of the Rock.

Linetski depicts the Land of Israel and Sinai, from Damascus in the north to the Red Sea in the south. Here too, the tribal allotments are signified by numbers that correspond to the key on the righthand side. Although the map's titles are in Russian and Yiddish, all the toponyms are written in Hebrew; only the Mediterranean Sea and Red Sea have also been translated into Yiddish. The Israelites' stations in the wilderness dot the southern half of the map. The heartland of Eretz Yisrael is teeming with Biblical place names as well as a few from the Mishnaic and Talmudic eras, such as Tiberias, Gamla, and Bethsaida (on the north shore of the Sea of Galilee).[92] In Hauran and by Damascus, there are a few Arabic and Greek place names, all of which are transcribed in Hebrew letters. These toponyms are yet another example of content from non-Jewish sources that penetrated the Hebrew map.

The symbols on Linetski's map distinguish between large cities, small cities, Levite cities, cities of refuge, and colonies ("*kalanin*"). The symbol for the latter is a small cottage. Among the Jewish settlements, the only ones that were included in the Galilee are Yesud haMa'ala and Rosh Pinna; and the same can be said for Zikhron Ya'akov in the Mount Carmel area. With respect to Judea, one can find the Hebrew colonies of Rishon LeZion, Petah Tikva, Qatra (i.e., Gedera), and Ekron. At any rate, the overall representation of these settlements is far from realistic. A particularly evident error is the placement of Rishon LeZion to the north of Petah Tikva. It stands to reason that this map came out in close proximity to the establishment of the first Jewish colonies. Just like Schottlaender's aforementioned tableau (see chapter 6), their very appearance on Linetski's map constitutes a noteworthy innovation.

Map of Jacob Goldzweig

In 1896, one Jacob Goldzweig published a booklet titled *Igeret le'keṣ ha'yamim* (an Epistle for the End of the Days).[93] According to its short introduction, the writer, a resident of Haifa, had already been living in Eretz Yisrael for forty years. However, the work itself was printed in Manchester and the cover page includes an English

---

[91] According to Josephus, Bezetha (or "New City") is the northeastern part of Jerusalem. Flavius Josephus, *History of the Jewish War against the Romans*, book V, chapter 4, 151.
[92] The inclusion of Bethsaida may attest to the use of a Christian source, for this city is mentioned in the New Testament.
[93] Jacob Goldzweig, *Igeret le'keṣ ha'yamim*, Manchester 1896 [Hebrew].

Fig. 127: Yitskhok Yoyel Linetski, "Die Carta Eretz Yisrael" map. 459 × 653 mm

title.⁹⁴ The lion's share of *Igeret le'keṣ ha'yamim* is taken up by a work on the gematria of Biblical verses that ostensibly proved that the redemption was near. This messianic-kabbalistic tract, which Goldzweig attributed to the famed kabbalist Rabbi Hayyim Vital, is followed by a couple of short works, among them "Several Points about the Settlement of Eretz Yisrael" and "These are the Names of the Colonies and the Number of their Inhabitants." This combination of topics demonstrates that, on the one hand, the author was a devout Jew who truly felt that the end of the days was imminent; and on the other hand, he was an unabashed supporter of the nascent settlement movement. It also stands to reason that Goldzweig saw a connection between the coming of the messiah and the first wave of Jewish colonization. What is more, the booklet praises various signs of modernization in the Land of Israel, such as the telegraph, the railroad, and the fact that the "boys and girls" in the Jewish colonies study "Torah and wisdom [i.e., general studies] in three languages."

It is against this backdrop that Goldzweig's Hebrew-English map should be understood. Printed on a rather large folio (430 x 571 mm) in 1893, this map combines old traditions with new trends.⁹⁵ Resembling a poster, the map was probably intended as a decoration for the home or sukkah. Perched between two Stars of David on the upper margin is a long English title: "A New and Original Biblical Map of the Holy Land." A bit lower is the understated Hebrew title—"Land of Canaan." The modest frame on the top-left corner contains lengthier English and Hebrew titles. The English version reads "Palestine or the Holy Land from Biblical Times to the Present," whereas the second Hebrew title is "Eretz Yisrael and the States that are around it as it was during the Time of Moses Our Rabbi, May Peace be upon Him, and Joshua and Solomon until the Destruction and the Present Day." In a nod to the Mizrah tradition, ornate pillars grace the folio's left and right side. The lower portion consists of seven compartments: the rightmost and leftmost frames are set aside for penciling in memorial days and birthdays, respectively; the second compartment to either side contains the map's Hebrew and English legends; the central frame depicts the Western Wall; to its left is Rachel's Tomb; and to its right the Cave of the Patriarchs in Hebron. The Wall and the Tomb are accompanied by relevant passages from the Bible. In the adjacent open space is a sketch of the Israelites' encampment in the wilderness. The camp's circular shape diverges from the standard rectangular template. These sort of elements, not least the illustrations of holy places, tie Goldzweig's map to the tableaux that were examined in the previous chapter.

This map of the Land of Israel and Sinai ranges from Egypt in the west to the wilderness of Trans-Jordan in the east, and from Beirut up north to the Sinai Peninsula's southern coast in the south. Goldzweig created an amalgam between new and

---

94 Nine years later, another edition of this booklet, which is attributed to R. Feibish Brauer, was published in Piotrków Trybunalski.
95 Jacob Goldzweig, "Palestine or the Holy Land from Biblical Times to the Present Day," London 1893.

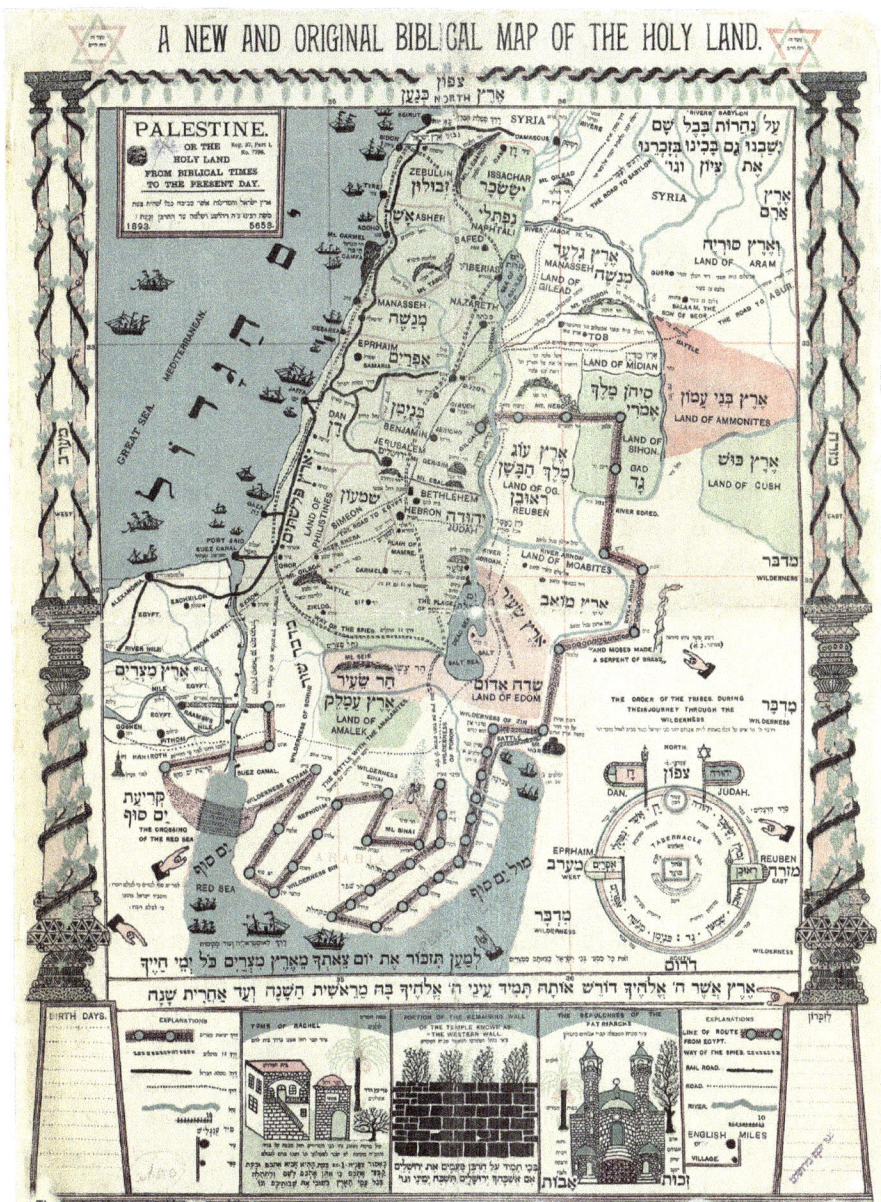

Fig. 128: Jacob Goldzweig, "A New and Original Biblical Map of the Holy Land." 430 x 571 mm

ancient, modern and traditional. From a geographical standpoint, most of this map is realistic. An obvious exception to this rule is the Nile River, which oddly undulates from west to east before spilling into the Mediterranean Sea. In addition, its mouth is presented in a distorted fashion, without any hint of the Nile Delta. The rivers of Babylon are compressed into the map's northeastern corner. Similarly, the rivers of Damascus, Amanah and Pharpar, spill into the Jordan, through the Jabbok, which also skews from its actual course and is connected also to Nahal Zered and the Arnon. Content wise, the map is by and large traditional. More specifically, it depicts toponyms from the Biblical epoch, and various sites are accompanied by verses from the Hebrew Bible and even the Passover Haggadah (i.e., "He who split the Sea of Reeds, His steadfast love is eternal"). Goldzweig's thoroughfares also derive from the Bible: "The Road from Paddan Aram," "The Roads from Babylon," "The Way of the 12 Spies," and the route of the Israelites' peregrinations. That said, the most accentuated track in the Land of Israel is the "Railroad," which is denoted by a thick black line. The Suez Canal is accompanied by a Yiddish caption reading "The *zuets canale* from the Great Sea." Near the Gulf of Sinai are the words "The Way to Australia and Other Places." In all likelihood, this alludes to the fact that Australia was a destination for Jewish immigration at the time. Even the accompanying illustrations reflect the author's vacillation between tradition and modernity. On the one hand, there are renderings of Moses on Mount Sinai and the brass serpent; and on the other hand, the smokestacks on some of the boats represent the modern technology of marine steam engines.

The "True Path" of Avigdor Malkov

In 1894, Avigdor ben Mordecai Malkov printed a map called *Mappah derekh ha'emet* ("The Map of the True Path"). Close to the bottom-left corner, beneath the Hebrew heading, is a long Russian title that offers both a transcription and translation of the Hebrew one: "Map of Palestine and the Israelites' Route from Egypt."[96] A year later, the author published Hebrew and Yiddish editions of a booklet, *Derekh emet* (True Path), which provides in-depth explanations of this map.[97] Facing eastward, the map was quite antiquated from a geographic standpoint. In contrast to, say, the Linetski and Goldzweig maps, Malkov's did not reflect the current geographic knowledge in the least. To begin with, the Mediterranean shoreline is rife with bays, the Sinai is under-sized, and the peninsula's coasts form round contours, rather than a triangular shape. Additionally, Babylon and its rivers are drawn to the north of the Hermon Mountain. The only modern elements are the Suez Canal and the shape of the Dead Sea.

---

[96] Mapa derekh emes t.e. Karta Palestiny i pravednoi dorogi vykhoda izraelityan Egipta. Sostavil Vigdor Malkov, Warsaw, 1894.
[97] Avigdor ben Mordecai Malkov, *Derekh emet* [True Path]: *An Explanation of the Map of the Land in which will be Explained the Voyages of the Israelites and the Borders of Eretz Yisrael with the Commentary* [...] *Divrei emet* [Words of Truth], Warsaw, 1895 [Hebrew]; Avigdor ben Mordecai Malkov, *Sefer derekh emet*..., Warsaw, 1895 [Yiddish]. Six years later, both versions netted a new edition.

Adorning the map's outer frame are several passages from the comfort prophecies. To the right, "Thus said the LORD: I accounted to your favor the devotion of your youth, your love as a bride—how you followed Me in the wilderness, in a land not sown" (Jeremiah 2:2). Up top, "You shall be a glorious crown in the hand of the LORD, and a royal diadem in the palm of your God: Nevermore shall your land be called desolate: But you shall be called [']I delight in her,['] and your land [']Espoused[']: For the LORD takes delight in you, and your land shall be espoused" (Isaiah 62:3–4). To the left, "And in that day, a great ram's horn shall be sounded and the strayed who are in the land of Assyria and the expelled who are in the land of Egypt shall come and worship the LORD on the holy mount in Jerusalem" (Isaiah 27:13). And below, "I the LORD have been your God ever since the land of Egypt[.] I will let you dwell in your tents again as in the days of old [Hosea 12:10]: I will show him wondrous deeds as in the days when You sallied forth from the l [and of] E[gypt] [Micha 7:15]: Forget not the love of Zion and adoration of Jerusalem forever [from *Selichot*]."⁹⁸

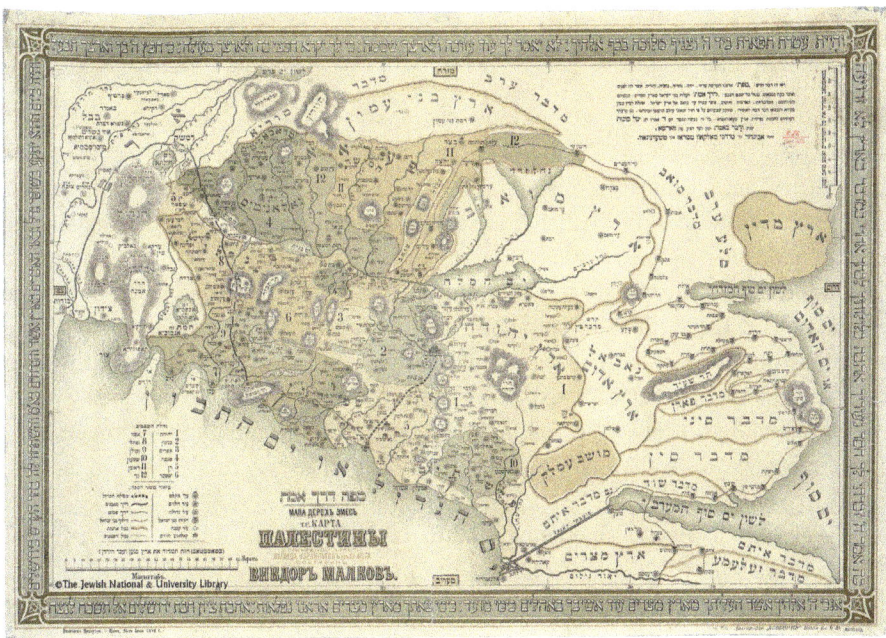

Fig. 129: Avigdor Malkov, "The Map of the True Path," 790 x 530 mm.

On the upper-right corner is a linear scale bearing the following caption: "(With this scale) you shall measure all the deserts and the land of Egypt, Moab, and

---

98 *Selichot* are a compendium of Jewish penitential poems and prayers that are customarily recited during the run-up to and the days between the High Holidays.

Ammon." To its left is an inscription that reveals the map's objective, its creator, and date of publication:

> Behold this novelty: "A map" of our sacred Land[,] its cities, its oceans, its rivers, its brooks, and its mountains, which existed in the past and which are currently to be found. The border of each and every tribe: "and the true path" of the Israelites' *aliyah* [ascent or immigration] from Egypt: their voyages and encampments; the deserts, the lands, and the seas that they traversed before arriving in the Land of Israel: I hope that you shall weigh with an open eye and confirm [the veracity] of item after item [on this map, which have been] refined sevenfold according to the sages of blessed memory and the world's geniuses[,] *rishonim* [and] *akhronim*.[99] I also closed the gap between their views and the wisdom of land surveying (*geographia*). All this was done and concluded on Wednesday[,] the day after the festival of Sukkot of the Hebrew year [...] here [in] Warsaw. By Avigdor son of Rabbi Mordecai Malkov from Surazh[,] the province of Chernigov.

The word "map" in the opening line merited large font and quotation marks, for the use of this term was seen as an innovation that the author took pride in.

On the bottom-lefthand side are a key of the tribal allotments and a legend of the map's symbols. Like the heading, these tools attest to a strong blend between tradition and modernity. For example, cities of refuge, Levite cities, and Israelite encampments appear side by side with "*Colonia yehudim*" (colony of Jews) and the railroad. On the traditional side of the equation, there is an emphasis on the period of the Bible. This is evident from, say, the toponyms and the tribal allotments. As in many of the earlier maps (see chapter 4), the route of the Israelites' peregrinations is circuitous. Aside for the Biblical names, there are also toponyms from the Mishna and Talmud, most notably Capolira (an appellation for a mountain or mountain-top city on the Land's northwestern border). In general, this term is used in the context of Mount Hor or Tor Amnon. With respect to the map at hand, Capolira sits atop a mountain on the Lebanese coast, from which a straight line darts southwards. This element is a cartographic expression of the discourse over Eretz Yisrael's western boundary. In the Sages' estimation, the border did not run along the coast, but a little further west. On the map, a sliver of the shoreline is actually outside the border.[100]

Among the modern elements herein are the Jaffa-Jerusalem railroad, which was inaugurated only two years before the map came out, and the new Hebrew colonies, which are symbolized by a house with a tiled roof. The colonies in the Galilee are Yesud haMa'ala, Rosh Pinna, Mishmar haYarden—the latter is erroneously located to the south of the Sea of Galilee—and Agudat ha'Elef (i.e., Ein Zeitim) near Safad. Further south, one can find Zikhron Ya'akov, Petah Tikva, Yehud, Mikveh Israel, Rishon LeZion, Nahalat Reuben (i.e., Ness Ziona), Ekron, Gedera, and Rehovot.

---

99 The *rishonim* and *akhronim* are the terms for the leading rabbis-cum-adjudicators who lived between the eleventh and fifteenth centuries and from the sixteenth century onwards, respectively.
100 The Babylonian Talmud, Gittin 8a.

Additional evidence of Malkov's familiarity with contemporaneous developments in the Land is a couple of Arabic names that are tied to the era's Jewish settlement, like *tanturot* (Tantura) and *wadi al khadr* (Hadera Stream). The map also bears witness to its maker's general geographic knowledge. For example, it includes distant sites that have no connection to either the Land of Israel or Jewish tradition, such as the city of Diyarbakır in eastern Turkey, Baghdad (which was erroneously located between the Tigris and the Euphrates, rather than on the Tigris), and the Suez Canal.

As above-mentioned, Malkov's *Derekh emet* is a detailed commentary of his map. Not surprisingly, then, it is informed by the same balance between tradition and innovation. While predicated on Biblical exegesis, his book frequently refers to Arabic toponyms using European languages. On occasion, the author provides Yiddish transcriptions of Russian names, such as "Arabiskei Zalibo" (i.e., the Arabian Gulf) and "Alanitski Zalibo" (the Eilanitian Bay (the Gulf of Eilat). In the two parts of *Derekh emet* that fall under the heading of "Small Geography," there is a description of the Land's geography, water sources, quarries, and more. The section on farming discusses the agricultural machines that were brought to the region by the "Ashkenazim," namely the German Templers.[101]

Matskevitsh's Borders of Zion

In 1898, Shabtai ben Yaakov Matsḳeyiṭsh (1832–1923), a Maskilic yeshiva scholar fluent in Russian,[102] printed a twenty-five page booklet in Vilnius titled *Gvulei ṣion* (Borders of Zion).[103] The work is, first and foremost, a Halakhic commentary on Eretz Yisrael's Biblical borders. For instance, the author expounded on the following issues: the seeming contradiction between the promised borders that extend "to the great river, the Euphrates" (Genesis 15:18) and those laid down in *Mas'ei* (Numbers 33–36); and the question of why the tribes of Gad, Reuben, and half of Manasseh settled down in Transjordan, outside the Land's boundaries. Accompanying *Gvulei ṣion* is a map bearing Hebrew and Russian headings.[104] Matskevitsh obviously saw the map as an integral part of this work, for it is mentioned on the book's cover, in both Hebrew and Yiddish, and on all three of its cover pages. Additionally, the map can be found right after the book's preliminary material.

---

[101] The Templers are a millennial German Protestant sect that began to settle the Holy Land in the late 1860s.
[102] Rejzen, *Leksikon fun der Yidisher Literatur*, vol. 5, p. 472.
[103] Shabtai ben Yaakov Matsḳeyiṭsh, *Borders of Zion* [...], Vilnius: Avraham Tsevi Ḳatsinelinboigen [Press], 1898 [Hebrew].
[104] "Map of the Holy Land with all the Places in its Vicinity, Accompanying the Book 'Gvulei Zion' by Shabtai Matsḳeyiṭsh." The Russian title reads "Karta drevnei Palestiny sb okrestnymi e mestnostiami. Sostavil Sh. Matskevich [Map of ancient Palestine with the countries around it. Compiled by Sh. Matskevich], Vilna: N. Matz, 1898."

Fig. 130: Shabtai ben Yaakov Matskevitsh, "Map of the Holy Land," 203 x 146 mm.

Printed on a regular page in the book, Matskevitsh's map is rather small. Nevertheless, it covers the entire Middle East, from Egypt to the Persian Gulf. The design is exceedingly plain, and there are no accompanying illustrations or special details. Facing north, the map is outfitted with geographical coordinates on its edges and a linear scale beneath the Russian title. Most of the captions are in Hebrew; only those few that pertain to non-Jewish sites, like the Arabian Desert, Egypt, Cyprus, and Asia Minor, also merited Russian names. A prominent line traces the Israelites' journey from Raamses, which the Russian caption identifies as Cairo, to Jericho. There are only a few sites within the boundaries of Eretz Yisrael: Jerusalem, the Dead Sea, Jericho, Jaffa, Mount Carmel, Acre, the Sea of Galilee, the Waters of Merom, Beirut, Mount Hor, and Turei Amnon (the last two are located in the north). Similar to his predecessors, Matskevitsh explicitly integrates traditional content with modern cartography. In other words, questions of Biblical exegesis are presented within the framework of a modern design that includes Maskilic trappings, like a Russian title, a scale and coordinates. Furthermore, there are references to decidedly non-Biblical places, such as Mesopotamia, Cypress, and Cairo.

The Map of Ephraim Michael Grover

Ephraim Michael Grover's "The Well-Researched and Precisioned Map of Eretz Yisrael" came out in Odessa a year after *Gvulei ṣion*.[105] I did not manage to find any biographical details about the author, so that the ensuing discussion concentrates entirely on the map itself. Facing north, the gray-green lithograph offers a modern cartographic account of the Land of Israel. In the upper-lefthand portion is a Star of David-topped cartouche bearing the map's title, author, and date of publication (1899) along with a linear scale in German Parasangs. The top-right corner is taken up by a similarly modern map of Jerusalem. Underneath this inset are two columns of Hebrew texts. The one to the right lists the post-exilic rulers of the Land, from the Roman conquest—in the year "3828" to "the Creation of the World" (i.e., 68 BCE)—until the return of the Ottomans in 1840. Titled "The Nature of the Land," the second column surveys the local weather and produce during each month of the Hebrew calendar. Stamped onto the bottom of the folio are the name of the Odessa-based print house, "Progress," and the Gregorian date of publication.

Grover's map ranges from Beirut in the north to the River of Egypt in the south, and from the Mediterranean Sea in the west to "The Wilderness of Syria" in the east. Eretz Yisrael is divided into the tribal allotments, which are denoted in large font by the first letter(s) of each tribe's name. The majority of toponyms are taken from the Bible, but there are a small handful of Arab names, such as *baḥr lot* (the Sea of Lot, namely the Dead Sea), Baalbek in Lebanon, and Nahar-Hasbeiya (the Litani River). The Hebrew colonies turn up on this map as well, where they are signified by a Star of David. With the following exceptions, the list of settlements is identical to that of Malkov: Mishmar haYarden is located in its correct place; the map at hand includes both Ein Zeitim and Agudat ha'Elef, even though both names refer to the same colony; Hadera is accurately located to the south of Zikhron Ya'akov; and Ness Ziona still goes by its Arab name—Wadi Chanin.

Grover's background and cultural milieu are reflected in the design and content: from a cartographic standpoint, the map is modern and very precise; and the accompanying texts provide scientific-cum-Maskilic information. The references to agriculture and the Jewish colonies tie the map to the Zionist movement, which was spreading its wings at the time—certainly in the author's city of Odessa. Moreover, the Arabic names point to firsthand knowledge of the region. The strong resemblance between this map and one that was subsequently published by Eliyahu Landa[106] indi-

---

[105] Ephraim Michael Grover, "The Well-Researched and Precisioned Map of Eretz Yisrael," Odessa, 1899 [Hebrew].
[106] Eliyahu ben Elazar Landa, "Map of Eretz Yisrael, Edited in Accordance to the Wisdom of the Great Traveler[,] the Rabbi [...] Y. Schwarz in his Book TvHa [*Tvu'ot ha'areṣ*] and with the Approval of the Architect[,] Dr. C. Schik in Accordance with the Reality that They Saw there [in the Land of Israel] with Their Eyes and was Drafted in Accordance with the Latest Maps of E.G. and P[inkas] Horowitz, Jerusalem 1912." Another edition came out in 1915.

Fig. 131: Ephraim Michael Grover, "The Well-Researched and Precisioned Map of Eretz Yisrael," 405 x 334 mm.

cates that the latter was predicated on the former. In addition, Landa noted in his preamble that one of his sources was "E.G.", probably Ephraim Grover.

Between the Jewish Enlightenment and Zionism

As discussed above, the Hebrew maps from the second half of the nineteenth century strike a balance between tradition and modernity. The former is evident in the Biblical topics and citations as well as from the contexts in which the maps were published. The traditional dimension stands out in the map of Goldzweig, who availed himself of the established Jewish design template, and the Matskevitsh map, which was incorporated into a standard rabbinic work of exegesis. Conversely, the increasing accuracy of the maps, the appearance of coordinates, multi-lingual texts, and Arabic toponyms, and references to contemporaneous reality are all emblematic of the Haskalah. In parallel to these two divergent trends, ever more attention was being paid to the colonies of the First Aliyah. The ways in which these sites were marked—with a Star of David or a modern house—testifies to the creators' positive attitude towards the settlement enterprise. This enthusiasm leads us to a discussion on maps that were drafted and printed within a full-fledged Zionist context.

As an ideological movement seeking to renew the Jewish hold over Eretz Yisrael, Zionism *ipso facto* viewed maps as a tool for advancing its territorial aims both within the movement's ranks and in its contacts with the outside world.[107] Given the objective of settling a tangible expanse, Zionism could no longer suffice with schematic or artistic maps. Embracing the ethos of modern cartography, the movement depicted the Land's borders, mountains, and valleys in a realistic fashion. While some Jewish mapmakers continued to follow in the footsteps of Rashi and the *Vilna gaon*, from here on realistic maps grounded on the fruits of modern research, surveying, and cartography were the prevalent form of Hebrew map. This was true of both Zionist maps and historical maps on the Biblical and Talmudic past, both of which would be robustly developed in conformance with scientific advances.[108]

The first writer to espouse this Zionist approach in a direct and lucid manner was Eliezer ben Yehudah. While mainly remembered for his efforts to rejuvenate the Hebrew language, Ben Yehudah also took great interest in the layout of the "ancestral land." As far back as 1883, a year after the beginning of the First Aliyah and the establishment of the maiden Hebrew colonies, he penned a book, *Sefer ereṣ yisrael*, on the Land's nature and geography.[109] Put differently, this work coincided with the embryonic stages of the Zionist movement's practical enterprise. In the introduction, the author reveals the primary impetus behind this book: improving his readers' knowledge of Eretz Yisrael and convincing them to make *aliyah*. "The last of all the Jewish people's hopes," Ben Yehudah averred, "are in this small land [...] There is only one land in the world where the Jewish people can live a life of peace and tranquility, in which he [i.e., the Jew] will not be told that he is a stranger and a foreigner." Despite the book's geographical emphasis, it does not include a map, so that we lack a tangible expression of Ben Yehudah's cartographic conception of the Land.

The Precious Land of Nahum Sokolow

Nahum Sokolow (1859–1936) was among the founders and most distinguished leaders of the Zionist movement.[110] From a very early stage in his life, Sokolow took an

---

[107] See Markus Kirchhoff, *Text zu Land: Palästina im wissenschaftlichen Diskurs, 1865–1920*.
[108] These developments inspired Hebrew researchers, like Benjamin Mazar, Michael Avi-Yonah, Yohanan Aharoni, and many later authors, to craft historical atlases and maps.
[109] Eliezer ben Yehudah, *Sefer ereṣ yisrael* [Book of the Land of Israel]: *On the Nature of this Land, its Seas and its Rivers, its Mountains and its Valleys and the Nature[,] Climate[,] and Flora and Fauna therein, its Cities and Villages*, Jerusalem: Yoel Moshe Salomon Press, 1883 [Hebrew].
[110] Florian Sokolow, *My Father, Nahum Sokolow* [Hebrew]; "Sokolow, Nahum," *Encyclopedia Judaica*.

interest in geography and would even write a Hebrew book on this topic, which drew on the methods of the American scholar Matthew Fontaine Maury.[111] An early proponent of Zionism who was aware of the connection between striving to acquire a territory and grasping its geography, Sokolow published articles about Eretz Yisrael's layout, vistas, and natural resources in *ha'Tsefirah* (The Call)—an influential Hebrew newspaper.[112] In 1885, he combined his expertise on these topics with his passion for Zionism by publishing *Ereṣ ḥemdah* (the Precious Land).[113] This short book is a systematic survey of Eretz Yisrael's geography and history. Its final chapter takes stock of the colonies of the First Aliyah. According to the introduction, Sokolow also intended to translate Lawrence Oliphant's *The Land of the Gilead*—a book that inspired many of the movement's earliest proponents—into Hebrew.[114]

Ereṣ ḥemdah was furnished with a color lithograph map. According to the preamble, the map was drafted "in the image" of Heinrich Kiepert's work.[115] It presents the Land of Israel in a northerly orientation, from the Sea to Transjordan, and from Beirut in the north to the ancient cities of Rehovot-in-the-Negev and Haluza in the south. Sokolow's modern outlook is reflected, first and foremost, in his use of a map by Kiepert that adhered to Edward Robinson's findings.[116] On the lefthand side of the map under review is a short list of the Arabic toponyms that the author interspersed to the west and north of Jerusalem, along with the Biblical sites that are identified with them. Further down is a glossary of basic Arabic geographical terms, like those for river and village. While all the map's captions and texts are written in German, its lone title (on the upper-right corner) is in Hebrew. A considerable portion of the sites are denoted using their Biblical-Hebrew names. However, quite a few are accompanied by their Arabic toponyms, albeit in smaller font. The more familiar sites are also identified by their appellations in Greek and Roman sources. As was customary in Kiepert's maps, all the names are transcribed in German letters. Glaring in their absence are the colonies of the First Aliyah. Since Sokolow's map came out in 1885 and given the precedence set by Linetski, one could have expected the Zionist leader to have included the new Hebrew settlements. Perhaps they were left out due

---

111 Nahum Sokolow, *Maṣukei areṣ* [Precipices of the Land]: *Or the Fundamentals of the Knowledge of Natural Geography [...]as it was Brought to Fruition [...]by the True Adherents of the Researchers and Following in the Footsteps of [...] M. P. Maury's Approach [...], Translated and Annotated with Enlightening Notes at the End of the Book* [...] Warsaw: Izaak Goldman Press, 1878 [Hebrew].
112 Ben-Arieh, "Developments in the Study of *Yediat Ha'aretz* in Modern Times, Up to the Establishment of the State of Israel" [Hebrew].
113 Nahum Sokolow, *The Precious Land: Including Information on the Regions of t[he] H[oly] L[and] according to the Greatest Travelers [...]*, Warsaw: Goldman Press, 1885 [Hebrew].
114 Oliphant, *The Land of the Gilead with Excursions in the Lebanon*.
115 "Map of the Holy Land[,] 'the Precious Land' by N. Sokolow," Berlin: Dietrich Reimer, 1885 [Hebrew]
116 I was unable to determine which of Kiepert's maps Sokolow's is based on. For more on Kiepert's maps, see Goren, "The Present as an Expression of the Past" [Hebrew].

Fig. 132: Nahum Sokolow, "Map of the Holy Land[,] 'the Precious Land,' Bearbeitet von H. Kiepert, Berlin 1885," 300 x 433 mm.

to their absence from Sokolow's source, the Kiepert map, which predated the new wave of Jewish immigration.

Sokolow also provided two inserts. A small map of the tribal allotments is located on the upper-left corner. Four different colors sort the tribes into the following groups: the progeny of Rachel, Leah, Zilpah, and Bilhah. The second insert is an exceedingly modern and accurate representation of Jerusalem, even if the title—"Ancient Jerusalem"—and its content reflect the city of the First and Second Temple Periods.

The Sapir and Krauze Map

Zionist activity at the turn of the twentieth century prompted a steady rise in maps of Eretz Yisrael. Given the large sample size, the ensuing discussion will concentrate on a few representative samples from this subgenre. [117] As noted above, most of these maps drew heavily on modern cartography. Therefore, they cannot be said to have broken new ground with respect to either form or content. A case in point is a map that was published in 1914 by the Zionist Committee in Berlin. Drafted by Eliyahu Sapir and Ephraim Krauze, this map also leaned on the Kiepert model.[118]

Eliyahu Sapir (1869–1922) was a native of Jerusalem and the son of a Maskilic family that was active in the *yishuv* (Jewish settlement in the Land of Israel). As a young man, he moved to Petah Tikva where he was among the first pedagogues to teach "Hebrew in Hebrew." Later on, he served as a clerk at the Jewish Colonization Association (JCA) and then worked as a branch manager of the Anglo-Palestine Bank in Jaffa.[119] Sapir's co-author, Ephraim Krauze, was a land surveyor and the less known brother of Eliahu Krause, the agronomist and educator.[120] Needless to say, Sapir and Krauze's educational background and Zionist enterprise undergirded their collaboration on the map.

Voluminous and highly detailed, the Sapir-Krauze map is a color lithograph with a fetching cartographic design. This Hebrew map presents the Land of Israel on both sides of the Jordan. From Jubayl (Arabic for Byblos) in the north, it extends eastwards to Damascus and then Hauran. The map's outer limits in the south and west are el-Audja (Nitzana) and the Mediterranean Sea, respectively. The area of Eretz Yisrael is brimming with toponyms, some of which are Biblical and some of which are Arabic. Hebrew colonies appear not only as points of settlement, but their land holdings are mottled in brown. As a result, the map is lightly peppered with sizable blocs of ter-

---

[117] Ben Arieh, "The Character of Hebrew Geographic Literature," p. 39 [Hebrew]. That said, only a smattering of books on the Land's geography came out during this same period.
[118] "Map of Eretz Yisrael (Palestine), in Accordance with the Kiepert Map, Prepared by E. Sapir and the Engineer E. Krauze, the Publishing House of the Zionist Committee (Berlin), 1913." Five years later, the map merited a second edition; "The Land of Israel, According to the Hebrew Map of E. Sapir-E. Krauze (the Zionist General Council Press) with Additions and Revisions, New York 1918".
[119] "Sapir, Eliyahu," *The Encyclopaedia Hebraica* [Hebrew].
[120] "Krauze, Eliyahu," *The Encyclopaedia Hebraica* [Hebrew].

Fig. 133: Eliyahu Sapir and Ephraim Krauze, "Map of Eretz Yisrael (Palestine)," 1370 x 1920 mm.

ritory, especially in places where there were a number of Jewish settlements in close proximity, like the Lower Galilee and the Carmel region.

The top-left corner houses an ancillary map of "The Tribes of Israel." As in Sokolow's version, the four colors herein represent the sons of Leah, Rachel, Zilpah, and Bilhah. Under this small inset is a linear scale (1: 200,000) and a legend of the map's symbols. Further down is "An Explanation on How to Read the Arabic Names," which indeed elaborates on the transcription and pronunciation of these toponyms, as well as a Hebrew glossary of the Arabic terms on this map. The lower right features a modern and accurate map of Jerusalem during the Great Revolt (or the First Jewish-Roman War), which ultimately led to the destruction of the Second Temple. In this map, colored lines delineate four of the city's walls: "The First Wall from the Days of Solomon," "The Second Wall from the Days of Hezekiah," "The Third Wall from the Days of Herod Agrippa," and "The Fourth Wall from the Days of Hadrian and Today." This account of the walls accords the early twentieth-century scholarship on this topic. Of course, this picture of Jerusalem no longer holds water.

In sum, Sapir and Krauze employed modern geographical tools, so that their map offers a realistic depiction of its subject matter. Their secondary maps represent the Land's past as the seat of Israelite/Jewish autonomy, while also broaching the topic of Jerusalem's destruction at the hands of the Romans—a national tragedy that culminated in the long years of exile. Conversely, the principal map depicts a territorial expanse that is ripe for an influx of Jews aspiring to revive their homeland.

The Land-of-Israel Republic

An extraordinary example of how cartography served as an outlet for Zionist enthusiasm is the "Map of the Land-of-Israel Republic," which came out in 1919. Towards the end of the Great War, Britain issued the Balfour Declaration recognizing Palestine as "a national home" of the Jews. In consequence, there was a sense in the Zionist camp that its dream of building a Jewish state was within reach. Published in Bucharest with the assistance of Poale Zion (a Zionist party), this Hebrew-Romanian map essentially proclaimed that the Hebrew republic was a done deal. What is more, the map features a picture of Louis Brandeis (the Jewish-American Supreme Court justice) that touts him as a worthy candidate for "President of the Hebrew Republic."[121] This endorsement did not come out of the blue. In 1919, Brandeis visited Eretz Yisrael in his capacity as honorary president of the Zionists of America. How-

---

[121] "Map of the Land-of-Israel Republic in Accordance with the Kiepert Map, Arranged and Brought to Print by D. Ben Menachem with the Assistance of the 'Poale Zion' Committee, Bucharest 1919." Alternatively, the Romanian title reads thus: "Harta Republicei Evreești, după Kipert desenat și editat de H. Brandes M. Lewin (Ben Menachem) și A. Israilovici cu concursul Comitetului 'Poale Zion,' București, 1919."

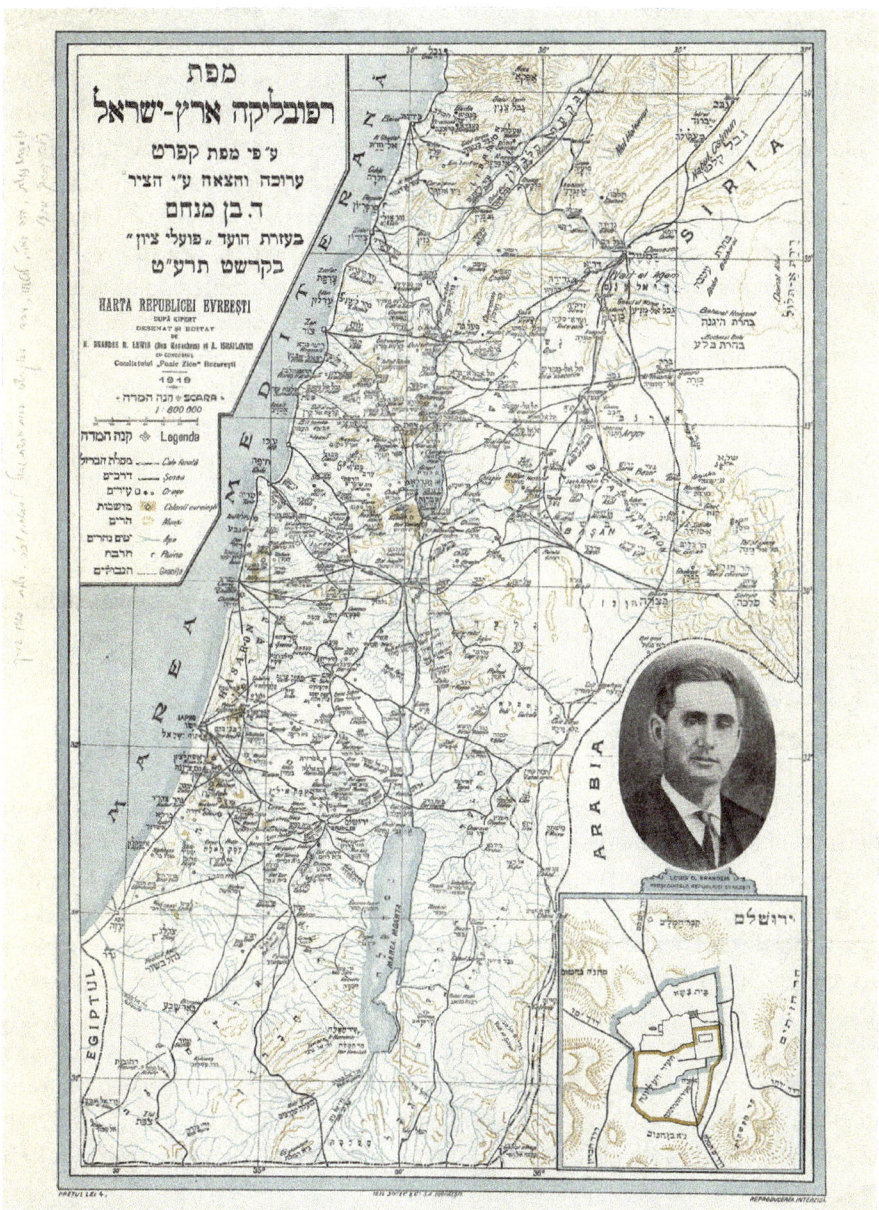

Fig. 134: D. Ben Menachem, "Map of the Land-of-Israel Republic," 413 × 600 mm.

ever, the following year a disagreement erupted between the judge and Chaim Weizmann, the leader of European Zionism. As a result, Brandeis distanced himself from the movement's establishment. In any event, presenting the Jewish luminary's image at this particular time—at the height of his influence as a Zionist activist—carried considerable weight.

This map is also grounded on those of Kiepert, as they are quite similar from a geographical standpoint. The "Brandeis map" ranges from Jubayl in the north to el-Audja (Nitzana) in the south, and from the Mediterranean Sea in the west to significantly beyond Damascus and Hauran in the east. Geographically and cartographically speaking, there are no remarkable innovations herein. A medley of Biblical and Arabic names is provided in both Hebrew and Latin letters. Jewish colonies are marked with a Star of David. The map showcases the Land's contemporaneous transportation routes, but the stretch of *rakevet ha'emek* (the Jezreel Valley Railway) between Haifa and Samakh, (a village on the southern shore of the Sea of Galilee) was inexplicably left out. On the bottom-right corner is a secondary map of Jerusalem. While most of its toponyms are from the Second Temple period, the compiler accurately depicted the roads leading to the city during the first decades of the twentieth century.

All told, the map's primary importance is its documentation of the Zionist rapture immediately following the British conquest of Palestine/Eretz Yisrael and the ensuing Balfour Declaration. These dramatic events raised hopes that the establishment of "the Land-of-Israel republic" was close at hand.

The Jabotinsky-Perlman Atlas

The so-called "Jabotinsky Atlas" is a synthesis between a decidedly Zionist view of the Land of Israel and modern geography with a global reach. As the book's editors, Ze'ev Jabotinsky (the charismatic leader of Revisionist Zionism) and Shmuel Perlman (a Hebrew newspaper and book editor), noted in their introduction, the atlas is designed "to serve Hebrew school pupils."[122] Content-wise, this illuminating book offers colored maps that are complimented by an abundance of written material. Most of the atlas is dedicated to general geography, but it also encompasses a great deal of "Jewish" information: the population of Jews in various cities and countries throughout the world, along with Jewish immigration flows over the centuries (map no. 9); a map of the ancient East (no. 43) and of the history of Eretz Yisrael (no. 44); and the history of Jerusalem, from its inception to the early twentieth century (nos. 47 and 48); *inter alia*.

As expected, the focus of our attention is on the map of Eretz Yisrael, which takes up a double folio (no. 14–15). On the top-lefthand corner are four cross-sections: the first tracks the descent of the Jordan River from the Hula to the Dead Sea; and the other three chart the elevation of the Galilee, Samaria, and Judea.

---

[122] Z. Jabotinsky and Dr. S. Perlman, *Atlas*, 1925, the Hebrew Publishing Company HA-SEFER, Ltd., London [Hebrew].

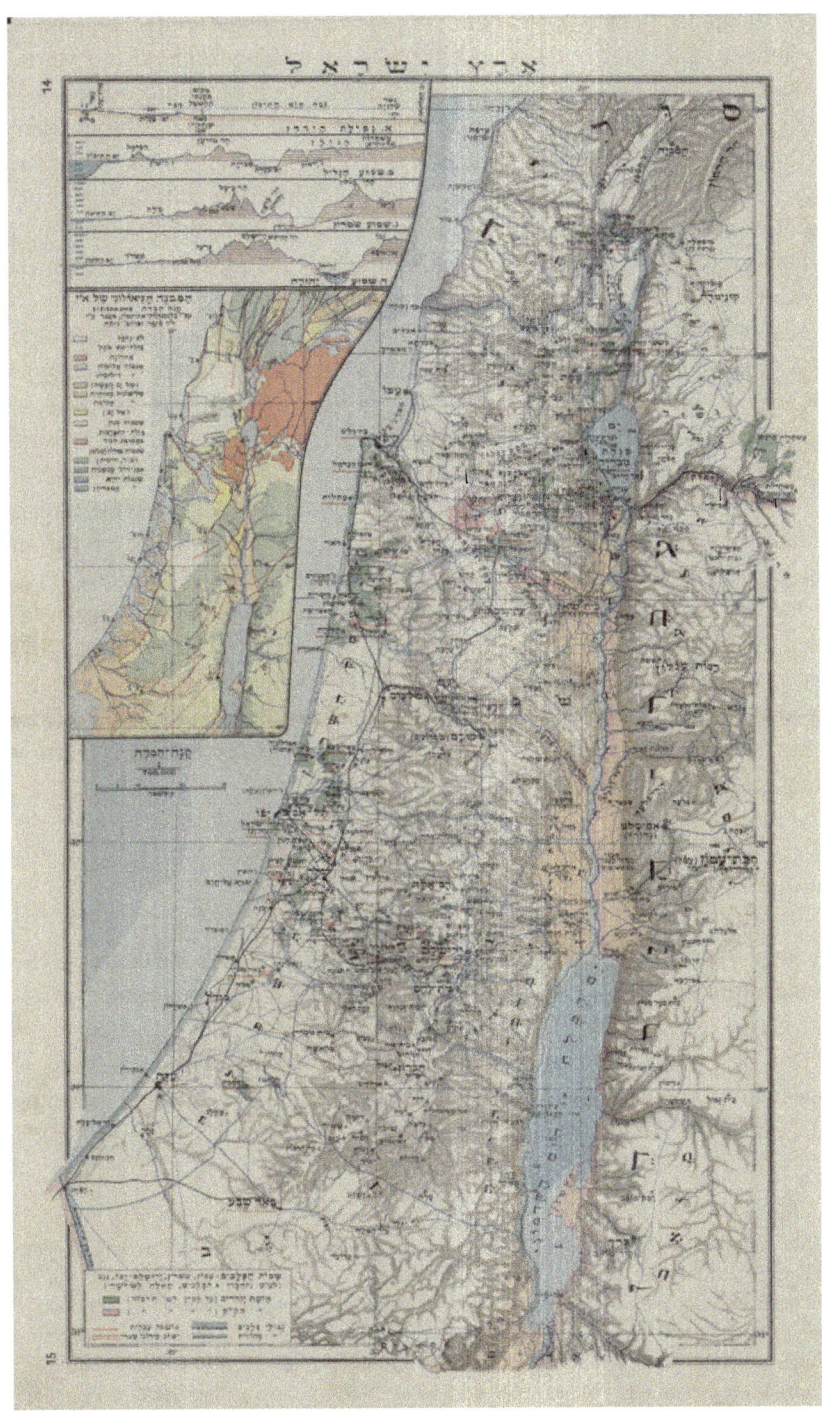

Fig. 135: Jabotinsky and Perlman, "Land of Israel" map, 1925, 240 x 420 mm.

Below these diagrams, there is a geological map that encapsulates Hans Fischer and Hermann Guthe's adaptation of the research conducted by Blanckenhorn and Aaronsohn.[123] The correlation between Guthe and Fischer's enterprise to this atlas is evident in the words of praise for the two in Jabotinsky and Perlman's introduction; the resemblance between the map in question and that of the German scientists; and the fact that the book was brought to press in conjunction with their Leipzig-based publisher, Wagner and E. Debes. Against this backdrop, there is no doubt that the Jabotinsky-Perlman map of the Land of Israel was profoundly influenced by Guthe and Fischer's findings.

The map itself extends from Sarepta (Sarafand) in Lebanon to the Negev, and from the Mediterranean Sea to Amman. Practically all the sites are Biblical toponyms or new Hebrew colonies. The latter are underlined in red, while cities are denoted with a wavy line. There are virtually no Arabic names, as even remote places were given Hebrew names. For example, Malhata and Ar'ara-in-the-Negev were dubbed "The Salt City" and "Aroer", respectively. Similarly, one can find Hebrew names like Adoraim, Shamir, Ein Rimon, and Yatir on Mount Hebron; and Kfar Bir'am and Yir'on were included in the Upper Galilee. Compared to the rest of Eretz Yisrael, there are few settlements in Samaria; among the exceptions are Dothan, Tirzah, Pirathon, and Timnath-serah. Territories that were classified as "The Land of Jews" and "The Land of the [J]NF" (Jewish National Fund) by the summer of 1924 are highlighted in green and red, respectively. In sum, Jabotinsky and Perlman undoubtedly viewed their subject through Zionist eyes, as they tethered Eretz Yisrael's history to a present and future of reviving the ancestral homeland.

Conclusion

In this chapter, I surveyed the Hebrew maps that were drafted over the course of the nineteenth century under the inspiration of the Haskalah, the attendant modernization of Jewish society, and the Zionist movement. The evolution of this subgenre is characterized by a gradual transition to maps that were predicated on the fruits of the European geographic research in all that concerned the Land of Israel, including the development of accurate maps on this topic.

---

[123] Professor Max Blanckenhorn (1861–1947) was a German geologist who is credited with pioneering research on Egypt and the Near East. Aaron Aaronsohn, a native of Zikhron Ya'akov, was among the first botanists in Eretz Yisrael. These two scholars took part in a research delegation on behalf of the Ottoman government. Herman Guthe (1849–1936) was also a leading German researcher of the Land of Israel. Among his numerous publications are a Biblical atlas and a Holy Land map. Hans Fischer (1860–1941) collaborated on the 1924 edition of the Guthe map; Hans Fischer, Palastina, Neue Ausgabe unter Mitwirkung von Prof. Dr. G. Dalman in Jerusalem, 1:400,000, Leipzig, Geograph. Anst. von Wagner, 1924, Hermann Guthe, *Bibelatlas mit einem Verzeichnis der alten und neuen Ortsnamen*, Leipzig, H. Wagner & E. Debes, 1911.

The Jewish Enlightenment prompted a synthesis between traditional Jewish knowledge, such as exegesis and Halakhic interpretation of the Land's borders, and general studies, not least the embracement of modern science. It is evident from the content of these particular Hebrew maps that their compilers availed themselves of non-Jewish sources, geographic and other forms of scientific knowledge, coordinates, modern scales, and European languages. In the maps from the latter stages of the 1800s, there are also references to the launching of the First Aliyah (Jewish colonization) and to signs of the region's fledgling modernization process, such as the Suez Canal and the railroad.

Zionist ideology called upon the Jewish nation "to return to history" by fulfilling its national aspirations via the obtainment of territorial autonomy.[124] With this objective in mind, the movement targeted a concrete geographical territory. From this point forward, Hebrew cartography could no longer make do with schematic maps and abstract artistic renderings, for the preoccupation with a specific place demanded a realistic map. Accordingly, Zionist activists viewed the expanse between the Sea and the wilderness and from Dan to Be'er Sheva as an actual realm. In consequence, it was incumbent upon the movement to cartographically represent this entity in terms of borders and scale, mountains and valleys, seas and rivers. To this end, Zionism indeed embraced the realistic map—a tool that steadily evolved over the course of the 1800s by various European factors that were not beholden to the Zionist cause. By coincidence, the London-based Palestine Exploration Fund (PEF) completed its above-mentioned topographic Survey of Western Palestine (SWP) in 1880, two years before the establishment of the maiden Hebrew colonies. From here on in, then, Zionist cartographers could rely on up-to-date modern maps. However, as champions of German education, the Zionist authors chose to base their maps on Kiepert's work, even though the more precise British maps already stood at their disposal. These developments also mark the stage in which Hebrew cartography ceased making unique contributions to the form and content of maps in general. For this reason, the Jewish-Hebrew map's evolution reaches its logical end on this note.

Be that as it may, the modern map of Eretz Yisrael would henceforth become a central Zionist tool on two different fronts: the movement's educational efforts within the Jewish community; and its public relations and diplomacy *vis-à-vis* the world at large. A fine example of the map's internal usage is its appearance in the Jewish National Fund's advertising campaigns and on the organization's little "blue box" (its iconic charity box).[125] As part of its endeavors to promote Hebrew and, in the

---

[124] The limited purview of this study precludes an in-depth look at the Jewish nation's "return to history." For a penetrating discussion on this concept, see Raz-Krakotzkin, "The Return to the History of Redemption (Or, What is the 'History' to Which the 'Return' in the Phrase 'The Jewish Return to History' Refers?)" [Hebrew].

[125] Bar-Gal, "'On the Wall Hangs a Map:' The Use of of the Jewish National Fund's Maps for Educational and Propaganda Needs" [Hebrew]. Bar-Gal, *Moledet and Geography in the Hundred Years of Zionist Education*, [Hebrew].

days to come, its Israeli image, the Zionist movement similarly placed an emphasis on Hebraicizing toponyms on its maps of the Land.[126] While the various manifestations of this process are beyond the scope of my book, this riveting topic deserves a comprehensive study.

---

[126] Maoz Azaryahu and Arnon Golan, "(Re)naming the Landscape: The Formation of the Hebrew Map of Israel 1949–1960," *Journal of Historical Geography*, Volume 27, Issue 2, April 2001, pp. 178–195.

Conclusion

Over the pages of this book, I unfolded the story of the Hebrew map and Jewish cartography. The corpus under review encompasses many and manifold graphical works that were created under different circumstances by a variety of authors from the second half of the eleventh century until the early 1900s. These maps and diagrams shed light on the manner in which conceptions of Eretz Yisrael—its sites and borders—were presented to the Jewish audience. With this in mind, I surveyed each of the maps on an individual basis, examining their content, classifying them into groups, and elaborating on their contexts and styles. In the process, the corpus' broad array of mapmakers—rabbis, Torah scholars, traditional exegetes, and latterly proponents of the Jewish enlightenment—came into view. Only a small portion of these figures were inhabitants of or paid a visit to the Land of Israel. Put differently, the vast majority never personally laid eyes on the Land's vistas. Whatever the case, every map reflects its author, his generation, and milieu as well as the ideas that he sought to convey. To some extent, the maps also expose the world of their readers.

Succinctly put, Hebrew cartography's evolution can be viewed as a chain of innovations, each of which triggered a change or development that influenced the ensuing maps. Rashi was the first Jew to draft geographic diagrams and maps. In so doing, he introduced a new medium, cartography, to Jewish civilization. By virtue of this step, Jewish exegetes were no longer limited to verbal explanations. Above all, the graphic dimension enabled them to better illustrate the relativity of directions and distances between various sites. Rashi's cartographic template—a schematic diagram *sans* illustrations—would become the traditional Jewish model, enduring into the nineteenth century. As part of his thoughts on Eretz Yisrael, Rashi chose to present its borders as per the portion of *Mas'ei* in the form of a map. He expressly stated that the "commandments dependent on the Land" obligate Jews to ascertain its precise boundaries. In all likelihood, the venerated rabbi included the map of the Land's borders in his exegesis for the sake of underscoring God's territorial promise to Abraham and his progeny. Against the backdrop of Judaism's ideological rivalry with Christianity and Islam, this message was a way of staking a claim to the Holy Land—a goal that is indeed articulated in the midrash with which Rashi opened his commentary on the Hebrew Bible.

The most perplexing question that arises from this study is thus: Why did the Jews refrain from making maps until the eleventh century? Not surprisingly, there are researchers who contend that there were, in fact, previous Jewish-authored maps. Menashe Har-El suggests that the mission chronicled in Joshua 18:4 ("Give out from among you three men for each tribe: and I will send them, and they shall rise, and go through the land, and describe it according to the inheritance of them; and they shall come again to me") should be understood as a survey that was undertaken for the purpose of drafting a map.[1] In addition, some scholars believe that the Book of Jubilees alludes to a lost map that may have been part of the original work.[2] These hypotheses notwithstanding, there are serious doubts as to whether Hebrew maps existed before Rashi's time. Even if they did, none have survived; and in any event, they had no impact on subsequent maps. Be that as it may, we are hard-pressed to explain why earlier Jewish sages did not compose graphical descriptions of the Land's borders for the purpose of elucidating, say, *Baraita de'tankhumin* (a discussion on the borders of the land that was left out of the Mishna), *shalosh artzot le'shevi'it*, and other related Halakhic issues. Even after Rashi broke new ground with his schematic diagrams, centuries would pass before Jews compiled maps describing the Israelites' exodus from Egypt and journey through the wilderness, the tribal allotments, and similar topics that pertain to the field of geography. This absence is all the more baffling given the fact that these sort of themes were prevalent in the Christian cartography since the Byzantine period. Our meticulous research on this corpus aside, we lack the resources to fully answer these questions. All that can be authoritatively said is that Rashi was also a pioneer in the field of cartography, where his influence would last for eight-hundred years.

Other Hebrew mapmakers would indeed follow in Rashi's footsteps. Disseminated in manuscripts, their maps' distribution increased several fold thanks to Elijah Mizrachi, who adapted the famed exegete's schemes. Primarily owing to the technology of printing, Mizrahi's output reached many Jews that studied the Bible using Rashi's commentary. The basic idea that stands at the heart of any map—analyzing the space with the help of graphical tools—became rather commonplace in Jewish society. While the early maps focused on the borders of Eretz Yisrael, the route of the Israelites' peregrinations was added to updated versions of the Mizrahi maps in the form of a simple broken line. In the centuries to come, this theme would be markedly enhanced in artistic Hebrew maps that were copied from Christian sources. Among the many innovations was the signification of the stations along the way.

At the vanguard of artistic, highly illustrated Jewish-authored maps was the mid-sixteenth-century Mantua Map by Joseph ben Jacob of Padua and Isaac ben Samuel Bassan—Italian Jews who propelled Hebrew cartography several steps forward. An analysis of this fetching map's content indicates that the motivation behind its creation was the rivalry with the era's prolific Christian cartographic output, not least the debut of maps covering the Exodus and the Land of Israel in Protestant and, later on, Catholic Bibles. Despite the Mantua Map's beauty, rich illustrations, and exegetical merit, to the best of our knowledge it did not attract a wide following, nor did it reverberate in subsequent maps—Hebrew or otherwise. The fact that there is but one extant copy of this map attests to this state of affairs.

---

1 Menashe Har-El, "Orientation and Mapping in Biblical Land" [Hebrew].
2 Alexander, "Notes on the 'Imago Mundi' of the Book of Jubilees"; Scott, *Geography in Early Judaism and Christianity: The Book of Jubilees.*

Avraham Zaddik, a member of Amsterdam's Jewish community in the 1600s, was the first cartographer to copy Christian van Adrichom's detailed map, translate it into Hebrew, and adapt it for a Jewish audience. Moreover, Zaddik described his entire venture in a text that appears on the map itself. In so doing, he blazed a cartographic path of copying and translating Christian maps for consumption in the Jewish world. The next Jewish mapmaker to embark on this course, Abraham bar Yaacov, also hailed from Amsterdam, but many of his successors towards the end of the eighteenth century and throughout the 1800s were Eastern Europeans. Among this wave were Hebrew cartographers who replicated Christian-authored maps and those that made secondary copies of Jewish adaptations of the former. Whatever the case, all these maps were garnished with rich artistic components.

Towards the end of the 1700s, a major wrinkle—the tribal allotments—was added to the traditional Hebrew map. However, the square contours of the Land's borders, as per Numbers 34, remained a defining element. From as far back as the fifth-century Madaba Map, the tribal allotments were a central topic in Christian Holy Land cartography. Besides the allotments, the eighteenth-century Jewish mapmakers phased in other topics: the rivers of Paradise, far-off lands ("from India and until Cush") that constituted the frontiers of the settled world, and toponyms from the Mishna and Talmud. A unique element of these particular maps is the cities belonging to Ephraim within the boundaries of Manasseh. The common denominator between the cartographic accounts of the Land's borders and the tribal allotments is their hidebound schematic design. In other words, the authors had no pretentions of representing geographical reality, such as the actual location of sites vis-à-vis others or the Land's mountains and waterways. Above all, the maps are devoid of artistry. The conservative leanings of these particular maps sharply diverged from their European contemporaries, which were replete with detailed topographic accounts and illustrations of human figures, animals, cities, towers, ships, raging waves, and monsters (both on the map itself and its margins).

The many immitations of the Rashi map testify to the importance of authority in the Jewish world, the inclination to ride on the coattails of preeminent rabbis, and a deep-seated aversion to changing or adding to their words. A case in point is Elijah Mizrachi and Simeon Halevi Aschaffenburg's attitude to Rashi's maps. Upon incorporating a dotted line signifying the route of the Israelites' peregrinations on Mizrachi's map, Aschaffenburg apologized and referred to the absence of this element on the original as a printer's error and not an oversight by his precursor (Mizrachi). Rashi was indeed beyond reproach. In fact, his erroneous location, in tractate Gittin, of Acre in the northeast continues to be printed in Orthodox publications to this day and age. Even authors who resided in Eretz Yisrael, like Shlomo Adeni, did not dare correct the savant; at the very most, they (e.g., Nahman Natan Koronil) tried to find excuses for this mistake. The power of tradition and the desire to associate oneself with the great scholars of yesteryear persevered into the nineteenth century. The authors of illustrated artistic maps that utterly deviated from the traditional variety nevertheless continued to attribute their cartographic positions to sages past.

For instance, Dov Baer Yozpa and Aharon ben Haim unwarrantedly claimed that their maps were based on those of the *Vilna gaon* and Rashi, respectively.

The traditional map takes an unambiguous stance, which is perhaps relevant to this day, regarding the borders of the Promised Land. In the Bible, these boundaries are spelled out in two disparate accounts: the Covenant of the Pieces promises a land running "from the river of Egypt unto the great river, the River Euphrates" (Genesis 15:18); and the above-cited portion of *Mas'ei* (Numbers 34:3–12) refers only to territory west of the Jordan, from south of Lebanon to the Negev-Sinai border. Biblical researchers have already taken stock of the disparity between these conceptions.[3] At any rate, the traditional maps of Eretz Yisrael, from Rashi on down, points to the Jewish exegetes' unwavering and express preference for the borders according to *Mas'ei*. In this respect, they sharply diverge from the Christian mapping tradition, which unhesitantly included the allotments of Gad, Reuben, and half of Manasseh in Transjordan within the borders of the Promised Land. Put differently, it was the Christian authors, of all people, who adopted more expansive borders.

As we have seen, there were two parallel tracks for creating Hebrew maps. Although the routes operated side by side, there was little in common between them. The traditional model focused on the borders of Eretz Yisrael, which were presented in a simple square template devoid of artistic drawings. Conversely, the second track produced an artistic Hebrew map that drew heavily on maps from the Christian-European cartographic accounts of the Holy Land (*Terra Sancta*) or Promised Land (*Terra Promissionis*). Like their Christian sources, the Hebrew maps included the following elements: the Israelites' journey; Eretz Yisrael was divided into the tribal allotments and encompassed territory on both sides of the Jordan; sites and related events from the Jewish scriptures; mountains, valleys, brooks, and rivers; fortified cities, towers, and tents; and historic scenes. Besides the geographical realm, these maps offer a rich array of images and drawings that comprise a set of symbols. Among the images are Jonah the Prophet and the huge fish; Joshua ben Nun; the personification of Egypt; the symbolization of "the land of milk and honey" in the form of cows and apiaries; and Moses and his brother Aaron. By interspersing a variety of graphical symbols and images, the Christian maps were adapted to the sensibilities of a Jewish readership.

The copying of maps, which was quite accepted in European cartography, was a difficult feat for Jewish authors, as they had to contend with the perpetual tension between the conservative approach to sacred texts and the material that had penetrated the community from outside the Jewish canon and even the Jewish world. More specifically, it was incumbent upon the Jewish copyists to suit the map to the palate of their co-religionists by, among other things, detecting and filtering out all the places, sites, and traditions that refer to the New Testament and other Christian texts.

---

3 Weinfeld, "The Extent of the Promised Land."

In the lengthy preamble to his map, Zaddik declared that steps were taken to ensure that everything in his map conformed to the Torah and "our rabbis of blessed memory." To this end, countless adjustments were made. For example, Mons Christi was converted to "fertile mountain" and Mount Precipice to "High Mountain." Yaakov Auspitz also stressed the importance of editing Christian material: "And every time I saw that the authors of these Roman plates veered away from this and did not go down the path of the exegesis of our Sages of blessed memory, I interpreted and duly inquired and modified [the content] here and there." The difference between the original Christian maps and their Jewish adaptations is evident in the regional breakdown of the Land. Many of the Christian maps provided two layers of division: representing the Biblical period were the tribal allotments; and the Hellenistic and Roman provinces—Judea, Samaria, Galilee, and Transjordan—which existed during the time of Jesus and the Apostles. Needless to say, the Jewish copyists sufficed with the division according to tribe.

The Christian-derived Hebrew maps triggered a second wave of replication, as Jewish authors further modified earlier copies. In all likelihood, they were unaware of the Christian source that was embedded into the Hebrew maps. A case in point is bar Yaacov's fingerprints on the maps of Dov Baer Yozpa. Similarly, the image of Joshua in Aharon ben Haim's map was taken from that of Auspitz, which the latter had borrowed from Lotter. While the initial copyist recognized the need for a comprehensive review of the source map, and thus omitted Christian sites, captions, or ideas on behalf of the Jewish audience, the secondary copyist felt as though he was on solid ground. More specifically, there was no reason to fear the existence of any material that was incommensurate with the Jewish faith. In any event, all the Jewish adapters sought to coat their maps with a veneer of conservatism. For instance, in the preamble to his illustrated map, Yozpa noted that he is predicating himself on the *Vilna gaon*.

The Jewish alteration of Christian sources touches on issues that go beyond the transferred knowledge or the manner in which maps were suited to the Jewish reader's tastes. Even in the absence of clear answers, the following questions demand our attention: In what type of settings were the maps copied? How were Jewish authors, including famous Torah scholars and rabbis, exposed to Christian-authored maps and books? And what were the intellectual circles in which information passed hands between Christian and Jew? In sixteenth-century North Italy and seventeenth century Amsterdam, this sort of cross-pollination is nearly to be expected. However, the existence of patently Christian sources in the work of Polish Jews, like Shlomo of Chelm and Yehonatan ben Yaakov, in the latter half of the eighteenth century—before the Haskalah—is bound to raise eyebrows. What is more, some of the authors hid the fact that their maps are based on Christian material, as this sort of openness was taboo among wide swaths of the Jewish reading public. Along with other documents, these maps perhaps shed light on the tension between openness and insularity within Jewish society, not least its East European branches.

Another group that straddles the fence between the traditional and artistic maps is a handful of small, unique tableaux featuring the sacred places of Eretz Yisrael. On the one hand, these maps are rooted in the Jewish pilgrimage literature and the attendant drawings of holy sites. From the standpoint of provenance and subject matter, then, these tableaux fall under the heading of the traditional maps. On the other hand, they are artistic works whose primary mode of representation is the figurative account. In this respect, they are closer in form to the artistic maps. Over the course of its development, from the first manuscripts in the 1830s to the late nineteenth century, this sub-genre underwent sweeping changes. The earlier tableaux only depict sacred sites along the traditional pilgrimage circuit. These shrines, especially the "tombs of the righteous" in the Galilee, where popular destinations during the late Middle Ages. Towards the end of the 1900s, though, the new settlements that were established by the First Aliyah—what the authors referred to as "the colonies of our Jewish brethren"—were added to the repertoire. As such, the later tableaux were on the same page as the contemporaneous Zionist maps, for both categories present the fledgling steps of the Zionist movement in Eretz Yisrael as well as the general modernization efforts throughout the area. For example, the Avigdor Malkov and Yitskhok Linetski maps denote the First Aliyah's maiden colonies using the symbol of a modern tile-roofed house. Likewise, Jacob Goldzweig signifies the railroad with a thick bold line.

The Jewish Enlightenment was accompanied by a spike of interest in geography, as Maskilim delved into the layout of Eretz Yisrael and its past. According to our findings, Hebrew geographic research dates back somewhat further than is customarily thought. By the *early nineteenth century*, champions of the Haskalah were apparently familiar with geography as a field of scientific enterprise, along with seventeenth- and eighteenth-century works on the historical geography of the Holy Land and its settlements. This literature underpinned Solomon Levisohn, Menachem Mendel Breslauer, and Jacob Kaplan's books. However, some of these authors endeavored to understate the influence of these Christian sources on their work, for reliance on "the sages of the nations" was at times the object of critique and derision. In any event, as the European Holy Land research made strides Jewish mapmakers increasingly availed themselves of the fruits of its labor. Modern geography's contribution to Jewish maps in the second half of the nineteenth century is evident. For instance, quite a few of these maps were multilingual and some were primarily in non-Hebrew tongues. At this stage, the tensions between tradition and innovation were once again pronounced and assumed new forms. The maps' representation of the Land was novel and modern, but a few of the established topics remained current.

Holy Land cartography reached its zenith with the publication of Heinrich Kiepert's maps and the highly accurate survey of the Palestine Exploration Fund (PEF). Kiepert's output served as the basis for the maps that were compiled by early Zionist proponents of cutting-edge geography. The Jewish movement used these maps to promote the land that it strove to repopulate. In this fashion, the Haskalah proceeded in lockstep with the European Enlightenment, as both advanced the research on and cartographic display of the Land of Israel.

The transition from solely covering Biblical sites and the tombs of the righteous to an emphasis on the new colonies reflect a marked shift in Jewish conceptions of the Holy Land. While on the topic, it also bears noting that the maiden settlements were presented in a positive light by, above all, symbols like modern tiled-roof buildings and Stars of David. This constituted a far-reaching change in both the perceived hierarchy of sanctified places and the Jewish agenda for Eretz Yisrael. Hebrew authors, including mapmakers, expressed this transformation in one of two ways: some presented the old alongside the new; and others preferred the Zionist efforts at renewal over the worship of graves and ruins, the modern over the old, and the settlement enterprise over pilgrimage, which they viewed as a passive undertaking. Whereas the new colonies symbolized hope and the future, the holy places stood for the past. The pilgrimage circuit was still important to Jews that espoused a traditional worldview, but was rejected or soft-pedalled by those with an inclination for mainstream Zionist ideology.[4] This approach manifests the itinerary of Mordechai Ben-Hillel Hacohen (a Zionist activist from Czarist Russia) for a tour through Eretz Yisrael: "From Jaffa to Rishon LeZion, and from there—to Gedera, Nahalat Reuben or Wadi Chanin [present-day Ness Ziona], and Be'er Tuvia or Kastina, and then back to Gedera; and from there to Ekron, and from Ekron by way of Rishon LeZion to Petah Tikva and Yehud. And then off to Jerusalem."[5] In the end, though, Hacohen also stressed his powerful attraction to Jerusalem and writes of the emotions that swelled up inside him during a visit to the ancient city.[6]

Much of the literature on maps concentrates on the mapmakers, while often ignoring their readers. Scholars are often hard-pressed to identify not only the creators, but the target audience and actual users. That said, a fair share of the maps under review clearly refer to the prospective audience. The traditional maps were meant to be used as a learning aid for Torah study. Consequently, they were integrated within Rashi-annotated volumes of the Pentateuch, Responsa collections, and commentaries on the following topics: Rashi's work, the Mishna, and the Talmud, *inter alia*. In all these cases, the author was reaching out to people who studied Torah, from young boys getting their feet wet with Rashi as their basic primer for the Pentateuch to seasoned Talmudists immersing themselves in complex topics, like the sabbatical year and writs of divorce. Aharon ben Haim, among others, averred that a graphic description would enhance the reader's comprehension of *Mas'ei*. Maps that were printed in Passover Haggadahs between the late seventeenth and nineteenth century naturally depict Egypt and the Land of Israel within the context of the Exodus. Needless to say, these maps give voice to the Seder rite's overarching theme—salvation.[7]

---

4 Bartal, "On Precedence" [Hebrew].
5 Hacohen, *The Travels of Mordecai Ben-Hillel Hacohen, 1899*, p. 654 [Hebrew].
6 Ibid, pp. 684–689.
7 The same can be said for drawings of Jerusalem and the Jewish Temple in Haggadahs.

Accordingly, they were intended to create the sense that the redemption is near, that readers will soon be able to leave their own "Egypt," travel to Eretz Yisrael, and celebrate Passover in the Temple. The above-mentioned tableaux were probably "marketed" as decorations for the home or sukkah. From a design standpoint, they are reminiscent of the Shiviti and Mizrah styles that graced houses and synagogues. The Maskilic maps by and large continued to fill Hebrew cartography's traditional role as a facilitator of study and exegesis, albeit with the help of modern scientific tools. Last but not least are the maps of Eretz Yisrael that were designed to promote Zionist ideals, especially the Jewish settlement of the Land.

Hebrew cartography's evolution from schematic diagram to modern map reflects the shifts in the way that Jews perceived the geographic expanse over the generations. Throughout the centuries of exile, the Land of Israel was an imagined realm. Though evoked on a daily basis in prayer, the Grace after the Meal, and ceremonies, Eretz Yisrael was presented in an abstract manner. So long as this was the prevailing attitude, the Promised Land could be grasped in a completely unrealistic geographical manner. For much of Jewish history, the descriptions of this realm entailed no maps whatsoever. In the eleventh century, Rashi introduced maps depicting the Land in a square frame, whose four sides consisted of the Mediterranean Sea, the River of Egypt, the Jordan River, and a line connecting Mount Hor, the entrance of Hamath, Zedad, and Ziphron (as noted, the locations of the last four sites have long been unknown). This paradigm, which is essentially an account of the Land's Biblical-Halakhic boundaries, was entirely sufficient as long as there was no functional need for these borders or for knowledge of the sites in between.

This theoretical and abstract outlook underwent a three-stage transformation over the course of the Hebrew map's development. First, due to the competition from Christian Holy Land maps, Jewish compilers adopted their rivals' concrete geographic approach whereby they portrayed the Land's vistas—its mountains and valleys, rivers and brooks, cities and villages. Second, Maskilim embraced the increasingly more scientific and realistic map. The disparities between Yehoseph Schwarz's first map, which he drafted while still in Germany, and the maps that accompanied the non-Hebrew editions of his *Tvu'ot ha'areṣ*, which were compiled in the Land of Israel and followed Robinson's-Kiepert's map, epitomizes the reification of the geographical expanse and the Jewish Enlightenment's acceptance of modern geographic and cartographic tools. Under the inspiration of Edward Robinson's book and Kiepert's maps, Jewish authors internalized the values of modern cartography.

The third and final phase of this transition was Zionism's enlistment of the accurate scientific map as a tool for education, enlightenment, and propaganda. Given its objective of returning the Jewish people to their "ancestral" home, it behooved the movement to refer to a palpable territory—a goal that was "right up the alley" of modern, precise, and tangible cartography. This approach is exemplified by the map in Sokolow's *Ereṣ ḥemdah* and, all the more so, "the Map of the Land-of-Israel Republic," which presents a clear political vision as well.

In summation, the Jewish attitude towards the Holy Land went through a metamorphosis between the eleventh and early twentieth centuries. From the object of prayers and dreams, Eretz Yisrael became a true destination for Jewish immigration—an expanse to be built and settled. In parallel, the Hebrew map became more realistic, as Zionist authors embraced the fundaments of modern cartography: accurate shape and scale, color, and graphic symbols. The Land of Israel was portrayed as an old-new realm in which Biblical toponyms were integrated with the new, late nineteenth-century settlements. It is at this juncture—when dream and reality meet, when the Land's theoretical-cum-abstract image is amalgamated with its modern scientific representation—that our voyage comes to a close.

Cartographic Material

Aaron ben Haim, "Map of the Borders of the Good Land," Warsaw: A. Bomberg Lithography, 1883, in idem, *Sefer moreh derekh*, Warsaw: Izaak Goldman Press, 1883 [Hebrew].

Christiaan van Adrichom, "Situs Terra Promissionis SS intelligentiam exacte aperiens," in idem, *Theatrum Terrae Sanctae*, Köln: Officina Birckmannica, 1590.

Aryeh Leib ben Isaac, "An Explanation of LC [the Land of Canaan] for All its Borders that are in the Bible," in Joshua Feivel ben Israel Teomim of Tarnogród, *Sefer kiṣvai areṣ*, Grodno: 1813 [Hebrew].

Yaakov Auspitz, "These are the Israelites' Journeys," in idem, *Be'er ha'luḥot*, Ofen: 1817 [Hebrew].

Yaakov Auspitz, "Map of the Israelites' Encampment," in idem, *Be'er ha'luḥot* [Hebrew].

Yaakov Auspitz, "Map of the Holy Land," in idem, *Be'er ha'luḥot* [Hebrew].

Avraham bar Jacob, "Map of the Holy Land," in *Passover Haggadah*, Amsterdam: Asher Anshel ben Eliezer and Issachar Ber ben Abraham Eliezer, 1695 [Hebrew].

D. Ben Menachem, "Map of the Land-of-Israel Republic," Bucharest: 1919.

Binder, M., "Charte worauf das iuedische Land vornehmlich, wie es zur Christi und der Apostel Zeiten gewesen ist, vorgestellet wird, Nach dem Entwurff des Herrn W.A. Bachiene," Norib, Erlang, 1772.

Willem Janszoon Blaeu, "Terra Sancta quae in sacris Terra Promissionis olim Palestina," Amsteldami, 1629.

Augustin Antoine Calmet, "Carte du Voïage des Israëlites dans le desert, depuis leur Sortie de l'Egipte (sic!) jusqu'au passage du Jourdain, Dressee par l'Auteur du Commentaire de l'Exode, et executée par Lièbaux Geografe," in idem, *Commentaire littéral sur tous les livres de l'Ancien et du Nouveau Testament*, vol. 1, pt. I, Paris: Emery, Saugrain, Pierre Martin, 1724–26.

Augustin Antoine Calmet, "Plan et Distribution de la Terre de Chanaan, Suivant la Vision d'Esechiel Chap. XLVIII Laquelle ne fut jamais executee a la lettre, depuis le retour de la Captivite," in idem, *Commentaire litteral sur tous les livres de l'Ancien et du Nouveau Testament*, Paris: Emery, T. VI., 1726, opposite p. 607.

Augustin Antoine Calmet, "A Geographical Map of the Old World according to the Division of it among Noah's Sons after the Dispersion of Them which Happened at Babel," in idem, *An Historical, Critical, Geographical [. . .] Dictionary of the Holy Bible*, vol. 2 (of 3) London: J. J. and P. Knapton, 1732.

Augustin Antoine Calmet, "Tabula Itineris, et Stationum Israelitarum in Deserto, delineata et incisa juxta Auctoris systema a P. Starck-man, Augustin Antoine Calmet," in idem, *Dictionarium historicum [...] Sacrae Scripturae*, vol. I, Augsburg: Martin Veith, 1738, p. 269.

Augustin Antoine Calmet, "Tabula Itineris, et Stationum Israelitarum in Deserto, delineata et incisa juxta Auctoris systema a P. Starck-man," in idem, *Dictionarium historicum...Sacrae Scripturae*, vol. I, Venezia: Sebastian Coleti, 1757, p. 235.

Claude Reignier Conder and Horatio Herbert Kitchener, "Map of Western Palestine in 26 sheets 1:63, 360," in *Survey of Western Palestine*, London: 1880.

Miles Coverdale, "The description of the lond of promes, called Palestina or the Holy lond", from the "Coverdale Bible", 1535.

Jean-Baptiste Bourguignon d'Anville, "Plan de la Ville de Jerusalem Ancienne et Moderne, Par le S.ʳ d'Anville, Ecrit par G. Delahaye, Jean Baptiste Bourguignon d'Anville," in idem, *Dissertation sur l'étendue de l'ancienne Jérusalem et de son temple, et sur les mesures hébraïques de longueur*, Paris: 1747.

Hans Fischer, "Palastina, Neue Ausgabe unter Mitwirkung von Prof. Dr. G. Dalman in Jerusalem," 1:400,000, Leipzig, Geograph. Anst. von Wagner, 1924, Hermann Guthe, Bibelatlas mit einem Verzeichnis der alten und neuen Ortsnamen, Leipzig: H. Wagner & E. Debes, 1911.

Juda Funkenstein, "Land of Canaan or the Promised Land," London: 1873 or 1874.

Meir Yekutiel Galgor, *"Mappa gevulot ha-areṣ, Aid for the Study of Holy Scripture,"* Vilna: Lithography of N. Matz, 1882.

Ephraim Michael Grover, "The Well-Researched and Precisioned Map of Eretz Yisrael," Odessa: 1899 [Hebrew].

Christoffer Heidmann, "Totius Terrae Sanctae Delineatio," Accurate recisa a Johanne a Felden. C. Buno fecit, in idem, *Palestina, sive, Terra Sancta: paucis capitibus distinctè ordineq[ue] explicata et / ab ipso autore olim in illustria Academia Iulia cum locorum indice publice iuris facta. Iam verò ob exemplarium raritatem unà cum quatuor tabulis ex Adrichomio et aliis collectis in studiosae iuventutis usum denuò edita*, Wolferbyti: Sumptibus Conradi Bunonis typis Iohannis Bismarci, 1655.

Johann Baptist Homann, *Iudaea seu Palaestina…hodie dicta Terra Sancta prout olim in Duodecim Tribus divisa separatis ab invicem Regnis Iuda et Israel collecta ex Tabulis Guil. Sansonij… a Joh. Baptista Homanno*, Norimbergae, 1707.

Heinrich Kiepert, "Map of Palestine, Chiefly from the Itineraries and Measurements of E. Robinson and E. Smith," in Edward Robinson and Eli Smith, *Biblical Research in Palestine, Mount Sinai and Arabia Petrea in 1838*, London: John Murray, 1841.

Kircher, Athanasius, "Tabula geographica divisionia gentium et populorum per tres filios Noe, Sem, Cham, Japhet," Amsterdam: Joannes Janssonius van Waesbergen, 1675.

Eliyahu ben Elazar Landa, "Map of Eretz Yisrael," Jerusalem: 1912.

Eliyahu ben Elazar Landa, "Map of Eretz Yisrael," Jerusalem: 1915.

Guy Michel Le Jay, "Sedes Filiorum Nohha vt legunt hebr. Genes. 10," Paris: 1629.

Jacob van Liesvelt, "Die ghelegentheit ende die palett des lants van Beloften," from a Dutch Bible, Antwerp: 1526.

Tobias Conrad Lotter, "Terra Sancta sive Palaestina," opera et studio Tobiae Conradi Lotter. Matth. Albrecht Lotter sculps. Aug. Vindel [Augsburg], 1759.

M. Löwi, "Palæstina nach der Vertheilung unter die 12 Stämme," ca. 1870.

Shabtai Matsḳeyiṭsh, "Map of the Holy Land with All the Places in its Vicinity, Accompanying the Book *Gvulei Ṣion*," Vilna: N. Matz, 1898 [Hebrew and Russian].

Gerard Mercator and Jodocus Hondius, "Situs Terra Promissionis S.S. Bibliorum intelligentiam exacte aperiens," in Gerard Mercator, *Atlas ou representation du monde universel et des parties d'icelui, faicte en tables et descriptions*, Amsterdam: Henricus Hondius, 1633.

Bendecitus Arias Montanus, "Tabula Terrae Canaan Abrahmae tempore et ante adventum filior. Israel cum vicinis et finitimis regionib," in *Biblia sacra Hebraice, Chaldaice, Graece, & Latine*, 8: *Liber Chanaan*, Antverpiae: Excudebat Christophorus Plantinus, 1572.

Joseph Moxon, "A Map of All the Earth and How After the Flood it was Divided among the Sons of Noah," London: 1671.

Salomon Munk, "The Land of Israel and its Division between the Twelve Tribes," in idem, *Palestine: Description géographique, historique, et archéologique*, tab. 69, Paris: 1845.

Salomon Munk, "The Route of the Hebrews' Journey upon Traversing the Desert," in idem, *Palestine: Description géographique, historique, et archéologique*, facing p. 122.

Sebastian Münster, "Signum C, Map no.7, Terra Sancta XVI nova tabula," in idem (ed.), *Claudius Ptolemaeus, Geographia*, Basel: Henricus Petri, 1542, map no. 44, Basel: Henricus Petri, 1542.

Abraham Ortelius, "Palestinae sive totivs Terrae Promissionis nova descriptio avctore Tilemanno Stella Sigenens," in idem, *Theatrum orbis terrarum*, Antwerp: A. Coppen van Diest, 1570–1571.

Abraham Ortelius, "Terra Sancta, A Petro Laicstain perlustrata, et ab eius ore et schedis à Christiano Schrot in tabulam redacta," Antwerp: Christopher Plantin, 1584.

Petrus Plancius, "Orbis terrarum typus de integro multis in locis emendates," Amsterdam, ca. 1594.

Richard Pococke, "A Plan of Jerusalem and the Adjacent Country," in T. Jefferys Sculp, *A Description of the East and Some Other Countries*, vol. 2, part 1, *Observations on Palestine, or the Holy Land, Syria, Mesopotamia, Cyprus, and Candia*, London: 1743–1745.

Claudius Ptolemaeus, "Tabula Moderna Terre Sancte," in idem, *Geographia*, Strassburg:
Johannes Schott, 1520.

Franciscus Quaresmius, "Imago transitvs filiorvm Israel per Mare Rvbrvm," in idem, *Historica theologica et moralis Terrae Sanctae elvcidatio*, vol. II, Antverpiae: Ex Officina Plantiniana Balthasaris Moreti, 1639, p. 973.

Nicolas Sanson, *Geographiae Sacrae ex Veteri et Novo Testamento desumptae, Tabula Secunda in qua Terra Promissa, sive Iudaea in suas Tribus partesq, Autore N. Sanson Abbavillaeo*, Paris: Pierre Mariette, 1662.

Eliyahu Sapir and Ephraim Krauze, "Map of Eretz Yisrael (Palestine), as per the Kiepert Map," Berlin: the Zionist Committee, 1913 [Hebrew].

Eliyahu Sapir and Ephraim Krauze, "The Land of Israel, in Accordance with the Hebrew Map of E. Sapir-E. Krauze," New York: the Zionist General Council Press, 1918 [Hebrew].

Yehospeh (Joseph) Schwarz, "Boundary of Palestine according to Numbers XXXIV 1–15, by Rabbi Joseph Schwarz, of Jerusalem," 1847, T. Sinclair Lithography, in idem, *A Descriptive Geography and Brief Historical Sketch of Palestine*, trans. from Hebrew, Isaac Leeser, Philadelphia: A. Hart, 1850.

Yehospeh (Joseph) Schwarz, "Map of Palestine, by J. Schwarz of Jerusalem 5607," 1847, T. Sinclair Lithography, in idem, *A Descriptive Geography and Brief Historical Sketch of Palestine*, 1850.

Matthaeus Seutter, "Palaestinae accurata descriptio geographica, ita adornata ut diversarum aetatum regna conditio et fata, in sacris oraculis indicata intelligi possint," in idem, *Grosser Atlas*, Augsburg: 1741.

Matthaeus Seutter, "Palaestina seu Terra a Mose et Iosua occupata et inter Iudaeos distributa per XII Tribus vulgo Sancta adpellata," a Ioanne Christophoro Hernberg [...] sculpsit Matth. Seutter, idem, *Atlas novus [...]*, Augsburg, ca. 1745.

Nahum Sokolow, "Map of the Holy Land[,] 'the Precious Land' by N. Sokolow," Berlin: Dietrich Reimer, 1885 [Hebrew].

Joshua Feivel ben Israel (Teomim) of Tarnogród, "Map of Eretz Yisrael," in idem, *Sefer kiṣvai areṣ*, Zholkva: Uri Feivish Segel, 1772 [Hebrew].

Joshua Feivel ben Israel (Teomim) of Tarnogród, "Map of the Tribal Allotments as per the Book of Ezekiel," in idem, *Sefer kiṣvai areṣ* [Hebrew].

Carel Willem Meredith van de Velde, "Map of the Holy Land Constructed by C. W. M. van de Velde from His Own Surveys in 1851 and 1852, from those Made in 1841 by Majors Robe and Rochfort Scott, Lieut. Symonds [...] and from the Results of the Researches Made by Lynch, Robinson, Wilson, Burckhardt, Seetzen," Engraved by Eberhardt and by Stichardt, Gotha: 1858.

Visscher family, "Die Gelegentheyt van t'Paradys ende t'Landt Canaan Mitsgaders de eerstbewoonde Landen der Patriarchen, door C.I. Visscher, Abraham van den Broeck fecit, C.I. Visscher Excu," in *Biblia*, map no. 1, 'S Graven-Hage 1645.

Visscher family, "Peregrinatie ofte Veertich-iarige Reyse, Der Kinderen Israels uyt Egipten [...] Door Claes Ianss Visscher," Hague: 1645.

Visscher family, "Tabula Geographica in qua Israelitarum, ab Aegypto ad Kenahanaeam usqve profectiones en rystplaetsen der Kindren Israels, from a Dutch Bible," Amsterdam: 1650.

Visscher family, "Die Gegend des iridischen Paradieses und des Landes Canaan [...] J. Sandrart sculpsit," *Biblia, Das ist: Die gantze Heilige Schrifft Altes und Neues Testaments*, [...] verteutscht von [...] Martin Luther, Nürnberg: In Verlegung Johann Andreä Endters seel, Erben, 1736.

Charles Wilson, *Ordnance Survey of Jerusalem, 1:2500*, Southampton: Ordnance Survey Office, 1864–1865.

Dov Baer ben Joseph Yozpa, "Map of the Holy Land," in *Bible, with Commentary*, Grodno and Vilnius: Simha ben Menahem Zymel, 1820 [Hebrew].

Dov Baer ben Joseph Yozpa "Map of the Holy Land," Warsaw: Peretz Feinroit Lithography [Hebrew].

Dov Baer ben Joseph Yozpa, and Moses Danzigerkron, "A List of the Surrounding Borders of Eretz Israel," Warsaw: Danzigerkron Lithography, 1881 [Hebrew].

Yaakov ben Avraham Zaddik (Justo), "A Drawing of the Situation of the Lands of Canaan," Amsterdam 1620.

Jacob Ziegler, "Septima tabula continet Tribum Iuda rursus, Idumaeam, Philistim, Extar Iordanem, Gentem Moab, Partem Aegypti, Arabiam petraeam, siue itinera Hebraeorum per desertum," in idem, *Terrae Sanctae . . . descriptio*, Strassburg: Petrus Opilius, 1532.

Bibliography

Adrichom, Christian van, *Iervsalem, et suburbia eius, sicut tempore Christi floruit [...] descripta per Christianum Adrichom Delphum*, Coloniae Agrippinae: Anno Christi, 584.
Adrichom, Christian van, *Theatrum Terrae Sanctae et Biblicarum historiarum cum tabulis georgraphicis aere expressis*, Köln: Officina Birckmannica, 1590.
Adrichom, Christian van, *Situs Terrae Promissionis SS Bibliorum inteliigentiam exacte aperiens*, a map in: *Theatrum Terrae Sanctae*.
Aiken, Edwin J., *Scriptural Geography: Portraying the Holy Land*, London: I. B. Tauris, 2010.
Albu, Emily, "Imperial Geography and the Medieval Peutinger Map," *Imago Mundi* 57/2 (2005): 136–148.
Alexander, Philip S., "Notes on the 'Imago Mundi' of the Book of Jubilees," *Journal of Jewish Studies* 33/1–2 (1982): 197–213.
Alexander Ziskind ben Moshe, *Yesod ve'shoresh ha'avodah*, Nowy Dwor: J. A. Krieger Printing House, 1782 [Hebrew].
Alfassi, Itzhak, "Chelm, Solomon ben Moses," *Encyclopaedia Judaica*, 2nd ed., vol. 4, Detroit: Macmillan Reference, 2007, p. 589.
Altschuler, Yehiel Hillel, *Sefer binyan ha'bayit*, Zholkva: 1775 [Hebrew].
Altschuler, Yehiel Hillel, *Sefer binyan ha'bayit*, Livorno 1782 [Hebrew].
Altschuler, Yehiel Hillel, *Bible: Book of Joshua, with Two Commentaries [....:] Miṣudat David [... and] Miṣudat Ṣion*, Livorno: E. Y. Kashtilo and A. Saadon Press, 1780–1782 [Hebrew].
Amram, David Werner, *The Makers of the Hebrew Books in Italy*, Philadelphia: J. H. Greenstone, 1909.
Aschaffenburg, Simeon Halevi, *Sefer devek tov*, Venice: Joan Di Gara Press, 1588 [Hebrew].
Auspitz, Yaakov, *Be'er ha'luḥot*, Ofen: 1817 [Hebrew].
Auspitz, Yaakov, *Be'er ha'luḥot*, Vienna: G. Halzingar Press, 1818 [Hebrew].
Avi-Yonah, Michael, *The Madaba Mosaic Map,* Jerusalem: Israel Exploration Society, 1954.
Avissar, Oded, *The Book of Tiberias: City, Region and its Settlements over the Generations*, Jerusalem: Keter, 1973 [Hebrew].
Azaryahu, Maoz and Golan, Arnon, "(Re)naming the Landscape: The Formation of the Hebrew Map of Israel 1949–1960," *Journal of Historical Geography*, Volume 27, Issue 2, April 2001, pp. 178–195
Babylonian Talmud, Oz V'Hadar, Gittin, Jerusalem 2006 [Hebrew].
Babylonian Talmud, Steinsaltz edition, Gittin, Jerusalem 2001 [Hebrew].
Babylonian Talmud with the Commentaries of Rashi, Tosafot and Piskai Tosafot and Rabeinu Asher and Piskei ha'Rosh and the Commentary on the Mishna of Maimonides, vol. VI, tractate Gittin, Amsterdam: Proops Press, 1752–1765 [Hebrew].
Bar-Gal, Yoram, "'On the Wall Hangs a Map:' The Use of the Jewish National Fund's Maps for Educational and Propoganda Needs," idem, Nurit Kliot, and Ammatzia Peled (eds.), *Erets Israel Studies—Aviel Ron Book*, Haifa: Haifa University, 2004, pp. 247–259 [Hebrew].
Bar-Gal, Yoram, *Moledet and Geography in the Hundred Years of Zionist Education*, Tel Aviv: Am Oved, 1992 [Hebrew].
Barber, Peter, "Visual Encyclopaedias: The Hereford and Other Mappae Mundi," *The Map Collector* 48 (1989): 3–8.
Barber, Peter, (ed.), *The Map Book*, New York: Walker, 2005.
Bardach, Judah, *Mazkir le'vnai reshef: [. . .] a List of the Collections and Manuscripts [. . .] from the Estate of [. . .] R. Simhah Pinsker [. . .]*, Vienaa: Y. Shlossberg, 1868 [Hebrew].
Barnai, Jacob, *Historiography and Nationalism: Trends in the Research of Palestine and its Jewish Yishuv (634–1881)*, Jerusalem: Magnes Press, 1995 [Hebrew].

Bartal, Israel, "To Forget and Remember: The Land of Israel in the Eastern European Haskalah Movement," in Aviezer Ravitsky (ed.), *The Land of Israel in Modern Jewish Thought*, Jerusalem: Yad Ben-Zvi, 1998, pp. 413–423 [Hebrew].

Bartal, Israel, "On Periodization, Mysticism, and Enlightenment—the Case of Moses Hayyim Luzzatto," *Simon Dubnow Institute Yearbook* 6 (2007): 201–214.

Bartal, Israel, "On Precedence: Time and Place in the First Aliya," in Yaffa Berlovitz (ed.), *Talking Culture: The First Aliya, an Interperiod Discourse*, Tel Aviv: Hakibbutz Hameuchad, 2010, pp. 15–24 [Hebrew].

Baruchson, Zipora, "On the Trade in Hebrew Books between Italy and the Ottoman Empire during the XVIth Century," *East and Maghreb* 5 (1986): 53–77 [Hebrew].

Bassan, Isaac ben R. Samuel, *Passover Haggadah with Illustrated Miracles and Wonders*, Mantua: Rufinalli, 1560 [Hebrew].

Beit Arie, Malachi, *Catalogue of the Hebrew Manuscripts in the Bodleian Library, Supplement*, Oxford: Clarendon Press, 1994.

Ben Aderet, Shlomo, and Koronil, Nahman Natan, *Piskei ḥallah*, Jerusalem: Elijah and Moshe Chai Sasson, 1876 [Hebrew].

Ben-Arieh, Yehoshua, "The First Surveyed Maps of Jerusalem," *Eretz-Israel*, vol. XI, Dunayevsky Memorial Volume, Jerusaslem: Israel Exploration Society, 1973, pp. 64–74 [Hebrew].

Ben-Arieh, Yehoshua, "The Catherwood Map of Jerusalem," *The Quarterly Journal of the Library of Congress* 31/3 (1974): 150–160.

Ben-Arieh, Yehoshua, *A City Reflected in its Times: New Jerusalem—the Beginnings*, Jerusalem: Yad Ben-Zvi, 1979 [Hebrew].

Ben-Arieh, Yehoshua, "The Character of Hebrew Geographic Literature about Eretz-Israel during the Nineteenth Century and until World War I," *Eretz-Israel*, vol. 22, David Amiran Volume, Jerusalem: the Israel Exploration Society, 1991 [Hebrew].

Ben-Arieh, Yehoshua, "Developments in the Study of *Yediat Ha'aretz* in Modern Times, up to the Establishment of the State of Israel," *Cathedra* 100 (August 2001): 305–338 [Hebrew].

Ben Bezalel, Haim, *Be'er mayim ḥayim: Explanations on Rashi's C[ommentary] on the Torah*, London: 1969 [Hebrew].

Ben Bezalel, Haim, *Be'er mayim ḥayim*, London 1993 [Hebrew].

Ben Bezalel, Haim, *Be'er mayim ḥayim*, Brooklyn 2011 [Hebrew].

Ben-Dor, Yehuda, "Leiner, Gershon Hanokh (Henikh) ben Jacob," *Encyclopaedia Judaica*, 2nd ed., vol. 12, 2007, p. 627.

Ben-Eli, Arie L. and Stieglitz, R. Raphael, *The National Maritime Museum, Haifa* London: 1972.

Ben Eliyhau, Eyal, *National Identity and Territory: The Borders of the Land of Israel in the Consciousness of the People of the Second Temple and the Roman-Byzantine Periods*, PhD thesis, Jerusalem: The Hebrew University, 2007 [Hebrew].

Ben-Gad Hacohen, David, "The Southern Boundary of the Land of Israel in the Tannitic Literature and the Bible," *Cathedra* 88 (1988): 15–38 [Hebrew].

Meir Ben Gedalia (MaHaRaM of Lublin), *Sefer meir einai ḥakhamim*, Venice: Bragadin Press, 1619 [Hebrew].

Meir Ben Gedalia (MaHaRaM of Lublin), *Sefer meir einai ḥakhamim*, Frankfurt: N. Weimann Press, 1709 [Hebrew].

Meir Ben Gedalia (MaHaRaM of Lublin), *Sefer meir einai ḥakhamim*, Fürth: R. Avraham Beng and Boneft Shene'ur Press, 1724 [Hebrew].

Meir Ben Gedalia (MaHaRaM of Lublin), *Sefer meir einai ḥakhamim*, Wilhermers dorf: Hirsch ben Haim of Fürth Press, 1737 [Hebrew].

Ben Haim, Aharon, *Sefer moreh derekh*, Grodno: Yaakov Yechezkel and Yechezkel Press, 1836 [Hebrew].

Ben Haim, Aharon, *Sefer moreh derekh*, Warsaw: Izaak Goldman Press, 1879 [Hebrew].

Ben Haim, Aharon, *Sefer moreh derekh*, Warsaw: Izaak Goldman Press, 1883 [Hebrew].
Ben Joseph Kass, Avraham, *Sefer toldot avraham*, Poland: 1749 [Hebrew].
Ben Meir (RaShBaM), Samuel, *Commentary on Exodus* [Hebrew].
Ben Yehudah, Eliezer, *Sefer ereṣ yisrael*, Jerusalem: Yoel Moshe Salomon Press, 1883 [Hebrew].
Ben Yehudah, Eliezer, *A Complete Dictionary*, Tel-Aviv: 1949 [Hebrew].
Benayahu, Meir, "Devotion Practices of the Kabbalists of Safed in Meron," *Sefunot* 6 (1962): 9–40 [Hebrew].
Benjamin, Israel Joseph, *Cinq années de voyage en Orient, 1846–1851*, Paris: 1856.
Benjamin, Israel Joseph, *Acht Jahre in Asien und Afrika: von 1846 bis 1855*, Hannover: 1858.
Benjamin, Israel Joseph, *Sefer masei yisrael*, trans. David Gordon, Lyck: Tsvi Hirsch Pettsall Press, 1859 [Hebrew].
Benjamin, Israel Joseph, *Eight years in Asia and Africa from 1846–1855*, Hannover: 1859.
Benkel, Abraham Szlomo, "R. Yehoszua Alter (Glicenstein), Chanoch (Henrik, Enrico) Glicenstein," in Eliezer Esterin (ed.), *Turek: A Memorial to the Jewish Community of Turek, Poland*, pp. 207–216 [Hebrew].
Berg, Johannes van den, and Ernestine G. E. van der Wall (eds.), *Jewish-Christian Relations in the Seventeenth Century, Studies and Documents*, Dodrecht and Boston: Kluwer Academic: 1988.
Berger, Pamela, *The Crescent on the Temple: the Dome of the Rock as image of the ancient Jewish sanctuary*, Leiden: Brill, 2012
Berik, Abraham, "Rabbi Shlomo of Chelm, the Author of *Merkevet ha'mishneh*," *Sinai* 61 (1967): 168–184 [Hebrew].
Berik, Abraham, "The Fate of a Manuscript of R. Shlomo of Chelm," *Sinai* 69/5–6 (1971): 258–271 [Hebrew].
Berliner, Avraham, *Selected Essays*, Jerusalem 1969, vol. 2, pp. 93–96 (Hebrew).
Berman, Isaac ben Phinehas, *Sefer ḥidushai yiṣhak*, Königsberg: A. Samter [Press], 1851 [Hebrew].
Bernheimer, Carolo, *Codices Hebraici Bybliothecae Ambrosianae*, no. 108, Florence: 1933.
Berry, Lloyd E. (ed.), *The Geneva Bible, A Facsimile of the 1560 edition*, Madison: University of Wisconsin Press, 1969.
Bick, Abraham (Shauli), *Rabbi Jacob Emden: The Man and His Thought*, Jerusalem: Mossad Harav Kook, 1974 [Hebrew].
Bloch, Joshua, *Venetian Printers of Hebrew Books*, New York: New York Public Library, 1932.
Bloch, Joshua, *Hebrew Printing in Riva di Trento*, New York: New York Public Library, 1933.
Bloch, Samson ben Isaac, *Shvilei olam*, vol. I, Zholkva: Meyerhoffer, 1822 [Hebrew].
Bloch, Samson ben Isaac, *Shvilei olam*, vol. II, Zholkva: Meyerhoffer, 1828[Hebrew].
Bloch, Samson ben Isaac, *Shvilei olam*, vol. III, Lwów: Grosman, 1855 [Hebrew].
Bloch, Samson ben Isaac, *Shvilei olam*, vol. III, Lwów: Grosman, 1855 [Hebrew].
Bloch, Samson ben Isaac, *Shvilei olam*, Salonica: Saadi Halevi Ashkenazi Press, 1853–1854 [Ladino].
Blücher, Ephraim Israel, *Grammatica aramaica in qua non solum de aramaismo biblico, sed etiam de illis idiomatibus, quae in variis Targumium et Talmud uti etiam in aliorum scriptorum vetero-rabbi-norum*, 1838.
Blücher, Ephraim Israel, *The Scroll of Ruth: With an Ashkenazic [German] Translation and Hebrew Commmentary*, Vienna: Franz Adler von Schmid Press, 1849.
Blücher, Ephraim Israel, *Die Synagogenfrage für deutsche Israeliten*, Wien 1860.
Blücher, Ephraim Israel, *Kannan & Nebenländer das heilige Land Palaestina in der ältern biblischen Zeit*, Grätz bei Posen, n.d..
Boer, Harem de, "Amsterdam as 'Locus' of Iberian Printing in the Seventeenth and Eighteenth Centuries," in Yosef Kaplan (ed.), *The Dutch Intersection: The Jews and the Netherlands in Modern History*, Leiden and Boston: Brill, 2008, pp. 87–110.

Bonfil, Robert (Reuven), *As by a Mirror: Jewish Life in Renaissance Italy*, Jerusalem: Zalman Shazar Center, 1994 [Hebrew].

Braslavi, Joseph, "The Jewish Community in Deir el Qamar and the Second Blood Libel (in the 19th Century)," in B. Z. Luria (ed.), *Zer Li'Gevurot: The Zalman Shazar Jubilee Volume*, Jerusalem: Kiryat Sefer, 1973, pp. 510–523 [Hebrew].

Brawer, Abraham J., "Schwarz, Yehoseph," *Encyclopaedia Judaica*, 2nd ed., vol. 18, 2007, p. 191.

Breslauer (Breza), Menachem Mendel, *Sefer glilot ereṣ yisrael*, Bratslav: Saltzbach Press, 1819 [Hebrew].

Brodsky, Harold, "The Seventeenth-Century Haggadah Map of Avraham bar Yaacov," *Jewish Art* 19–20 (1993–1994): 148–157.

Brodsky, Harold, "Ezekiel's Map of Restoration," in idem (ed.), *Land and Community: Geography in Jewish Studies*, Bethesda: University Press of Maryland, 1997, pp. 17–29.

Brodsky, Harold, "The Hebrew Holy Land Map of Avraham bar Yaacov, Amsterdam, 1695," *The Portolan* 51 (Fall 2001): 15–21.

Buber, Salomon, *A Lofty Town, is the City of Zholkva*, Krakow: 1903 [Hebrew].

Busi, Gulio, *Libri Ebraici a Mantova*, vol. 1, Firenze: Edizioni Cadmo, 1996.

Calmet, Augustin Antoine, *Commentaire litteral sur tous les livres de l'Ancien et du Nouveau Testament: Les trois premiers livres des Rois*, Paris: Pierre Emery, 1711.

Campbell, Tony, *The Earliest Printed Maps 1472–1500*, London: The British Library, 1987.

Shlomo (Solomon) of Chelm, *Merkevet ha'mishneh*, vol. III, Salonica: 1785 [Hebrew].

Shlomo (Solomon) of Chelm, *Ḥug ha'areṣ*, manuscript, n.d [Hebrew]..

Shlomo (Solomon) of Chelm, *Hug Ha-Aretz ha-Shalem*, ed. S. D. Rosenthal, Jerusalem: Frank Institute, 1988 [Hebrew].

Chorin (Choriner), Áron ben Kalman, *Sefer emek ha'shaveh*, Prague: Alzenwagner Press, 1803 [Hebrew].

Cohen, Isaac of Ostrowo (Isaac Naftali Hertz Buking), *The Abridged Book of Mizrachi*, Prague: 1604 [Hebrew].

Cohen, Menachem (ed.), *Mikraot Gedolot 'Haketer': Ezekiel*, Ramat Gan: Bar Ilan University Press, 2000 [Hebrew].

Cohen, Mortimer J., *Jacob Emden, a Man of Controversy*, Philadelphia: Dropsie College, 1937.

Cohen, Richard I., *Jewish Icons: Art and Society in Modern Europe*, Berkeley: University of California Press, 1998.

Cohen, Tova, *From Dream to Reality: Descriptions of Eretz Israel in Haskalah Literature*, Ramat Gan: Bar Ilan University Press, 1982 [Hebrew].

Conder, C.R. and Kitchener, H.H., *The survey of Western Palestine*, London: Committee of the Palestine Exploration Fund, 1881–1889

Connolly, Daniel K., "Imagined Pilgrimage in the Itinerary Maps of Matthew Paris," *Art Bulletin* 81 (1999): 598–622.

Cosgrove, Denis, "Mapping New Worlds: Culture and Cartography in Sixteenth Century Venice," *Imago Mundi* 44 (1992): 65–89.

David, Abraham, "Lipschutz, Israel ben Gedalia," *Encyclopaedia Judaica*, 2nd ed., vol. 13, 2007, p. 74.

Delano-Smith, Catherine, "Maps in Bibles in the Sixteenth Century," *The Map Collector* 39 (1987): 2–14.

Delano-Smith, Catherine, "Maps as Art and Science: Maps in Sixteenth Century Bibles," *Imago Mundi* 42 (1990): 65–83.

Delano-Smith, Catherine, "Maps and Religion in Medieval and Early Modern Europe," in David Woodward, Catherine Delano-Smith and Cordell D. K. Yee (eds.), *Plantejaments I Objectius D'Una Historia Universal de la Cartographia/Approaches and Challenges in a Worldwide History of Cartography*, Barcelona: Institut Cartogràphic de Catalunya, 2001, pp. 179–200.

Delano-Smith, Catherine, "The Intelligent Pilgrim: Maps and Medieval Pilgrimage to the Holy Land," Rosamund Allen (ed.), *Eastward Bound: Travel and Travellers, 1050–1550:* Manchester: Manchester University Press, 2004: 107–130.
Delano-Smith, Catherine, "The Exegetical Jerusalem: Maps and Plans for Ezekiel Chapters 40–48," Lucy Donkin and Hanna Vorholt (eds.), *Imagining Jerusalem in the Medieval West*, Oxford: Proceedings of The British Academy, 175 (2012), p. 41–75.
Delano-Smith, Catherine, and Mayer I. Gruber, "Rashi's Legacy: Maps of the Holy Land," *The Map Collector* 59 (1992): 30–35.
Delano-Smith, Catherine, and Elizabeth M. Ingram, *Maps in Bibles: 1500–1600: An Illustrated Catalogue*, Geneva: Librairie Droz, 1991.
Delano-Smith, Catherine, and Roger J. P. Kain, *English Maps: A History*, London: British Library, 1999
Delano-Smith, Catherine, "La Carte de Palestine," in Marcel Watlet (ed.), *Gérard Mercator Cosmographe: le temps et l'espace*, Antwerp: Fonds Mercator, 1994, pp. 268–283.
Gatta, Giovanni Francesco Della, *Palaestinæ sive Terre Sancte descriptio,* Romae: Apud Ioannem Franciscium vulgo Della Gatta, 1557.
Demsky, Aaron, "From Kzib unto the River near Amamanh (Mish. Shebi'it 6:1; Halla 4:8): A Clarification of the Northern Border of the Returnees from Egypt," *Shnaton—An Annual for Biblical and Ancient Near Eastern Studies* 10 (1986–1989): 71–81 [Hebrew].
Dilke, Oswald Ashton Wentworth, *Greek and Roman Maps*, Baltimore: John Hopkins, 1998.
Edson, Evelyn, *Mapping Time and Space: How Medieval Mapmakers Viewed Their World,* British Library Studies in Map History, vol. 1, London: The British Library 1997.
Eidelberg, Shlomo, "Lublin, Meir ben Gedaliah," *Encyclopaedia Judaica*, 2nd ed., vol. 13, 2007, p. 245.
Eidels, Samuel (Shmuel) Eliezer ha-Levi (MaHaRShA), *Sefer ḥiddushei halakhot*, Fürth: Reuven Fürsth Press, 1706 [Hebrew].
Eidels, Samuel (Shmuel) Eliezer ha-Levi (MaHaRShA), *Sefer ḥiddushei halakhot* Wilhelmsdorf: Hirsch ben R. Haim of Fürth, 1720 [Hebrew].
Eidels, Samuel (Shmuel) Eliezer ha-Levi (MaHaRShA), *Sefer ḥiddushei halakhot*, Berlin: 1756 [Hebrew].
Eidels, Samuel (Shmuel) Eliezer ha-Levi (MaHaRShA), *Sefer ḥiddushei halakhot*, Nowy Dowr: J. A. Krieger, 1792 [Hebrew].
Eidels, Samuel (Shmuel) Eliezer ha-Levi (MaHaRShA), *Ḥidushai halakhot*, Slavita: Moshe Shapira, 1816 [Hebrew].
Elbaum, Jacob, *Openness and Insularity: Late Sixteenth Century Jewish Literature in Poland and Ashkenaz*, Jeruslaem: Magnes Press, 1990 [Hebrew].
Elijah ben Shlomo Zalman, *Sefer ṣurat ha'areṣ le'gvuloteha*, Shklov: Aryeh Leib ben Shene'ur Feivish and Shabbatai ben Ben-Zion, 1802 [Hebrew].
Elijah ben Shlomo Zalman, *Aderet Eliyahu, Commentary on the Prophets and the Hagiographa*, Jerusalem: Elijah Landau, 1905 [Hebrew].
Elliot, James, *The City in Maps—Urban Mapping to 1900*, London: The British Library, 1987.
Emden, Jacob, *Sefer leḥem shamayim*, part 2, vol. 4, Altona: 1768 [Hebrew].
Emden, Jacob, *Mor ve'kṣiah*, Altona: 1761, (repr. New York 1952–1953) [Hebrew].
Encyclopaedia Hebraica, Jerusalem and Tel-Aviv: Encyclopaedia Publishing Co.,
  "Avraham ben Avigdor," vol. 1, p. 292 [Hebrew].
  "Hezekiah ben Manoah," vol. 17, 1965, p. 269 [Hebrew].
  "Falklish, Eliezer," vol. 27, 1975, p. 891 [Hebrew].
  "Krauze, Eliyahu," vol. 30, 1978, p. 32 [Hebrew].
  "Sapir, Eliyahu," vol. 26, 1974, p. 228 [Hebrew].
  "Yoffe, Mordecai," vol. 20, 1971, pp. 61–62 [Hebrew].

"Ziskind, Alexander," vol. 3, pp. 632–633 [Hebrew].
Encyclopaedia Judaica, 2nd ed., Detroit: Macmillan Reference, 2007
"Abraham ben Jacob," vol. 1, p. 294.
"Adeni, Solomon Bar Joshua," vol. 1, p. 390.
"Altschuler, R. David ben Aryeh Leib," vol. 3, p. 533.
"Samuel Eliezer ben Judah Ha-Levi," vol. 6, pp. 143–144.
"Elijah ben Solomon Zalman," vol. 6, pp. 341–346.
"Falklish, Eliezer," vol. 27, p. 891.
"Munk, Solomon," vol. 14, p. 615.
"Rashi," vol. 13, pp. 1558–1565.
"Sokolow, Nahum," vol. 18, pp. 747–749.

Etkes, Immanuel, "On the Question of the Haskalah's Forerunners in Europe," in idem (ed.), *The East European Jewish Enlightenment*, Jerusalem: Zalman Shazar Center, 1993, pp. 25–44 [Hebrew].

Etkes, Immanuel, (ed.), *The East European Jewish Enlightenment*, Jerusalem: Zalman Shazar Center, 1993 [Hebrew].

Etkes, Immanuel, *The Gaon of Vilna—The Man and His Image*, Jerusalem: Zalman Shazar Center, 1998 [Hebrew] (trans. into English, Jeffrey Green, Berkeley: University of California Press, 2002).

Eusebius, *Das Onomastikon, der Biblischen Ortsnamen*, ed. Erich Klostermann, Hildesheim: G. Olms, 1966 [Leipzig: 1904].

Eusebius, *The Onomastikon of Eusebius*, vol. XIX and XXI, trans. E. Z. Melamed, Jerusalem: Tarbiz, 1966 [Hebrew].

Feiner, Shmuel, *Haskalah and History: The Emergence of a Modern Jewish Historical Consciousness*, Oxford: Littman Library of Jewish Civilization, 2002.

Feiner, Shmuel, *The Jewish Enlightenment in the Eighteenth Century*, Jerusalem: Zalman Shazar Center, 2002 [Hebrew].

Feiner, Shmuel, *The Jewish Enlightenment*, Philadelphia: University of Pennyslvania Press, 2004.

Feiner, Shmuel, *Haskalah and History: The Emergence of a Modern Jewish Awareness of the Past*, Jerusalem: Zalman Shazar Center, 1995 [Hebrew].

Feiner, Shmuel, *The Jewish Enlightenment in the 19th Century*, Jerusalem: Carmel Press, 2010 [Hebrew].

Feiner, Shmuel and Israel Bartal (eds.), *The Varieties of Haskalah*, Jerusalem: Magnes Press, 2005 [Hebrew].

Feiner, Shmuel and David Sorkin (eds.), *New Perspectives on the Haskalah*, London: Littman Library of Jewish Civilization, 2001.

Fiorani, Francesca, *The Marvel of Maps, Art, Cartography and Politics in Renaissance Italy*, New Haven: Yale University Press, 2005.

Fischer, Yona, *Art and Artisanship in Nineteenth-Century Eretz Yisrael*, Jerusalem: Israel Museum 1979 [Hebrew].

Flusser, David (ed.), *The Josippon (Josephus Gorionides)*, Jerusalem: Bialik Institute, vol. 1, 1978, vol. 2, 1980 [Hebrew].

Freudenthal, Gad, "Hebrew Medieval Science in Zamosc ca. 1730. The Early Years of Rabbi Israel ben Moses Halevy of Zamosc," in Resianne Fontaine, Andrea Schatz, and Irene E. Zwiep (eds.), *Sepharad in Ashkenaz: Medieval Learning and Eighteenth-Century Enlightened Jewish Discourse*, Amsterdam, 2007.

Friedberg, Bernhard, *History of Hebrew Typography in Italy, Spain-Portugal, Turkey and the Orient from its Beginning and Formation about the Year 1472*, Antwerp: 1932.

Friedberg, Bernhard, *History of Hebrew Typography in Poland: From the Beginning of the Year 1534 and its Development up to Our Days*, 2nd ed., Tel Aviv: 1950 [Hebrew].

Frumkin, Aryeh Leib, *The Annals of the Sages of Jerusalem*, Jerusalem: Salomon Press, 1928–1930 [Hebrew].
Frumkin, Aryeh Leib, *The Annals of the Sages of Jerusalem*, vol. III, Jerusalem: Or Hahohma, 2003 [Hebrew].
Fuller, Thomas, *A Pisgah-Sight of Palestine and the Confines Thereof*, London: John Williams, 1650.
Funkenstein, Amos, *Styles in Medieval Biblical Exegesis: An Introduction*, Tel Aviv: Ministry of Defence, 1990 [Hebrew].
Gaon, Moshe David, *Heart Lockets: The Different Versions and Fate of the Book that was Read for Generations in the Ladino Tongue*, Jerusalem: Ma'arav Press, 1933 [Hebrew].
Gaon, Moshe David, *Oriental Jews in Eretz-Israel (Past and Present)*, Jerusalem: 1938 [Hebrew].
Garel, Michel, "La premier Carte de Terre Sainte en Hebreu (Amsterdam, 1620/21)," *Studia Rosenthaliana* 21/2 (1987): 131–139.
Geiger, Ari, *The Commentary of Nicholas of Lyra on Leviticus, Numbers and Deuteronomy: A Study of the Jewish Sources of the Literal Commentary on Leviticus-Deuteronomy and the Moral Commentary on Deuteronomy*, PhD thesis, Ramat-Gan, Israel: Bar-Ilan University, 2006 [Hebrew].
Gelber, Nathan, "Tarnogrod," *Encyclopaedia Judaica*, 2nd ed., vol. 19, 2007, p. 516.
Gelis, Jacob, *Among the Greats of Jerusalem*, Jerusalem: Sifriyat Rishonim, 1967 [Hebrew].
George, Wilma, *Animals and Maps*, Berkeley: University of California Press, 1969.
Ginsburg, Saul, "Individuals and Generations: A Hundred Years of the Annals of One Family," in idem, *Historical Writings*, Tel Aviv: Dvir, 1944, pp. 215–253 [Hebrew].
Ginzberg, Louis, *The Legends of the Jews*, Baltimore: Johns Hopkins University Press, 1998.
Glick, Shmuel (ed.), *Zekhor Davar le-Avdekha: Essays and Studies in Memory of Dov Rappel*, Ramat-Gan, Israel: Bar-Ilan University, 2007 [Hebrew].
Godlewska, Anne, *The Napoleonic Survey of Egypt, a Masterpiece of Cartographic Compilation and Early Nineteenth-Century Fieldwork*, Cartographica Series, vol. 25, nos. 1 and 2, monograph no. 38–39, Toronto: University of Toronto Press, 1988.
Godlewska, Anne, "The Idea of the Map," in Susan Hanson (ed.), *Ten Geographical Ideas that have Changed the World*, New Brunswick, NJ: Rutgers University Press, 1997, pp. 15–39.
Goldzweig, Jacob, *Sefer igeret le'kaiṣ ha'yamin*, Manchester, 1896 [Hebrew].
Goren, Haim, "An Eighteenth Century Geography: 'Sefer Yedei Moshe' by Rabbi Moshe Yerushalmi," *Cathedra* 34 (January 1985): 75–96 [Hebrew].
Goren, Haim, *'Go View the Land': German Study of Palestine in the Nineteenth Century*, Jerusalem: Yad Ben-Zvi, 1999 [Hebrew].
Goren, Haim, "The Present as an Expression of the Past: The Holy Land in the Work of the Historical Cartographer Heinrich Kiepert," *Horizons in Geography* 51 (1999): 99–118 [Hebrew].
Graetz, Heinrich, "Die talmudische topogrpahie," *Monatsschrift für Geschichte und Wissenschaft des Judentums* 2 (1853): 106–113.
Grossman, Abraham, "Marginal Notes and Addenda of R. Shemaiah and the Text of Rashi's Biblical Commentary," *Tarbiz* 60 (1990–1991): 67–98 [Hebrew].
Grossman, Abraham, "Eretz Israel in the Thought of Rashi," *Shalem* 7 (2001): 15–31 [Hebrew].
Grossman, Abraham, *Rashi*, Jerusalem: Zalman Shazar Center, 2006 [Hebrew].
Grossman, Abraham, "The Wording of Rashi's Commentary on the Hebrew Bible and the Jewish-Christian Polemic," *Sinai* 137 (2006): 32–48 [Hebrew].
Grossman, Abraham, *Rashi: Religious Beliefs and Social Views*, Alon Shvut: Tvunot, 2008 [Hebrew].
Grossman, Abraham, "Revolution and Sense of Commitment—the Contours of Rashi's Image," in idem and Japhet, Sara (eds.), *Rashi: The Man and His Work*, Jerusalem: Zalman Shazar Center, 2008, [Hebrew].

Grossman, Abraham and Benjamin Z. Kedar, "Rashi's Maps of the Land of Israel and their Historical Meaning," *The Israel National Academy of Sciences Newsletter* 25 (2003): 26–29 [Hebrew].

Gruber, Mayer I., "What Happened to Rashi's Pictures?" *Bodleian Library Record* 14/2 (1992): 111–124.

Gruber, Mayer I., "Light on Rashi's Diagrams from the Asher Library of Spertus College of Judaica," *The Solomon Goldman Lectures,* vol. VI, Chicago: The Spertus Institute of Jewish Studies Press, 1993, pp. 73–85.

Gruber, Mayer I., "Notes on the Diagrams in Rashi's Commentary to the Book of Kings," *Studies in Bibliography and Booklore* 19 (1994): 29–41.

Gruber, Mayer I., "The Sources of Rashi's Cartography," in Norman Simms (ed.), *Letters and Texts of Jewish History*, Hamilton, New Zealand: Outrigger, 1998, pp. 61–67.

Gruber, Mayer I., "Rashi's Map Illustrating His Commentary on Judges 21:19," *Proceedings of the Rabbinical Assembly* 65 (2004): 135–141.

Guthe, Herman, *Bibelatlas mit einem Verzeichnis der alten und neuen Ortsnamen*, Leipzig, H. Wagner & E. Debes, 1911.

Haag, Hans Jacob, "Die vermutlich älteste bekannte hebräische Holzschnittkarte des Heiligen Landes (um 1560)," *Cartographica Helvetica* 4 (1991): 23–26.

Haag, Hans Jacob, "Hebräische Karte des Heiligen Landes um 1560," in *Zentralbibliothek Zürich, Schätze aus vierzehn Jahrhunderten*, Zürich: Verlag Neue Zürcher Zeitung, 1991.

Haag, Hans Jacob, "Elle mas'e vene Yisra'el asher yatz'u me-eretz Mitzrayim: eine hebräische Karte des Heiligen Landes aus dem 16. Jahrhundert," in Klaus Herrmann, Margarete Schlüter, and Giuseppe Veltri (eds.), *Jewish Studies between the Disciplines, Judaistik zwischen den Disziplinen: Papers in Honor of Peter Schäfer on the Occasion of His 60th Birthday*, Leiden: Brill, 2003, pp. 269–278.

Hacker, Joseph, "Mizrahi, Elijah," *Encyclopaedia Judaica*, 2nd ed., vol. 14, 2007, pp. 393–395.

Hacohen, Haim Isaiah, *Sefer tshu'ot ḥayim*, Lublin: Shneidermasser and Hirschenharen Press, 1894 [Hebrew].

Hacohen, Mordechai Ben-Hillel, *The Travels of Mordecai Ben-Hillel Hacohen, 1899*, in Abraham Ya'ari (ed.), *Travels to Eretz Yisrael*, Tel-Aviv: 1945–1946, p. 654 [Hebrew].

Hailperin, Herman, *Rashi and the Christian Scholars*, Pittsburgh, University of Pittsburgh: 1963

Halper, Jeff, *Between Redemption and Revival: The Jewish Yishuv of Jerusalem in the Nineteenth Century*, Oxford: Westview, 1991.

Haparchi, Ishtori, *Kaftor va'ferech*, Jerusalem: Luntz, 1976 [Hebrew].

Har-El, Menashe, "Orientation and Mapping in Biblical Land," *Israel—People and Land* 1/19 (1983–1984): 157–168 [Hebrew].

Harley, J. Brian, "Silences and Secrecy: The Hidden Agenda of Cartography in Early Modern Europe," *Imago Mundi* 40 (1988): 57–76.

Harley, J. Brian, "Deconstructing the Map," *Cartographica* 26/2 (1989): 1–20.

Harley, J. Brian, "Maps, Knowledge and Power," in Denis Cosgrove and Stephen Daniels (eds.), *The Iconography of Landscape*, Cambridge: Cambridge University Press, 1989.

Harley, J. Brian, *The New Nature of Maps: Essays in the History of Cartography*, ed. Paul Laxton, Baltimore: 2001.

Harley, J. Brian and David Woodward, *The History of Cartography*, vol. I-III, Chicago: University of Chicago Press, 1987–2007.

Harvey, David, *The Condition of Postmodernity*, Oxford: Blackwell, 1990.

Harvey, Paul Dean A., *Medieval Maps*, London: The British Library, 1991.

Harvey, Paul Dean A., *Mappa Mundi: The Hereford World Map*, Toronto: University of Toronto Press, 1996.

Harvey, Paul Dean A., *Medieval Maps of the Holy Land*, London: The British Library, 2012.

Heller, Marvin J., *The Sixteenth Century Hebrew Book: An Abridged Thesaurus*, Leiden: Brill, 2004.
Heller, Marvin J., *The Seventeenth Century Hebrew Book: An Abridged Thesaurus*, Leiden: 2011.
Hock, Simon, *Die Familien Prags. Nach den Epitaphien des Alten Jüdischen Friedhofs in Prag*, Pressburg: Druck von Adolf Alkalay, 1892.
Hopkins, I. W. J., "Nineteenth-Century Maps of Palestine: Dual-Purpose Historical Evidence," *Imago Mundi* 22 (1968): 30–36.
Horovitz, Haim ben Dov Dvrish, *Sefer ḥibat yerushalayim*, Jerusalem: Y. Baeck Press, 1844 [Hebrew].
Ilan, Zvi, *Attempts at Jewish Settlement in Trans-Jordan, 1871–1947*, Jerusalem: Yad Ben-Zvi 1984 [Hebrew].
Ilan, Zvi, *Tombs of the Righteous in the Land of Israel*, Jerusalem: Kana, 1997 [Hebrew].
Immerzeel, Mat, Waldemar Deluga, and Magdalena Łaptaś, "Proskynetaria- Inventory," *Series Byzantina* 3 (2005): 25–31.
Ingram, Elizabeth M., "A Map of the Holy Land in the Coverdale Bible: A Map of Holbein?" *The Map Collector* 64 (1993): 26–33.
Ingram, Elizabeth M., "Maps as Readers' Aids: Maps and Plans in Geneva Bibles," *Imago Mundi* 45 (1993): 29–44.
Ish-Shalom, Michael, *Holy Tombs: A Study of Traditions concerning Jewish Holy Tombs in Palestine*, Jeruslaem: Kook Foundation and Palestine Institute of Folklore and Ethnology, 1948 [Hebrew].
Jacotin, Pierre, "Carte topographique de l'Egypte et de plusieurs parties des pays limitrophes," *Description de l'Egypte*, XX, Paris 1818.
Jabotinsky, Ze'ev and Perlman, Shmuel, *Atlas*, 1925, the Hebrew Publishing Company HA-SEFER, Ltd., London [Hebrew].
Jahn, Johann, *Biblische Archaeologie*, Wien: C.F. Wappler, 1, 1797.
Jerusalem Judaica (catalog of a public auction held in April 2008).
Josephus, Flavius, *History of the Jewish War against the Romans*, trans. Lisa Ullman, Jeruslaelm: Carmel, 2010 [Hebrew].
Juhasz, Esther, *The 'Shiviti-Menorah': A Representation of the Sacred—between Spirit and Matter*, PhD thesis, Jeruslaem: The Hebrew University, 2004 [Hebrew].
Kaczynski, Bernice M., "Illustrations of Tabernacle and Temple Implements in the *Postilla in Testementum Vetus* of Nicolaus de Lyra," *The Yale University Library Gazette* 48/1 (July 1973): 1–11.
Kadmon, Naftali, "Cartography," *Encyclopaedia Hebraica*, vol. 30, 1978, pp. 131–135 [Hebrew].
Kaplan, Jacob, *Eretz Kedumin, Mekh'karei ereṣ*, Vilnius 1839, see: Solomon Löwisohn (Levisohn)
Kaplan, Yosef, "*Gente Politica*: The Portuguese Jews of Amsterdam vis-a-vis Dutch Society," Cahya Brasz and idem (eds.), *Dutch Jews as Perceived by Themselves and by Others*, Leiden, Boston, and Köln: Brill, 2001, pp. 22–40.
Kaplan, Yosef, *From New Christians to New Jews*, Jerusalem: Zalman Shazar Center, 2002 [Hebrew].
Karp, Abraham J. (ed.), *From the Ends of the Earth: Judaic Treasures of the Library of Congress*, Washington D.C.: Library of Congress, 1991.
Katz, Ben Zion, *Rabbinate, Hasidism and Enlightenment*, vol. 1, Tel-Aviv: Dvir, 1956, vol. 2, Tel-Aviv: Dvir, 1958 [Hebrew].
Kedar, Benjamin Z., "Rashi's Map of the Land of Canaan, ca. 1100, and its Cartographic Background," in Richard J. A. Talbert and Richard W. Unger (eds.), *Cartography in Antiquity and the Middle Ages: Fresh Perspectives, New Methods*, Leiden: Brill, 2008, pp. 155–168.
Kedar, Benjamin Z., "Rashi's Map of the Land of Canaan and its Cartographic Background," in Joseph R. Hacker, Yosef Kaplan, and B. Z. Kedar (eds.), *From Sages to Savants: Studies Presented to Avraham Grossman*, Jerusalem: Zalman Shazar Center, 2010, pp. 111–128 [Hebrew].

Kirchhoff, Markus, *Text zu Land: Palaestina im wissenschaftlichen Diskurs 1865–1920*, Schriften des Simon-Dubnow-Instituts, 5, Goettingen: Vandenhoeck & Ruprecht, 2005.

Klausner, Israel, Samuel Mirsky, David Derovan, and Menahem Kaddari, "Elijah ben Solomon Zalman," *Encyclopaedia Judaica*, 2nd ed., vol. 6, 2007, pp. 341–346.

Klausner, Jospeh, *The History of the New Hebrew Literature*, vol. 1–2, Jerusalem: Achiasaf, 1952 [Hebrew].

Klausner, Jospeh, "R. Alexander Ziskind of Grodno—the Hasid amongst Mitnagdim," in Umberto Cassuto, idem, and Joshua Gutmann (eds.), *Book of Asaf*, Jerusalem: Mossad Harav Kook, 1953, pp. 427–432 [Hebrew].

Klausner, Jospeh, "Alexander Susskind ben Moses of Grodno," *Encyclopaedia Judaica*, 2nd ed., vol. 1, 2007, pp. 630–631 [Hebrew].

Koronil, Nahman Natan, *Piskei ḥallah [...] Our Rabbi the RaShBA [R. Shlomo ben Aderet]*. Printed one time [...] in Constantinople [in the] year 1518 [...] and at the end I affixed two rulings on customs and conventions regarding the laws of ḥallah from the genius Rabbi Yaacov [so]n of Tzahal among the sages of the generation before me and more from me [...],Jerusalem, Elijah and Moshe Chai Sasson [Press], 1876.

Kortüm, Hans-Henning, "Der Pilgerzug von 1064/5 ins Heilige Land," *Historische Zeitschrift* 277/3 (2003): 561–592.

Krieger, Pinchus, *Parshan-Data: Supercommentaries on Rashi's Commentary on the Pentateuch*, Monsey, NY: 2005 [Hebrew].

Kupfer, Ephraim, "Jaffe, Mordecai ben Abraham," *Encyclopaedia Judaica*, 2nd ed., vol. 11, 2007, pp. 67–68.

Laor, Eran, *Maps of the Holy Land, a Cartobibliography of Printed Maps 1475–1800*, New York and Amsterdam: Alan R. Liss & Meridian, 1986.

Leiner, Gershon Ḥanokh Henikh, *Sefer sidrei tohorot*, vol. 1, tractate Keilim, Józefów: Shlomo and Baruch Zetser and Yechezkel Raner Press, 1873 [Hebrew].

Leiner, Gershon Ḥanokh Henikh, *Sefer sidrei tohorot*, vol. 2, tractate Oholot, Piotrków: Shlomo Belchatovski Press, 1908 [Hebrew].

Leon, Jacob Judah (Templo), *Afbeeldinghe vanden Tempel Salomonis*, Middelburg: 1642.

Leon, Jacob Judah (Templo), *Retrato del Templo de Selomo*, Middelburg: 1642.

Leon, Jacob Judah (Templo), *Retrato del Tabernaculo*, Amsterdam: Gillis Joosten, 1654.

Levy-Rubin, Milka, "From Eusebius to the Crusader Maps: The Origin of the Holy Land Maps," in Bianca Kühnel, Galit Noga-Banai and Hanna Vorholt (eds.), *Visual Constructs of Jerusalem*, Turnhout: Brepols, 2014, 253–263.

Liberman, Haim, *Rachel's Tent*, vols. 1–3, New York: 1980–1984 [Hebrew].

Lichtenstädter, Benjamin Wolf, *Amtaḥat binyamin*, Fürth: Zürendorffer & Sommer, 1846 [Hebrew].

Lichtenstädter, Benjamin Wolf, *Amtaḥat binyamin*, 1848 [Hebrew].

Linetski, Yitskhok Yoyel, *Die Kurze geographia fon Palestina*, Odessa: P. A. Zeleni Press, 1882 [Yiddish].

Löwisohn (Levisohn), Solomon, *Mekh'karei ereṣ*, Vienna: George Holzinger, 1819 [Hebrew].

Löwisohn (Levisohn), Solomon, *Biblische Geographie*, Wien 1821.

Löwisohn (Levisohn), Solomon and Jacob Kaplan, *Eretz Kedumin, Mekh'karei ereṣ*, Vilnius: Menachem Mann and Simcha Zimel Press, 1839 [Hebrew].

Magid, Shaul, *Hasidism in Transition: The Hasidic Ideology of Rabbi Gershon Henoch of Radzin in Light of Medieval Jewish Philosophy and Kabbala*, Ann Arbor, University of Michigan, 1994.

Magid, Shaul, *Hasidism on the Margin: Reconciliation, Antinomianism and Messianism in Izbica/Radzin Hasidism*, Madison, WI: University of Wisconsin Press, 2003.

MaHaRShA, See: Samuel (Shmuel) Eliezer ben Judah ha-Levi Eidels

Maimon, Yehuda Leib (ed.), *The Book of Rashi*, Jerusalem: Rav Kook Institute, 1956 [Hebrew].

Maimon, Yehuda Leib (ed.), *Annals of the Vilna Gaon*, Jerusalem: 1970 [Hebrew].

Malkov, Avigdor ben Mordecai, *Sefer derekh emet...*, Warsaw, 1895 [Yiddish].
Mann, Vivian B. and Bilski, Emily D., *The Jewish Museum: New York*, London: Scala Books 1993.
Marcus, Ahron, *Hasidism*, trans. from German, Moshe Shenfeld, Bnei Brak: Netzach Press: 1980 [Hebrew].
Margalioth, Mordechai (ed.), *Encyclopedia of Great Men in Israel*, vol. IV, Tel Aviv: Joshua Chachik, n.d [Hebrew]..
Marton, Yehouda, Paul Schveiger, and Radu Ioanid, "Arad," *Encyclopaedia Judaica*, 2nd ed. vol. 2, 2007, p. 331.
Maser, Edward A. (intro., trans., and commentaries), *Baroque and Rococo Pictorial Imagery: The 1758–60 Hertel Edition of Ripa's 'Iconologia,'* New York: Dover 1971.
Matskeyitsh, Shabtai ben Yaakov, *Gvulei șion* [Borders of Zion], Vilnius: Avraham Tsevi Katsinelinboigen, 1898 [Hebrew].
Merian, Matthaeus, *Figures de la Bible, demonstrans les principales histoires de la Saincte Escriture, Biblische Figuren, Icones Biblicae*, Amsterdam: N. Visscher, 1659?.
Me'am lo'ez, part 4, the Book of Numbers, Constantinople: 1764 [Ladino and Hebrew].
Me'am lo'ez, part 4, Salonica: M. Nachman and D. Israeligia, 1796–1803 [Ladino].
Me'am lo'ez, part 4, the Book of Numbers, ed. Yitzhak Magriso, Salonica: 1815 [Ladino].
Me'am lo'ez, part 4, the Book of Numbers, Livorno: Nachman Sa'adon Press, 1822–1823 [Ladino].
Me'am lo'ez, part 4, Numbers, Salonica: Saadi Halevi Ashkenazi Press, 1863–1867 [Ladino].
Me'am lo'ez, part 4, the Book of Numbers, ed. Yitzhak Magriso, Izmir: 1872 [Ladino].
Mellinkoff, Ruth, *The Horned Moses in Medieval Art and Thought*, Berkeley: University of California Press, 1970.
Mellinkoff, Ruth, "More about Horned Moses," *Journal of Jewish Art* 12–13 (1986–1987): 184–198.
Mercator, Gerard and Hondius, Jodocus, *Atlas ou Représentation du monde universel et des parties d'icelui, faicte en tables et descriptions*, Amsterdam: 1633.
Metzger, Thérèse, "Le manuscript enluminé Cod. Hebr. 5 de la Bibliothèque d'Etat à Munich," in *Études de Civilisation Médiévale (IXe-XIIe siècles), Mélanges offerts à Edmond-René Labande*, Poitiers: C.É.S.C.M., 1974, pp. 537–552.
Meyer, Michael Albert, *Response to Modernity: A History of the Reform Movement in Judaism*, New York: Oxford University Press, 1988 [Hebrew].
Michael, Reuven, *Solomon Löwisohn: A Selection of His Works*, Jerusalem: Dinur Centre, 1984 [Hebrew]
Miklish, Yoseph ben Issachar, *Yosef da'at*, Prague: Gershom Katz, 1609 [Hebrew].
Mintz-Manor, Limor, "Symbols and Images in the Spanish-Portuguese Congregation of Amsterdam in the 17th and 18th Centuries," *Pe'amim (Studies in Oriental Jewry)* 120 (Summer 2009): 9–60 [Hebrew].
Mintz-Manor, Limor, *The Discourse on the New World in Early Modern Jewish Culture*, PhD thesis, Jerusalem: The Hebrew University, 2011 [Hebrew].
Mishna [...] with the Commentary of Our Rabbi Obadiah of Bertinoro and the Primary Portion of the Tosafot Yom Tov [...] and a Commentary [...] [...] by the name Tiferet Yisrael [...],by Our Rabbi Yisrael Lipschitz Son of Our Rabbi Gedalia [...] with Numerous Additions [...] Warsaw: Y. Goldman, 1873 [Hebrew].
Mishna, the Order of Zeraim, with All the Exegetes and Many Additions as Explained in the Second Title Page, Vilnius: the Widow Romm and Sons Press, the publishing house of the rabbi our rabbi Haim Noah Eisenstadt may his candle illuminate in Warsaw, year [..., 1878] [Hebrew].
Mishna [...] with the Commentary of Our Rabbi Obadiah of Bertinoro and with the Primary Portion of the Tosafot yom tov [...] and with the Commentary Tiferet yisrael, Authored by [...] Our Rabbi Yisrael Lipschitz son of Our Rabbi Gedalia [...], Vilnius: the Romm Press, 1913 [Hebrew].

Mishna – Six Orders of the Mishna with the Commentaries of the Rishonim and Aḥaronim, photographic print as per the Romm Publisher 1909, Jerusalem 1955–1958, Order Zeraim, tractate Shevi'it [Hebrew].
Mizrachi, Elijah ben Abraham, *Commentary on Rashi on the Torah*, Venice: Daniel Bomberg, 1545 [1527] [Hebrew].
Mizrachi, Elijah ben Abraham, *Sefer (melekhet) ha'mispar*, Constantinople: Soncino, 1533 [Hebrew].
Mizrachi, Elijah ben Abraham, *The Five Books of the Torah with the Translation and Com[mentary] of Rashi and the Abridged Mizrachi [Supercommentary]and the Five Scrolls with the Com [mentary] of Rashi and the Supercommentary of Ba'al ha'akedah R. Isaac Arama [...]* Riva di Trento, 1521 [Hebrew].
Mizrachi, Elijah ben Abraham, *Elijah Mizrahi [on] the Commentary of Our Rabbi Shlomo Yitzhaki*, Krakow: Issac ben Aaron of Prostitz, 1595 [Hebrew].
Mizrachi, Elijah ben Abraham, *Gur aryeh, Levush ha'orah, Siftai ḥakhamim*, part 4, Numbers, Warsaw: Isaac Rathauer Print, 1864 [Hebrew].
Mohr, Abraham Menachem Mendel, *Shvilei olam*, Lemberg: Shernetzel, 1856 [Hebrew].
Montanus, Benedictus Arias, *Biblia sacra Hebraice, Chaldaice, Graece, & Latine*, vol. 8, Antverpiae: Christophorus Plantinus, 1572.
Morse, Victoria, "The Role of Maps in Later Medieval Society: Twelfth to Fourteenth Century," in David Woodward (ed.), *The History of Cartography*, vol. 3, book 1, *Cartography in the European Renaissance*, Chicago and London: Chicago University Press, 2007, pp. 25–52.
Moscrop, John James, *Measuring Jerusalem: The Palestine Exploration Fund and British Interests in the Holy Land*, London: Leicester University Press, 2000.
Munk, Salomon, *Palestine: Description géographique, historique, et archéologique*, Paris: 1845.
Munk, Salomon, *Palästina: geographische, historische und archäologische Beschreibung dieses Landes, und kurze Geschichte seiner hebräischen und jüdischen Bewohner*, nach dem Französischen von S. Munk bearbeitet von M. A. Levy, Leipzig 1871–1872.
Namenyi, Ernest M., "The Illustration of Hebrew Manuscripts after the Invention of Printing," in Cecil Roth (ed.), *Jewish Art*, Jerusalem: Massada, 1971, pp. 149–162.
Narkiss, Betzalel, "Rashi's Maps," in Eli Shiller (ed.), *Zev Vilnay Jubilee Volume*, Jerusalem: Ariel, 1984, pp. 435–439 [Hebrew].
Nebenzahl, Kenneth, *Maps of the Holy Land*, New York: Abbeville Press, 1986.
Nebenzahl, Kenneth, "Zaddiq's Canaan," in T. C. Van Uchelen, K. van der Horst and G. Schilder (eds.), *Theatrum orbis librorum: liber amicorum Presented to Nico Israel on the Occasion of His Seventieth Birthday*, Utrecht: HES, 1989, pp. 39–46.
Ne'eman, Yehuda, "Is Acre Part of Eretz Israel?" *Sinai* 77 (1991): 28–41 [Hebrew]
Neher, Andre, "New Material concerning David Gans as Astronomer," *Tarbiz* 45/1–2 (1975–1976): 138–147 [Hebrew].
Neher, Andre, *David Gans (1541–1613) and His Times: Jewish Thought and the Scientific Revolution of the Sixteenth Century*, Jerusalem: Rubin Mass, 2005 [Hebrew].
Neubauer, Adolf, *Catalogue of the Hebrew Manuscripts in the Bodleian Library*, Oxford: 1886.
Ofer, Joseph, "The Ḥazzekuni Commentary on the Bible and its Different Versions," *Megadim* 8 (1989): 69–83 [Hebrew].
Ofer, Joseph, "The Maps of the Land of Israel in Rashi's Commentary on the Torah and the Status of MS Leipzig 1," *Tarbiz* 76/3–4 (2007) [Hebrew].
Offenberg, A. K., "Bibliography of the Works of Jacob Jehuda Leon (Templo)," *Studia Rosenthaliana* 12 (1978): 111–132.
Offenberg, A. K., "Jacob Jehuda Leon (1602–1675) and His Model of the Temple," in Johannes van den Berg and Ernestine G. E. van der Wall (eds.), *Jewish-Christian Relations in the Seventh Century, Studies and Documents*, Dordecht, Boston, and London: Kluwer, 1988, pp. 95–115.

Oliphant, Lawrence, *The Land of the Gilead with Excursions in the Lebanon*, New York: 1881.
O'Loughlin, Thomas, "Map as Text: A Mid-Ninth Century Map for the Book of Joshua," *Imago Mundi* 57 (2005): 7–22.
O'Loughlin, Thomas, "Map Awareness in the Mid-Seventh Century: Jonas' Vita Columbani," *Imago Mundi* 62 (2010): 83–85.
Ortelius, Abraham Ortelius, *Theatrum orbis terrarum*, Antwerp: A. Coppen van Diest, 1570–1571.
Meir and Joseph ben Jacob of Padua, Psalms […] in] the Ashkenazic Tongue, Mantua: 1562.
Pearl, Chaim, *Rashi*, London: Peter Halban, 1988.
Peled, Haviva, "Seven Artists," in Yona Fischer, *Art and Artisanship in Nineteenth-Century Eretz Yisrael*, Jerusalem: Israel Museum, 1979 [Hebrew].
Pelli, Moshe, "Rabbi Áron Chorin's Ideological and Halakhic War on Behalf of
Religious Reform in Judaism," *Hebrew Union College Annual* 39 (1968): 184–185 [Hebrew].
Penkower, Jordan S., "Bomberg's First Bible Edition and the Beginning of his
Printing Press," *Kirjath Sepher* 58/3 (1983): 586–604 [Hebrew].
Penkower, Jordan S., "A Replica of a Diagram of the Temple and its Courts that Rashi Sent to R. Samuel of Tzavyareh," *Mikraot Gedolot HaKeter, the Book of Ezekiel*, Ramat Gan, Israel: Bar-Ilan University Press, 2000 [Hebrew].
Piccirillo, Michele and Alliata, Eugenio (eds.), *The Madaba Map Centenary 1897–1977*, Jerusalem: Studium Biblicum Francescanum, 1999.
Pilarczyk, Krzysztof, "Hebrew Printing Houses in Poland against the Background of their History in the World," *Studia Judaica* 7/2 (2004): 201–222.
Pliny the Elder, *Historia Naturalis*, book VIII, Loeb Classical Library, no. 38, Cambridge, MA: Harvard University Press, 1967.
Prawer, Joshua, *A History of the Latin Kingdom of Jerusalem*, vol. 1, Jerusalem: Bialik Institute: 1984 [Hebrew].
Praz, Mario, *Studies in Seventeenth-Century Imagery*, Rome: Edizioni di Storia e Letteratura, 1964.
Preschel, Tovia, "More on Yehiel Hillel Altschuler," *Sinai* 66 (1970): 169–172 [Hebrew].
Ptolemaeus, Claudius, *Geographia*, Lugduni: apud Hugonem à Porta, 1541. Quaresmius, *Historica theologica et moralis Terrae Sanctae elvcidatio*, 2 vols, Antwerp: Balthasar Moreti, 1639.
Raz-Krakotzkin, Amnon, "The Return to the History of Redemption (Or, What is the 'History' to Which the 'Return' in the Phrase 'The Jewish Return to History' Refers?)," in eds. S. N. Eisenstadt and Moshe Lissak, *Zionism and the Return to History: A Reappraisal*, Jerusalem: Yad Ben-Zvi 1999, pp. 249–276 [Hebrew].
Raz-Krakotzkin, Amnon, *Censorship, Editing and the Text: Catholic Censorship and Hebrew Literature in the Sixteenth Century*, Jerusalem: Magnes Press, 2005 [Hebrew].
Ratzaby, Yehuda, "R. Shlomo Adeni and His Work *Melekhet shlomo*," *Sinai* 106 (1990): 243–244 [Hebrew].
Rée, Peta, "Shaw, Thomas (1694–1751)," *Oxford Dictionary of National Biography*, Oxford: Oxford University Press, 2004.
Reiner, Elchanan, "From Joshua to Jesus—the Transformation of a Biblical Story to a Local Myth (a Chapter in the Religious Life of the Galilean Jew," *Zion* 61/4 (1986): 281–317 [Hebrew].
Reiner, Elchanan, *Pilgrims and Pilgrimage to Eretz Yisrael, 1099–1517*, PhD diss., Jeruslaem: The Hebrew University, 1988 [Hebrew].
Reiner, Elchanan, "'Since Jerusalem and Zion Stand Separately:' The Jewish Quarter of Jerusalem in the Post-Crusade Period," in Yossi ben Artzi, Israel Bartal, and Elchanan Reiner (eds.), *Studies in Geography and History in Honour of Yehoshua Ben-Arieh*, Jerusalem: Magnes Press, 1999, pp. 277–321 [Hebrew].
Reiner, Elchanan, "Traditions of Holy Places in Medieval Palestine—Oral versus Written," in Rachel, Sarfati, *Offerings from Jerusalem*, Jerusalem: Israel Museum, 2002, pp. 9–17.

Reiner, Elchanan, "'Oral versus Written:' The Shaping of Traditions of Holy Places in the Middle Ages," in Yehsoua Ben-Arieh and Elchanan Reiner (eds.), *Studies in the History of Eretz Israel, Presented to Yehuda Ben Porat*, Jerusalem: Yad Ben-Zvi, 2003, pp. 308–345 [Hebrew].

Reisen (Rejzen), Zalman, *Leksikon fun der yidisher literatur, prese, un filologye*, Vilna: 1926–1929 (New York: 1960) [Yiddish].

Reland, Hadrian, *Palaestina ex monumentis veteribus illustrata* (Trajecti Batavorum), Utrecht: Broedelet, 1714.

Riemer, Jack and Berenbaum, Michael, "Leeser, Isaac," *Encyclopaedia Judaica*, 2nd ed., vol. 12, 2007, pp. 600–601.

Ripa, Cesare, *Iconologia*, Roma: 1598.

Ripa, Cesare, *Iconologia*, Padua 1611 (repr. New York 1976).

Riqueti, Joseph Shalit ben Eliezer, *Ḥokhmat ha-mishkan*, Mantovah: 1675/6 [Hebrew].

Ritter, Michael, "Seutter, Probst and Lotter: An Eighteenth-Century Map Publishing House in Germany," *Imago Mundi* 53 (2001): 130–135.

Robinson, Arthur H. and Petchenik, Barbara B., *The Nature of Maps*, Chicago: University of Chicago Press, 1976.

Robinson, Edward, *Palästina und die südlich angrenzenden Länder: Tagebuch einer Reise im Jahre 1838 in Bezug auf die biblische Geographie unternommen von E. Robinson und E. Smith*; mit neuen Karten und Plänen in fünf, Halle: Blättern, 1841–1842.

Robinson, Edward and Eli Smith, *Biblical Researches in Palestine, Mount Sinai and Arabia Petrea in 1838*, London: John Murray, 1841.

Robinson, Edward, *Biblical Researches in Palestine*, London: John Murray, 1856.

Röhricht, Reinhold, *Bibliotheca Geographica Palaestinae*, Berlin: 1890 (repr. Jerusalem: 1963).

Rosenau, Helen, "The Architecture of Nicolaus de Lyra's Temple Illustrations and the Jewish Tradition," *Journal of Jewish Studies* 25/2 (1974): 294–304.

Rosenau, Helen, *Vision of the Temple: The Image of the Temple of Jerusalem in Judaism and Christianity*, London: Oresko Bks, 1979.

Rosenthal, Avraham, "When did Daniel Bomberg Begin to Print?" *Sinai* 78 (1976): 186–191 [Hebrew].

Rosenthal, Judah, "The Antichristian Dispute," in Simon Federbush (ed.), *Rashi: His Teachings and Personality*, New York: World Jewish Congress, 1958, pp. 45–59 [Hebrew].

Roth, Cecil (ed.), *The Casale Pilgrim: A Sixteenth-Century Illustrated Guide to the Holy Places*, London: Soncino Press, 1929.

Rubin, Rehav, "Iconography as Cartography: Two Cartographic Icons of the Holy City and its Environs," in G. Tolias and D. Loupis (eds.), *Eastern Mediterranean Cartography*, Athens: Institute of Neohellenic Research, 2004, pp. 347–378.

Rubin, Rehav, "One City, Different Views: A Comparative Study of Three Pilgrimage Maps of Jerusalem," *Journal of Historical Geography* 32 (2006): 267–290.

Rubin, Rehav, "Mapping a Myth: The Cartographic Image of the Overthrowing of Sodom and Gomorrah," paper presented at the International Conference on the History of Cartography, Copenhagen, July 2009.

Rubin, Rehav, "A Sixteenth-Century Hebrew Map from Mantua," *Imago Mundi* 62/1 (2010): 325–340.

Rubin, Rehav, "A Chronogram Dated Map of Jerusalem," *The Map Collector* 55 (Summer 1991): 30–31.

Rubin, Rehav and M. Levy-Rubin, "An Italian Version of a Greek-Orthodox Proskynetarion," *Oriens Christianus* 90 (2006): 184–201.

Rubin, Rehav and M. Levy-Rubin, "The Early Cartographic Tradition of the Holy Land and the Origins of the Crusader Maps of Jerusalem," in *Eretz Israel*, vol. 28, Teddy Kollek Volume, Jerusalem: The Israel Exploration Society, 2007 [Hebrew with English abstracts].

Ruderman, David B., *The World of a Renaissance Jew: The Life and Thought of Abraham ben Mordecai Farissol*, Cincinnati: Hebrew Union College Press, 1981.
Rudy, Kathryn M., *Virtual Pilgrimages in the Convent: Imagining Jerusalem in the Late Middle Ages*, Turnhout, Belgium: Brepolis, 2011.
Sabar, Shalom, "Messianic Aspirations and Renaissance Urban Ideals: The Image of Jerusalem in the Venice Haggadah, 1609," Bianca Kühnel (ed.), *The Real and Ideal Jerusalem in Jewish, Christian and Islamic Art*, Jerusalem: The Hebrew University, 1998, (=*Jewish Art*, 23–24 (1997) pp. 295–312.
Sabar, Shalom, "Herlingen, Aaron Wolff (Schreiber) of Gewitsch," *Encyclopaedia Judaica*, 2nd ed., vol. 9, 2007, p. 23.
Sabar, Shalom, "Seder Birkat Ha Mazon—Wien 1720, the earliest Illuminated Manuscript of the artist Aaron Wolff Schreiber Herlingen of Gewitsch," in Shmuel Glick (ed.), *Zekhor Davar le-Avdekha: Essays and Studies in Memory of Dov Rappel*, Jerusalem 2007, pp. 455–427 [Hebrew].
Sabar, Shalom, "From Amsterdam to Bombay, Baghdad, and Casablanca: The Influence of the Amsterdam Haggadah on the Haggadah Illustration among the Jews in India and the Lands of Islam," Yosef Kaplan (ed.), *The Dutch Intersection: The Jews and the Netherlands in Modern History*, Leiden and Boston: Brill, 2008, pp. 279–299.
Salomons, Reuven, "Tracking the Vanished Mapmaker," *Et-mol* 215 (2011): 10–13 [Hebrew].
Salway, Benet, "The Nature and Genesis of the Peutinger Map," *Imago Mundi* 57/ 2 (June 2005): 119–135.
Samet, Moshe, "Emden, Jacob," *Encyclopaedia Judaica*, 2nd ed., vol. 6, 2007, pp. 392–394.
Sanudo, Marino, *Liber Secretorum Fidelium Crucis*, Hanoviae: 1611 (repr. Jerusalem: Massada, 1972).
Sanudo, Marino, *Part XIV of Book III of Marino Sanuto's Secrets for True Crusaders to Help them to Recover the Holy Land*, The Library of the Palestine Pilgrims Texts Society, vol. 12, trans. Stewart Aubrey, London: Committee of the Palestine Exploration Fund, 1896.
Sanudo, Marino, *The Book of the Secrets of the Faithful of the Cross* (*Liber secretorum fidelium crucis*), trans. Peter Lock, Farnham, Surrey: Ashgate, 2011.
Sarfati, Rachel, "The Illustrations of Yiḥus ha-Avot: Folk Art from the Holy Land," in idem (ed.), *Offerings from Jerusalem: Portrayals of Holy Places by Jewish Artists*, Jerusalem: Israel Museum, 2002, pp. 21–30.
Sasson, Avraham, *Hebrew Authors and Scholars in the Nineteenth Century and Their Contribution to the Historical-Geographical Study of the Land of Israel in the Nineteenth Century*, PhD. thesis, Ramat-Gan, Israel: Bar-Ilan University, 1998 [Hebrew].
Sasson, Avraham, "Solomon Levisohn and His Contribution to the Study of the Land of Israel in the Nineteenth Century," *Jerusaelm & Eretz-Israel* 1 (2003): 153–172 [Hebrew].
Sasson, Avraham, "The Life and Enterprise of Rabbi Yehoseph Schwarz, Author of *Tvu'ot ha'areṣ*," *Sinai* 134 (2005): 181–195 [Hebrew].
Scafi, Alessandro, *Mapping Paradise: A History of Heaven on Earth*, Chicago: Univeristy of Chicago Press, 2006.
Schacter, Jacob J., *Rabbi Jacob Emden: Life and Major Works*, PhD diss., Cambridge, MA: Harvard University, 1988.
Schattner, Isaac, "Review of Gustav Hölscher, *Drei Weltkarten*, 1949," *Imago Mundi* 8 (1951): 116.
Schattner, Isaac, "Maps of Ereẓ Israel," *Encyclopaedia Judaica*, 2nd ed., vol. 13, 2007, pp. 500–505.
Schattner, Isaac, "Maps of Eretz Yisrael," *Encyclopaedia Hebraica*, vol. 6, 1957, pp. 1160–1171 [Hebrew]..
Schattner, Isaac, *The Map of Eretz Yisrael and its Annals*, Jerusalem: Bialik Institute, 1951.

Schubert, Ursula, *Die Jüdische Buchkunst*, trans. Yedidya Peles, Tel Aviv: Hakibbutz Hameuchad, 1994 [Hebrew].
Schwarz, Yehoseph (Joseph), *Sefer tvu'ot ha'areṣ: The Book of Joseph's Words*, Jerusalem: Y. Bek Press, 1845 [Hebrew].
Schwarz, Yehoseph (Joseph), *Das Heilige Land*, Frankfurt am Main: J. Kaufmann, 1852 [Hebrew].
Schwarz, Yehoseph (Joseph), *Sefer tvu'ot ha'areṣ*, Lemberg: Y. M. Shtand Press, 1865 [Hebrew].
Schwarz, Yehoseph (Joseph), *Sefer tvu'ot ha'areṣ*, Jerusalem: Abraham Moses Luncz Press, 1900 [Hebrew].
Schwarz, Yehoseph (Joseph), *Sefer tvu'ot ha'areṣ*, Jerusalem: the Committee to Publish the Writing of R[abbi] J[ospeh] Schwarz, 1978 [Hebrew].
Schwarz, Yehoseph (Joseph), *Sefer tvu'ot ha'areṣ*, Jerusalem: Ariel, 1979 [Hebrew].
Schwarz, Yehoseph (Joseph), *Sefer tvu'ot ha'areṣ*, Jerusalem: the M. Schwarz Family, 1998 [Hebrew].
Schwarz, Yehoseph (Joseph), *A Descriptive Geography and Brief Historical Sketch of Palestine* (trans. from the Hebrew by Isaac Leeser), Philadelphia: A. Hart, 1850.
Schwarz, Yehoseph (Joseph), *In the Gates of Jerusalem: Documents on the Annals of Jerusalem and its Inhabitants*, Jerusalem: The Writings of Rabbi Yehoseph Schwarz Press, 1979 [Hebrew].
Scott, James M., *Geography in Early Judaism and Christianity: The Book of Jubilees*, Cambridge: Cambridge University Press, 2002.
Sed-Rajna, Gabrielle, "Some Further Data on Rashi's Diagrams to His Commentary of the Bible," *Jewish Studies Quarterly* 1 (1993/4): 149–157.
Sed-Rajna, Gabrielle, *Les Manuscrits Hébreux Enluminés des Biblithèques de France*, Leuven: Peters, 1994.
Seder Haggadah shel pesakh [Arrangement of the Passover Haggadah] with Illustrated Miracles and Wonders [...] by Isaac ben R. Samuel Bassan, Mantua, Rufinalli [Press], 1560 [Hebrew].
Seder Haggadah shel Pesakh [Arrangement of the Passover Haggadah], Venice: Yo'ani Ḳalioni, 1629 [Hebrew].
Seder Haggadah shel Pesakh [Passover Haggadah]: with a Beautiful Interpretation and Drawings [...] on Copper Plates / by the lad Abraham bar Yaacov from the family of Abraham our Patriarch, Amsterdam: the printing press of Asher Anshel ben Eliezer and Issachar Ber ben Abraham Eliezer, 1695 [Hebrew].
Sefer yiḥus ha'ṣadikim ha'nikbarim be'eretz yisrael u'be'yerushalayim [The Book of the Pedigree of the Righteous Buried in Eretz Yisrael and in Jerusalem] [...], brought to print by Gershom ben (Moses) Asher, Mantua: Press of Jacob ben Naphtali Hacohen of Gazolo, 1561 [Hebrew].
Seiffe, Y., "Mosheh Fainkind," in Eliezer Esterin (ed.), *Turek: A Memorial to the Jewish Community of Turek, Poland*, Tel Aviv: 1982, pp. 203–206 [Hebrew].
Seutter, Matthaeus, *Atlas novus sive Tabulae geographicae totius orbis*, Augustae Vindelicorum, M. Seutter, ca. 1757.
Sezgin, Fuat, *The Contribution of the Arabic-Islamic Geographers to the Formation of World Map*, Frankfurt am Main: Institute for the History of Arabic-Islamic Science, 1987.
Sezgin, Fuat, *Mathematical Geography and Cartography in Islam and their Contribution in the Occident*, Frankfurt am Main: Institute for the History of Arabic-Islamic Science, 2000–2007.
Shachar, Isaiah, *Jewish Tradition in Art: The Feuchtwanger Collection of Judaica*, trans. R. Grafman, Jerusalem: Israel Museum, 1981.
Shalev, Zur, "Sacred Geography, Antiquarianism and Visual Erudition: Benito Arias Monatno and the Maps of the Antwerp Polyglot Bible," *Imago Mundi* 55 (2003): 55–80.
Shalev, Zur, *Sacred Words and Worlds Geography, Religion and Scholarship, 1550–1700*, History of Science and Medicine Library, vol. 21, *Scientific and Learned Cultures and their Institutions*, vol. 2, Leiden and London: Brill, 2012.

Shapira, Natan (Neta) ben Shimshon, *Bai'urim al ha'eshel ha'gadol Rashi* (*Explanations on Rashi the Great Tamarisk of Blessed Memory*), Venice: the Matteo Zanetti and Comino Presigno Press, 1593 [Hebrew].

Sharif, Sharif, *Hidden Places: Wall and Ceiling Paintings in Ottoman Palestine (1856–1917)*, Tel Aviv: Eretz Israel Museum, 2002.

Shaw, Thomas, *Travels, or Geographical, Physical and Miscellaneous Observations Relating to Several Parts of Barbary and the Levant*, Oxford: Printed at the Theatre, 1738.

Shemerok, Chone and Bartal, Israel, "'Tla'ot Moshe': The First Geographical Description of Eretz Israel in Yiddish," *Cathedra* 40 (July 1986): 121–137 [Hebrew].

Shereshevsky, Esra, *Rashi, the Man and His World*, New York: Sepher Hermon Press, 1982.

Shilhav, Yosef, "Interpretations and Misinterpretations of Jewish Territoriality," *Studies in the Geography of Israel* 12 (1986): 142–150 [Hebrew].

Shirley, Rodney W., *The Mapping of the World: Early Printed World Maps, 1472–1700*, London: Holland Press, 1984.

Shisha Ha-Levy, Avraham, "Miṣudat David and Miṣudat Zion," *Yeshurun* 3 (1997): 617–48 [Hebrew].

Shisha Ha-Levy, Avraham, "Mezudat David and Mezudat Zion," *Yeshurun* 6 (1999): 661–677 [Hebrew].

Shisha Ha-Levy, Avraham, *Jewish Life in Renaissance Italy*, New York: Ogen Press, 1955 [Hebrew].

Shulvass, Moses Avigdor, *Rome and Jerusalem: The Annals of Italian Jewry's Attitude to the Land of Israel*, Jerusalem: Mosad ha-Rav Kook, 1944 [Hebrew].

Sieber, Franz Wilhelm, *Reise von Cairo nach Jerusalem und wieder zurück […]*, Prag and Leipzig 1823.

Simonshon, Shlomo, *History of the Jews in the Duchy of Mantua*, Jerusalem: Kirjath Sepher, 1977.

Skalova, Zuzana, "A Holy Map to Christian Tradition, Preliminary Notes on Painted Proskynetaria of Jerusalem in the Ottoman Era," *Eastern Christian Art* 2 (2005): 93–103.

Smalley, Beryl, *The Study of the Bible in the Middle Ages*, Oxford: Blackwell, 1952 (repr. South Bend, IN: University of Notre Dame Press, 1964).

Sokolow, Florian, *My Father, Nahum Sokolow*, Jerusalem: The Zionist Library, 1970 [Hebrew].

Sokolow, Nahum, *Maṣukei areṣ*, Warsaw: Izaak Goldman Press, 1878 [Hebrew].

Sokolow, Nahum, *Ereṣ ḥemdah* [*The Precious Land*], Warsaw: Izaak Goldman Press, 1885 [Hebrew].

Sotheby's, *Magnificent Judaica and Manuscripts: Including Property Formerly in the Furman Collection* (catalog of an auction held on Tuesday, December 12, 2000), New York: Sotheby's, 2000.

Spitz, Jonas, *Eretz ṣvi Das biblische Palästina: zum bessern Verstaendnisse der heiligen Schrift in geographischer Beziehung fuer Schul-und Privat-Gebrauch*, Prag: M. J. Landau, 1853.

Stein Kokin, Daniel, "Entering the Labyrinth: On the Hebraic and Kabbalistic Universe of Egidio da Viterbo," in Ilana Zinguer, Avraham Melamed, and Zur Shalev (eds.), *Hebraic Aspects of the Renaissance: Sources and Encounters*, Leiden and Boston: Brill, 2011, pp. 27–42.

Stern, David, "Mapping the Redemption: Messianic Cartography in the 1695 Amsterdam Haggadah," *Studia Rosenthaliana* 42–43 (2010–2011): 43–63.

Sussmann, Yaakov, "The 'Boundaries of Eretz-Israel,'" *Tarbiz* 45/1–2 (1975–1976): 213–257 [Hebrew].

Teomim, Joshua Feivel ben Israel of Tarnogród, *Sefer kiṣvai areṣ*, Zholkva: The Grandsons of Uri Feivish Segel, 1772 [Hebrew]..

Teomim, Joshua Feivel ben Israel of Tarnogród, *Sefer kiṣvai areṣ*, Grodno: Simcha Zimel ben Menachem Nachum Press, 1813 [Hebrew].

Tishby, Ariel (ed.), *Holy Land in Maps*, Jerusalem: Israel Museum and New York: Rizzoli, 2001.

Tishby, Ariel (ed.), *Holy Land in Maps*, Jerusalem: Israel Museum, Jerusalem 2001 [Hebrew].

Tobias, Alexander and Derovan, David, "Ḥayyim ben Bezalel," *Encyclopaedia Judaica*, 2nd ed. vol. 8, 2007, pp. 480–481.

Tolkes, Jerucham, "Chorin, Aaron," *Encyclopaedia Judaica*, 2nd ed., vol. 4, pp. 667–668.

Torah (Pentatuch), *The Five Books of the Torah with the Translation and Com[mentary] of Rashi and the Abridged Mizrachi [Supercommentary] and the Five Scrolls with the Com[mentary] of Rashi and the Commentary of Ba'al ha-Akedah*, Riva di Trento: 1521 [Hebrew].

Torah (Pentatuch), *Five Books of the Torah: [. . .] Adorned and Arrayed in Ten Raiments of Light and Commentaries of Precious Glory*, volume 4, Book of Numbers, Frankfurt an der Oder: P. P. Dadre, 1746 [Hebrew].

Torah (Pentatuch), *Ḥumash makor ḥayyim*, Berlin: 1831–1833 [Hebrew].

Touitou, Elazar, "What Motivated Rashi to Write a Commentary on the Pentateuch," in Avraham Grossman and Sara Japhet (eds.), *Rashi: The Man and His Work*, vol. 1, Jerusalem: Zalman Shazar Center, 2008, pp. 33–62 [Hebrew].

Tsafrir, Yoram, "The Provinces of Eretz Yisrael—Names, Borders and Administrative Jurisdictions," in Zvi Bras et al. (eds), *Eretz Israel from the Destruction of the Second Temple to the Muslim Conquest*, Jerusalem: Yad Ben-Zvi, 1982, pp. 350–386 [Hebrew].

Tzartza, Samuel, *Sefer mekor ḥayim: An Explanation on the Torah*, Mantua: 1559 [Hebrew].

Vilna Gaon = Elijah ben Shlomo Zalman, *Sefer ṣurat ha'areṣ le'gvuloteha*, Shklov: Aryeh Leib ben Shene'ur Feivish and Sabbatai ben Ben-Zion, 1802 [Hebrew].

Vilna Gaon = Leib ben Shene'ur Feivish and Sabbatai ben Ben-Zion, *Sefer aderet eliyahu, a Commentary on the Prophets and the Hagiographa*, by Elijah of Vilnius[,] a Righteous Person of Blessed Memory, published by Elijah son of Rabbi Eliezer Landau[,] grandson of the Genius Rabbi Elijah, Jerusalem (1905).

Vilnay, Zev, "Maps of Palestine in Rabbinical Literature," *Eretz-Israel*, vol. 2 (dedicated to the memory of Zalman Lif), Jerusalem: Israel Exploration Society, 1953, pp, 89–92 [Hebrew].

Vilnay, Zev, *The Hebrew Maps of the Holy Land: A Research in Hebrew Cartography*, 2nd ed., Jerusalem: Ahiever, 1968 [Hebrew].

Wajntraub, Eva and Wajntraub Gimpel, *Hebrew Maps of the Holy Land*, Wien: Brüder Hollinek, 1992.

Wajntraub, Eva and Wajntraub Gimpel, "Hebrew Map Showing the Dispersion of the Sons of Noah," in Harold Brodsky (ed.), *Land and Community: Geography in Jewish Studies*, Bethesda MD: University Press of Maryland, 1997, pp. 31–35.

Warhaftig, Itamar, "The Borders of Eretz Yisrael according to Maimonides," *Techumin* 2 (1981): 398–411 [Hebrew].

Wazana, Nili, *All the Boundaries of the Land: The Promised Land in Biblical Thought in Light of the Ancient Near East*, Jerusalem: the Bialik Institute, 2007 [Hebrew].

Weinberg, Joanna, *Azariah de' Rossi's Observations on the Syriac New Testament*, London: Warburg Institute and Turin: Nino Aragno Editore, 2005.

Weinberg, Raphael Shimon, "Yoseph ben Yehoshua Cohen and His Book *Maziv gevulot amim*," *Sinai* 72 (1973): 333–364 [Hebrew].

Weinfeld, Moshe, "The Extent of the Promised Land: Two Points of View," *Cathedra* 47 (1988): 3–16 [Hebrew].

Williams, John, *The Illustrated Beatus—a Corpus of the Illustrations of the Commentary on the Apocalypse*, vol. 4, London: Harvey Miller, 2002.

Frederick de Wit, "Terra Sancta, sive Promissionis, olim Palestina, recens deline, et in lucem edita per Fredericum de Wit," Tot Amsterdam: Gedruckt by Frederick de Wit, ca. 1680.

Wood, Denis, "How Maps Work," *Cartographica* 29/3–4 (1992): 66–74.

Wood, Denis, *The Power of Maps*, London: Guilford Press, 1992.

Woodward, David, "Reality, Symbolism, Time and Space in Medieval World Maps," *Annals of the American Association of Geography* 75 (1985): 510–521.

Woodward, David, "Medieval Mappaemundi," in John B. Harley and idem, *History of Cartography*, vol. I, Chicago: University of Chicago Press, 1987, pp. 286–370.

Woodward, David, *Maps as Prints in the Italian Renaissance, Makers, Distributors and Consumers*, London: The British Library, 1996.

Woodward, David, *The History of Cartography*, vol. III, part 2, *Cartography in the European Renaissance*, Chicago and London: University of Chicago Press, 2007.

Ya'ari, Abraham, "Miscellaneous Bibliographical Notes 1: On the History of the Hebrew Printing at Nowy Dwor," *Kirjath Sepher* 10 (1933): 372–374 [Hebrew].

Ya'ari, Abraham, "Miscellaneous Bibliographical Notes 25: Illustrations of Jerusalem and the Temple as Decoration for Hebrew Books," *Kirjath Sepher* 15 (1938–1939): 377–382 [Hebrew].

Ya'ari, Abraham, "Miscellaneous Bibliographical Notes 37: The Drawing of the Seven Walls of Jericho in Hebrew Manuscripts," *Kirjath Sepher* 18 (1941–1942): 179–181 [Hebrew].

Ya'ari, Abraham, "Miscellaneous Bibliographical Notes 41: Hebrew Map of Palestine Printed at Nowy Dwor 1784," *Kirjath Sepher* 19 (1942–1943): 204–217 [Hebrew].

Yerushalmi, Yosef Hayim, *Haggadah and History*, 2nd ed., Philadelphia: Jewish Publication Society of America, 1975.

Yisraeli, Yael, *To the Light of the Menorah*, Jerusalem: Israel Museum, 1998 [Hebrew].

Yoffe, Mordechai, *Levush ha'orah*, Prague: Haim ben Yaacov ha-Cohen, 1604 [Hebrew].

Yudlov, Isaac (ed.), *The Haggadah Thesaurus: Bibliography of Passover Haggadot from the Beginning of Hebrew Printing until 1960*, Jerusalem: Magnes Press, 1997 [Hebrew].

Zalkin, Mordechai, *A New Dawn: The Jewish Enlightenment in the Russian Empire—Social Aspects*, Jerusalem: Magnes Press, 2000 [Hebrew].

Zalkin, Mordechai, "Scientific Literature and Cultural Transformation in Nineteenth-Century East European Jewish Society," *Aleph* 5 (2005): 249–271.

Ziegler, Jacob, *Quae intus continentur Syria, [...] Palestina, [...] Arabia, [...] Aegyptus [...]*, Argentorati, Apud Petrum Opilionem, Strassburg: [1532]).

Ziskind, Alexander, *Yesod ve'shoresh ha'avodah*, Grodno: Baruch ben Yosef, 1795 [Hebrew].

Ziskind, Alexander, *Yesod ve'shoresh ha'avodah*, Yehuda Lima ben Aryeh Leib Segal and Simcha Zimel ben Yechezkel, 1810 [Hebrew].

Ziskind, Alexander, *Yesod ve'shoresh ha'avodah*, Grodno and Vilnius: Menachem Mann and Simcha Zimel Press, 1817 [Hebrew].

Zögner, Lothar (Hg.), *Antike Welten Neue Regionen: Heinrich Kiepert 1818–1899*, Staatsbibliothek zu Berlin—Preußischer Kulturbesitz, Ausstellungskataloge, Neue Folge 33, Berlin: 1999.

Zunz, Leopold, "On the Geographical Literature of the Jews from the Remotest Times to the Year 1841," A. Asher (trans. and ed.), *The Itinerary of Rabbi Benjamin of Tudela*, New York: 1927, pp. 393–448.

Index

Aaron 7, 28, 41, 62, 77, 114, 116, 121, 146, 231 f., 235, 280, 286
Abraham XX, 38, 45, 47, 99, 113 f., 137, 168 f., 209, 222, 230, 283, 285
Abu Ghosh 213
Achzib 1, 14, 133, 215
Acre 1, 14, 69–72, 74 f., 87, 133 f., 151, 164, 200, 204, 206, 232, 245, 257, 268, 285
Adeni, Shlomo 73–75, 200, 285
Adoraim 280
Adrichom, Christian van 65, 87, 102–104, 108–115, 119, 125, 129–133, 135, 173, 195, 223, 232, 285
Adullam 208
Ahab 111
al-Aqsa 207
al-Arish 131, 227, 238, 244
Alfalo, Shalom 205
Allotments of the tribes, see: Tribal allotments
Alma 37
Altschuler, David ben Aryeh 181, 182
Altschuler, Yehiel Hillel 22, 151, 178, 181–184, 186, 202, 236
Amalek 97, 196, 251
Ammon 30, 51, 83, 147, 174, 187, 191, 199, 266
Ammonites 24, 34, 83, 194, 196
Ammuqa 208
Amsterdam 65, 72, 75, 102 f., 108, 110, 113 f., 116–121, 124, 155, 171, 285, 287
Anglo-Palestine Bank 274
Antioch 35, 83, 156, 194
Antipatris 238, 251
Antwerp 38, 87, 97, 142, 173
Apameas 176
Apolonia 133
Apostles 135, 287
Arabah 245, 257
Arabia 140, 167, 173, 226 f., 236 f., 251, 267
Arabian Desert 140, 226, 268
Arabic 20, 37, 168, 237 f., 243, 245, 259 f., 267, 269 f., 272, 274, 276, 278, 280
Arad 77, 84
Aram 28, 191, 194, 264
Aram-Naharaim 28, 153, 194, 198 f.
Aram-Zobah 194, 196
Arama, Isaac ben Moses 42

Archelaus 251
Arias Montano, Benedictus 87
Arnon River 34, 83, 113, 124 f., 140, 187, 194
Aroer 187, 194, 280
Ascent of Akrabbim 4, 7, 24, 41, 50, 84, 189, 195, 208
Aschaffenburg, Simeon ben Isaac Halevi 42–45, 47, 49, 285
Ashdodian 28
Asher 1, 26, 65, 72, 100, 103, 113, 131, 166, 176, 178, 180, 182, 190, 192, 195
Ashkelonian 28
Asphalt, sea of, see: Dead Sea
Assyria 176, 194, 265
Ataroth Addar 27 f.
Auspitz, Yaakov 140–152, 159, 170 f., 184, 287
Austro-Hungarian Empire 77
Avraham ben Avigdor 49
Azariah de' Rossi, 99
Azmon 4, 24, 41, 50, 84, 175, 189, 195, 208
Ḥazzekuni 23–25, 42

Baal Zephon 82
Baalbek 207, 254, 269
Balfour Declaration 276, 278
Banyas, see Pamyas
Barbier, Nicolas 97
Bassan, Isaac ben Samuel 90, 92, 93, 96, 284
Bashan 24, 83, 133, 147, 187, 191, 194, 199, 245, 254
Basra 251
Bass, Shabbethai 53 f.
Bay of Haifa, see: Haifa Bay
Beatus maps 19
Be'er Sheva 36, 42, 45, 133 f., 247, 281
Be'er Tuvia 217, 289
Behar, Nissim 216
Beirut 37, 206, 208, 210, 215, 238, 252, 262, 268 f., 272
Beit Guvrin 36
Beit She'an 207, 215
Beit Shemesh 190
Ben Haim, Aaron 158 f.
Ben Hillel, Mordechai 289
Ben Isaac, Aryeh Leib of Sejny 194 f.
Ben Yaacov, Yehonatan 137–140, 170, 171
Ben Yehuda, Eliezer 271

Benjamin, Israel Joseph 248f.
Benjamin, tribe 27f., 65, 76, 85, 87, 131f., 136, 139, 164, 173, 178f., 182, 190, 195, 271
Berlin 13, 30f., 34, 69, 75f., 182, 226, 243, 272–274
Berman, Isaac ben Pinchas 196–198, 202
Bertius, Petrus 222
Beth Aven 195
Beth Hogla 27
Beth She'an 133, 226
Bethany 111
Bethel 13, 42, 45, 69, 192, 237
Bethoron 37
Bethphage 111
Bethsaida 260
Betulia 251
Bible 1f., 17, 20, 23, 26, 30, 35, 37f., 54, 78, 80, 82, 84, 86f., 96f., 100–103, 108, 110f., 118f., 124–126, 128f., 133, 135, 139, 141–143, 147, 151f., 157, 166–169, 173f., 176, 178f., 182, 184, 194, 202f., 212, 215, 223f., 227, 231, 235–238, 240, 257, 262, 264, 266, 269, 283f., 286
Bilhah 274, 276
Binder, M. M. 146, 159f., 171
Black Sea 224
Blanckenhorn, Max 280
Bloch, Samson 38, 92, 99, 222f.
Blücher, Ephraim Israel 251–255
Bnei Brak 254
Bnei Yehuda 217
Bomberg, Daniel 38f.
Bonn 243
Book of Jubilees 284
Brahe, Tycho 49
Brandeis, Louis 276–278
Breslau 214f., 218, 220
Breslauer, Menachem Mendel 223f., 226–229, 232, 236, 288
Buda, 140
Budapest 140, 251
Byblos 274

Caesarea 150, 251, 255
Cairo 167, 245, 268
Caleb ben Jephunneh 45
Calmet, Augustin Antoine 22, 142, 144, 185f., 232
Calvari 136
Campbell, Tony XVII

Canaan 4, 7, 10, 18f., 42, 45, 65f., 82, 87, 106, 108, 133, 135, 145, 161–163, 168f., 174, 192f., 243, 253, 257f., 262
Canaanites 18
Carmel, see: Mount Carmel
Catholic maps XXIII, 133, 146
Cave of the Patriarchs 86, 95, 155, 262
Chelm, Shlomo ben Moshe of 125–136, 171, 202, 287
Choriner, Aaron 75–78
Christian IV, King of Denmark 110
Cities of refuge 254
Colmar 71
Cologne 102
Constantinople 38, 54, 58, 74, 207, 249
Copenhagen 214
Courteau, Thomas 97
Coverdale, Myles 97
Crémieux, Adolphe 243
Crescas, Elisha ben Abraham 84
Crusades 17, 96
Culi, Yaakov 54
Cush 174, 176, 179, 191, 194, 285
Cyprus 198, 224, 245, 268

Damascus 113, 164, 166, 178, 194, 215, 227, 231, 248, 251f., 257, 260, 264, 274, 278
Dan 19, 26, 65, 103, 124, 130–132, 136, 166, 173, 176, 178, 180, 182, 190, 192, 195, 197, 245, 281
Danzig 71
Danzigerkron, Moshe 153, 157
Dead Sea 4, 7, 24, 30, 34f., 41, 50, 66f., 83, 103, 110f., 115, 121, 124f., 132, 140, 147, 149, 151, 159, 164, 175, 178f., 182, 195, 197, 199f., 203f., 208f., 227f., 231f., 234, 238, 245, 251f., 264, 268f., 278
Deir el Qamar 207f., 210, 212f.
Delano-Smith, Catherine 8, 20, 97, 184
Delft 102
Della Gatta, Giovanni Francesco 97
Dessau 71
Diglath 176
Dome of the Rock 85, 96, 207, 212, 260
Dor 59, 64, 146, 163f.
Dothan 130, 280
Dubno 138, 226

Ebal 195
Ebstorf map 116

Eden, see also Paradise 28, 167–169, 175, 194
Edom 4, 24, 28, 30, 34, 41, 50, 66, 84, 89, 146, 176, 187, 191, 195 f., 245, 257
Edrei 194
Egypt 4 f., 24, 29, 32, 41, 50, 57, 66, 74, 80–82, 84, 88, 90, 97, 101, 111–113, 115 f., 118 f., 121, 132, 139, 144–146, 152 f., 155, 158, 160–162, 167, 176, 188, 191, 196, 198, 200, 224, 231 f., 234, 243, 248, 257, 262, 264–266, 268, 280, 284, 286, 289 f.
Eidels, Samuel (Shmuel) Eliezer ha-Levi 69–71, 76–78
Eilat, see Ezion Geber
Ein Rimon 280
Ein Zeitim 266, 269
Ekron 24, 164, 260, 266, 289
Ekronian 28
el-Audja (Nitzana) 274, 278
Eleazar the Priest 148
Eli, Tomb of 86, 95 f., 209
Elijah ben Shlomo Zalman, see the Vilna Gaon
Elim 82, 126, 195
Em al-Gamal 217
Em al-Tut 217
Emden, Jacob 198–201
Emir Bashir 212
En Gannim 36
En Rogel 190
Enlightenment 69 f., 126, 135, 163, 186, 203, 221, 224, 229, 247, 270, 281, 288–290
Ephraim 26 f., 65, 129–131, 136, 139, 159, 166, 176, 178 f., 182, 195, 252, 270, 285
Eretz Israel 3, 14, 17, 69, 186, 221 f.
Esau 45
Eskrich, Pierre 184
Etham 41, 49, 82, 88, 187, 195 f.
Ezion Geber 166, 195, 257
Euphrates 28–30, 153, 158, 168 f., 175 f., 191, 194, 196, 199 f., 203, 257, 267, 286
Eusebius of Caesarea XXII, 19, 86, 173, 223, 236, 237
Exodus 23, 34, 42, 54, 82, 88, 94, 114, 118, 124, 144 f., 153, 179, 284, 289
Ezekiel 1 f., 8, 13, 15, 20–22, 26, 30, 82, 108, 116, 142, 151 f., 174, 177 f., 183–187, 202

Fainkind, Mosheh ben Pinhas 209–212
Falklish, Avraham 121, 124 f.
Farissol, Abraham 99, 222

Feinroit, Peretz, 153
First Aliyah 215 f., 270–272, 281, 288
Fischer, Hans 214, 219, 280
Frankfurt an der Oder 66
Fredrik Henry, Prince of Orange 110
Froschauer 97
Fuller, Thomas 125, 129, 223
Funkenstein, Juda 20, 257–259

Gad 26, 28, 65, 82 f., 126, 131, 139, 153, 166, 178, 183, 187, 194, 196, 199, 267, 286
Galgor, Meir Yekutiel 161 f.
Galilee 37, 76, 83, 96, 133, 136, 199, 204, 207, 210, 215, 217, 219, 232, 247, 251, 256, 260, 266, 276, 278, 280, 287 f.
Gamla 260
Ganbash, Moshe 205–210, 212, 218–220
Gans, David 49
Gaza 36, 124, 176, 190, 192, 205, 212, 257
Gazan 28
Gedera 217, 260, 266, 289
Geneva Bible 97 f., 143
Gerar 176, 195
Gergesa 111, 136
Gershon 62, 146, 163, 165
Geshuri 191
Gibraltar 57
Gihon River 169
Gilead 83, 151, 159, 179, 187, 194, 196, 199, 238, 272
Gilezer 83
Gilgal 84, 155, 251
Gittin 7, 14, 17, 28, 32, 69–72, 200, 266, 285
Godlewska, Anne 234
Goldzweig, Jacob 216, 260, 262–264, 270, 288
Goos, Abraham 108
Goshen, the land of 32, 81, 143
Graetz, Heinrich 240, 242
Great Sea, see also Mediterranean 4, 7, 20, 23, 24, 28, 35, 41, 66, 72, 75, 81, 88, 113, 124, 143, 145, 147, 151, 153, 158, 178, 179, 182, 194, 196, 200, 228, 254, 264
Gregory VII 17
Grodno 152 f., 156 f., 159, 161 f., 178 f., 192 f.
Grover, Ephraim Michael 269 f.
Günther of Bamberg 17
Gush Halav 37, 208
Guthe, Hermann 280

Habakkuk 108
Hadad-Rimmon 236
Hadera 217, 267, 269
Hadid 132
Haifa Bay 103, 164, 232
Haim ben Bezalel 50–52, 78
Halacha XIX, XX
Halberserg, Haim Isaiah Hacohen 65 f.
Haluza 272
Hamath 6 f., 23, 35, 41, 49, 51, 83, 156, 178, 187, 194, 290
Haran 208, 215, 257
Harley, Brian 102
Haskala, see also Jewish enlightenment 126, 172, 202, 221, 222, 229, 230, 242, 247, 270, 280, 287, 288
Hauran 178, 260, 274, 278
Hazar Addar 4, 24, 37, 41 f., 50, 84, 208
Hazar Enan 6 f., 23, 41, 51, 74, 83, 175, 178, 187
Hebrew 1–3, 10, 14, 17, 19–23, 26, 28–30, 34 f., 37 f., 41 f., 45, 47, 49 f., 52–55, 57, 65 f., 69–71, 73, 75–86, 88 f., 92 f., 95, 97, 99–101, 108–111, 113 f., 119–121, 124, 126 f., 129, 132 f., 135–142, 145 f., 151–153, 155, 157, 159, 161–164, 166 f., 169–174, 176, 178, 182, 184, 186, 190–194, 196, 198, 200, 202–206, 208–219, 221–224, 226 f., 229–232, 235, 237 f., 240, 242 f., 245–249, 251 f., 254 f., 257, 259 f., 262, 264, 266–272, 274, 276, 278, 280–291
Hebron 57, 73, 75, 86, 124, 155, 204, 207 f., 215, 220, 245, 260, 262, 280
Heidmann, Christoffer 139 f.
Herlingen, Aaron wolf Shreiber 121 f., 124 f.
Hermon, Mountain 115, 187, 194, 264
Herodium 133
Heshbon 83, 187, 208
Hezekiah ben Manoah 23 f., 26
Hinnom Valley 190
Hippos 133
Hirsch, Meir Isaac 69 f., 158, 161, 248, 252
Hirsch, Joseph 252
Hittites 24
Holy Land 8, 14, 17, 19, 50, 79 f., 84, 86, 97, 100, 102 f., 108, 113, 117, 119, 121, 137, 146–149, 166, 170 f., 173, 186, 190 f., 202, 204, 209–216, 219 f., 222, 227, 232–234, 237 f., 245, 247, 249, 262 f., 267 f., 280, 283, 285 f., 288–291

Hondius, Jodocus 114–117
Hormah 254
Hukkok 176
Hula Lake, Hula Sea 103, 124, 149, 164, 166, 182, 231, 278
Humboldt, Alexander von 248

Ibn Hawqal 20
India 100, 119, 174, 176, 179, 194, 248 f., 285
Ior 19, 103, 124
Ireland XVIII
Ishmael 176, 194
Ishtori Haparchi 35, 85 f.
Israel Museum 214, 219, 226
Israelites' peregrinations VII, XV, XXIII, 3, 11, 23, 34, 42, 80, 82, 87, 89, 90, 101, 111, 119, 121, 127, 132, 139, 142, 143, 157, 196, 227, 229, 231, 243, 245, 252, 264, 266, 284, 285
Issachar 65, 113, 131, 178, 180, 182, 195, 210
Istanbul 37 f., 125
Italy 38, 41 f., 80, 92, 96, 99 f., 120, 171, 287
Izhbitza 163, 171
Izmir 57, 60, 63, 125

Jaazer, sea of 124, 125, 194
Jabbok River 83, 187, 194, 238
Jabotinsky, Ze'ev 278–280
Jacob 23, 42, 45–47, 50, 88–90, 92 f., 97, 99 f., 124, 163, 192, 198, 224, 284
Jacob's ladder 23, 42, 45–47, 50
Jacotin, Pierre 232, 234
Jaffa 124, 207, 217, 227, 231, 234, 245, 257, 260, 266, 268, 274, 289
Janssonius, Jan 120
Jarmuth 208
Jebus 192, 195, 198
Jenin 207
Jericho 27 f., 30, 83–85, 87, 97, 111, 115, 124, 132, 143, 155, 179, 188, 192, 195 f., 198, 206, 208, 227, 268
Jerome 19 f., 173, 223, 229, 236
Jerusalem 8, 10, 13, 17, 19 f., 28, 32, 34, 72–77, 85 f., 95–97, 100, 103, 111, 118, 124, 126, 137, 151, 155, 162, 176–178, 182, 187–190, 192, 195 f., 198, 202, 204, 207 f., 210, 213, 215–217, 219 f., 226, 230 f., 235 f., 238, 243, 245, 247–249, 254, 257, 259 f., 265 f., 268 f., 271 f., 274, 276, 278, 280, 289

Jesus   84, 103, 111, 135 f., 232, 287
Jewish colonies XVIII, 215 – 217, 220, 260, 262, 266, 269 – 272, 274, 278, 280, 281, 288, 289
Jewish Colonization Association   274
Jewish enlightenment, see also: Haskalah XV, 126, 163, 171, 186, 221, 224, 229, 247, 270, 281, 283, 288, 290
Jezreel   36, 199, 226, 251, 278
Johann Boemus   99
Jonah   117, 286
Jordan River   4, 19, 23, 34, 50, 66, 84, 103, 115, 124, 132, 151, 153, 166, 178, 184, 194, 210, 227, 231, 256, 278, 290
Joseph ben Jacob of Padua   90, 92, 93, 99, 100, 284
Joseph ben Joshua Hacohen   99
Josephus   110, 131, 260
Joshua   13, 19, 24, 27 f., 34, 45, 73, 84, 86 – 88, 124, 126 – 129, 148, 150 f., 153, 157, 159, 161, 163, 166, 173 – 176, 178 – 182, 186, 190, 192, 195, 202, 208, 231, 236, 243, 262, 283, 286 f.
Joshua Alter   209, 211
Joshua of Sikhnin   18
Joshua son of Nun   86
Josippon, Book of   109 f., 236
Judah   14, 26, 28, 57, 65, 69, 76, 85, 87, 130 – 133, 136, 145, 164, 173, 176, 178 – 180, 182, 190, 195, 198 f.
Judah Loew ben Bezalel   50
Judas Iscariot   130, 136

Kadesh Barnea   4, 7, 24, 41, 50, 84, 175, 195, 245, 257
Kaftor va'ferech   35 – 37, 86
Kain, Roger   XVIII
Kaplan, Jacob   102, 120, 223 – 225, 229, 288
Kara, Joseph   26 f.
Karakh   37
Kastina   289
Kepler, Johannes   49
Keturah   169, 191
Kfar Bir'am   280
Kfar Cana   136
Kfar Hananya   37
Kfar Saba   217
Kiepert, Heinrich   171, 238, 242, 247, 272 – 274, 276, 278, 281, 288, 290
Kiryat Yearim   28

Kishon River   103, 121, 125
Kohath   62, 146
Koriat, Jehuda   57
Koronil, Nahman Nathan   74 f., 78, 285
Krakow   41, 43, 45, 69 f.
Krauze, Ephraim   274 – 276
Krieger, Johann Anton   37, 69, 137 f., 178, 180
Krup[?], Jacob   88 – 90
Kziv   69, 72, 74 f., 200

Ladino   54, 56 f., 62, 65, 223
Laicstain, Peterus   125
Lake Bardawil   134, 144, 164, 166
Land of Havilah   175, 176, 194
Land of Sihon   4, 24, 28, 34, 41
Latakia   37
Leah   274, 276
Lebanon Mountains   108, 115, 207, 231
Lebonah   69
Leiner, Gershon Chanoch (Henich)   163 – 167, 171
Lemberg   125, 223, 230, 251 f.
Leon, Jacob Jehuda   65
Leshem   166, 176, 182, 190, 194, 198
Levisohn, Solomon   223 f., 229, 288
Levites 13, 151, 177, 178, 227, 231
Levy, Rabbi   19, 86, 173, 243
Library of Congress   245
Lichtenstaedter, Beniamin   167 – 169
Liesveldt, Jacob van   173
Linetski, Yitskhok Yoyel   259 – 261, 264, 272, 288
Lipschitz, Yisrael ben Gedalia   71 f.
Litani River   227, 269
Livorno   57, 59 f., 62, 64, 182, 184
López de Gómara   100
Lot   208, 215, 269
Lotter, Tobias Conrad   135, 149 – 151, 287
Löwi, M.   254 – 256
Lublin   65 f., 69 – 71
Luria, Haim Shlomo   205, 206, 209, 212, 218 f.
Luria, Isaac   205
Luria, Solomon 71
Lydda   132, 251
Lyra, Nicholas of   20 f., 178

Ma'avarata   176
Madaba Map   19, 86, 173, 285
Magriso, Yitzhak   54, 56 f.
Mahanayim,   217

MaHaRaL = Judah Loew ben Bezalel 50, 52, 53
MaHaRaM = Meir Ben Gedalia 70,–72, 76
MaHaRSha = Eidels, Samuel (Shmuel) Eliezer ha-Levi 69–71, 75–78, 137
Maimonides 14, 72, 74, 125, 215, 243
Malkov, Avigdor 264–267, 269, 288
Manasseh 26, 28, 65, 87, 131, 136, 139, 153, 159, 166, 178–180, 182 f., 187, 194–196, 199, 267, 285 f.
Mantua 80 f., 84 f., 87–97, 99–101, 118, 146, 284
Marah 82, 145 f., 195
Mas'ei, Torah portion 1, 3, 8, 10, 18, 21–23, 25, 29, 35, 42, 50, 54, 57, 65, 66, 78, 87, 101, 139, 153, 157, 174, 176, 179, 180, 188, 194, 195, 202, 231, 267, 283, 286, 289
Matskevitsh, Shabtai 267 f., 270
Mattanah 24
Maury, Matthew Fontaine 272
Mazkeret Batya 217
Me'am Lo'ez 54–65, 146, 147,
Mecca XXII
Media 174, 176
Mediterranean, see also Great Sea XXII, 4–7, 14, 20, 23, 26, 28, 30, 51, 72, 74, 81, 83, 99, 108, 115, 124–126, 138–140, 142, 153, 158, 159, 164, 166, 175, 184, 188, 190, 195, 199, 200, 203, 204, 206, 209, 222, 228, 231, 238, 244, 251, 252, 257, 260, 263, 264, 269, 274, 278, 280, 290
Megiddo 36, 226, 237
Meir Ben Gedalia, see MaHaRam
Meiron 215
Memphis 167
Menorah 93, 206
Merari 62, 146
Mercator, Gerard 99, 114–117
Merian, Matthaeus 118 f.
Mesopotamia 198, 208, 245, 268
Metula 217
Midian 174, 191, 194, 196
Midrash 18, 87, 179, 187, 207
Migdol 82, 88, 167, 187
Miklish, Yosef ben Issachar 52 f., 78
Mikveh Israel 217, 266
Mintz, Moshe 100, 120
Miriam 60, 62, 65, 87, 147
Mishmar haYarden 217, 266, 269
Mishna 14, 69, 71–73, 78, 194, 200, 203, 251, 266, 284 f., 289

Miṣudat David 182
Miṣudat Ṣion 181
Mizrachi, Elijah 10, 28, 34, 37–43, 45, 47, 51, 53–55, 66, 80, 83, 89, 93 f., 101, 176, 179, 194 f., 284 f.
Mizrah 219 f., 262, 290
Moab 4, 24, 28, 30, 34, 41, 50, 66, 82 f., 143, 146 f., 165, 174, 183, 187, 191, 194, 196, 199, 227, 245, 265
Mohr, Abraham Menachem Mendel 223
Montefiore, Moses 213, 243
Moses 62, 77, 82 f., 97, 100, 114, 116, 124, 126 f., 146, 153, 162, 178, 199, 222, 230–232, 235, 262, 264, 286
Motza 216
Mount Abarim 164, 196, 199, 227
Mount Ararat 167
Mount Carmel 124, 204, 207, 257, 260, 268
Mount Gerizim 195
Mount Hermon 147, 153, 158, 176, 207, 231
Mount Hor 6 f., 23, 26, 28, 32, 34 f., 40 f., 51, 67, 82 f., 153, 164, 175, 187, 194–196, 266, 268, 290
Mount Horeb 192, 196
Mount Lebanon, 215
Mount Moriah 192
Mount Nebo 83, 227
Mount Seir 24, 49, 82, 192, 196
Mount Tabor 111, 190, 195
Moxoene 176
Munk, Salomon 211–213, 243–246
Munk Lithography Studio 211
Muslim 17–20, 112, 210, 260

Nablus 12 f., 69, 124, 198, 207 f., 210, 215
Nahalat Reuben 217, 266, 289
Nahaliel 24
Naphtali 65, 100, 131, 176, 178, 180, 190, 192, 195, 208
Napoleon 221, 232, 257
Nazareth 111, 136
Near East 167, 280
Negev 68, 84, 195, 252, 254, 272, 280, 286
Nephtoah 136, 189
Nile Delta 108, 112, 115, 121, 124, 139, 143 f., 263
Nile River 263
Nisin she-bayam XVIII, 28, 29, 34, 35, 49, 81
Nowy Dwor 69, 137 f., 140, 178, 180

Numbers 33   1–6, 17, 23f., 34, 54, 66, 80, 82, 87f., 116, 143, 145, 157, 163f., 188, 231, 267
Numbers 34   2, 4, 7–9, 23, 26, 41, 50, 66f., 80, 82–84, 124, 156, 164, 238, 285f.

Obadiah   71f., 111
Oboth   24
Odessa   259, 269
Og   4, 24, 28, 34, 41, 83, 161, 169, 179, 191, 194, 254
Oliphant, Lawrence   272
Ono   132
Onomasticon   19, 173
Ophrah   237
Ortelius, Abraham   125
Ostrog   69
Ottoman Empire   38

Palestine   57, 65, 86, 100, 125, 129, 137, 162, 208, 212f., 215, 218f., 221, 227, 232, 234, 236–238, 242–247, 250, 252, 259, 262, 264, 267, 274–276, 278
Palestine Exploration Fund (PEF)   247, 281, 288
Palmyra   169
Pameas   166, 255
Pamyas, Paneas, Banyas, 7, 8, 166, 194, 198, 210
Paradise, see also Eden   XXII, XXIII, 109, 168, 169, 175, 179, 194, 285
Paran Desert   147, 195
Paris   2–4, 6f., 9, 13, 15, 20f., 26f., 34, 186, 234, 243, 245
Passover Haggadah   90, 93, 113, 138, 264, 289
Pathros   167
Peniel   194
Pentapolis   111
Peqʿin   207
Peregrinations of the Israelites, see Israelites' peregrinations
Perlman, Shmuel   278–280
Persia   167, 174, 176
Persian Gulf   168, 224, 268
Petah Tikva   217, 260, 266, 274, 289
Petchenik, Barbara B.   XVII
Petermann, August   248
Pharaoh   81, 97, 116

Philistines   4, 24, 28, 34, 41, 51, 74, 86, 145, 153, 190, 196f., 227f., 251
Phoenice   133
Pi Hahiroth   82, 187, 196
Pinia, Chaim Salomon   212, 220
Pirathon   280
Pishon River   168, 169, 175, 194
Pithom   32, 158
Port Katz, Isaac   90
Prague   26, 43, 47, 49, 52, 54, 76, 121, 124, 223f., 229, 252
Pressburg   121
Promised Land   1, 3f., 7, 10, 12, 23, 32, 37, 41, 49, 73, 83f., 97, 113, 147, 151, 163, 173, 186, 188, 199, 257, 286, 290
Prostitz, Isaac of   41
Protestant maps   119
Provincia Arabia   83
Provincia Palaestina   83
Provinciali, Abraham   99
Provinciali, David   99
Ptolemaios   99
Ptolemy   125

Qaqun   37
Qarantal   111
Quaresmius, Franciscus   118, 146f., 159, 171

Raamses   3, 24, 32, 41, 49, 81, 158, 167, 187, 196, 268
Rabbat of Ammon   194, 196
Rachel   138, 262, 274, 276
Rachel's Tomb   262
Radzin   163
Rafah   37
Rakkath   176, 195f.
Ramah   251
Ramla   210, 212f.
Raphiaḥ   37
RaShBam, Rabbi Samuel ben Meir   2, 66, 68, 114
Rashi   1–24, 26, 28–30, 32, 34f., 37–43, 45, 47, 49–55, 57, 65–70, 72–74, 78, 80, 82f., 86, 89, 94f., 101, 109f., 114, 152, 156f., 159, 161f., 171, 173, 175–179, 184, 187, 189, 191, 195f., 203, 226, 238, 271, 283–286, 289f.
Red Sea   19, 30, 32, 34, 41f., 46, 52, 66, 81f., 88f., 93, 108, 118, 121, 139, 144–147, 153,

158 f., 164, 166 f., 171, 187, 192, 194–196, 199, 203, 227, 231, 249, 260
Rehovot 266,
Rehovot in the Negev 272
Reland, Hadrian 129, 223, 226, 229, 236 f.
Rephidim 82, 97
Reuben 26, 28, 65, 83, 129, 131, 153, 165 f., 178, 183, 187, 194, 196, 199, 267, 286
Reuveni, David 100
Rhinocorura, see also al-Arish
Riblah 6 f., 23, 41, 50, 83, 175
Ripa, 116
Riqueti, Joseph Shalit ben Eliezer 120
Rishon LeZion, 217, 260, 266, 289
Ritter, Carl 248
Riva di Trento 42, 92
River of Egypt 32, 41, 50 f., 72 f., 84, 151, 175, 178, 182, 189, 196, 199 f., 227, 238, 254, 269, 290
Robinson, Arthur M. XVII
Robinson, Edward 171, 236, 237 f., 242, 245, 247, 254, 257, 272, 290
Rome 100, 116
Rosh Pinna 217, 260, 266
Rothschild, Baron 217, 257
Royal Cities 254

Sabbioneta 92
Sachem Golan 217
Safad 73–75, 87, 204 f., 207 f., 210, 212, 215, 219 f., 266
Salekah 194
Salonica 54, 56, 126 f., 223
Salustius maps 19
Samaria 111, 133, 150, 198, 215, 217, 232, 251, 256, 278, 280, 287
Samuel, Tomb of 2, 69, 86, 88, 90, 93, 95, 114, 129, 216, 284
Samuel of Tzavyareh 2, 17
Sana'ah 73
Sanson, Guillaume 149
Sanson, Nicholas 149
Sanudo, Marino 86, 103, 125, 158, 173
Sapir, Eliahu 274–276
Saulcy, Louis Félicien de 258
Schattner, Isaac XV
Schottlaender Lithography Studio 214
Schubert, Gotthilf Heinrich von 113, 236
Schwarz, Yehoseph (Joseph) 75, 230–243, 249, 269, 290

Scriptures, Holy 79, 97, 109, 129, 162, 221, 224, 229, 247, 251
Sea of Galilee 7, 23, 28, 30, 41 f., 50, 66, 83 f., 87, 103, 111 f., 115, 121, 124 f., 131, 140, 149, 151, 158, 164, 175 f., 179, 182, 186, 194 f., 197, 199 f., 204, 209 f., 231 f., 238, 260, 266, 268, 278
Seetzen, Ulrich Jasper 227, 245, 247
Serpent of brass 82, 264
Seutter, Matthaeus 133–135, 150, 164, 166
Shalit, Aaron ben Haim ha-Levi 89 f., 92 f.
Shalosh Artzot le-shevi'it XIX 73, 74, 127, 200, 284
Shamir 280
Shapira, Natan ben Shimshon 45–47, 57, 71, 87, 94 f.
Sharon 111
Shaw, Thomas 40–42
Shechem 12 f., 69, 124, 198, 207 f., 210
Shemaiah of Soissons 2
Shepham 6 f., 23, 41, 51, 83, 194
Shfeya 217
Shiloh 13, 27, 37, 69, 86, 96, 124, 210
Shklov 186, 188
Showbread table 30
Shunem 36
Sibarim 178
Sidon 36, 124, 147, 164, 180, 190, 195, 199, 203, 206, 208, 210, 231, 251, 257
Sihon 24, 66, 83, 161, 187, 194
Sihor 24, 227
Simeon 26, 42, 44, 65, 131 f., 173, 176, 178, 180, 182, 190, 197, 207, 210, 215 f., 285
Sinai, Desert XVIII, 30, 41, 80 f., 101, 108, 111 f., 115, 137, 139, 143, 147, 164, 166, 170, 195, 227, 245, 251 f., 254, 257, 260, 262, 264, 286
Sinai, Mount XVIII, 32, 34, 49, 82, 97, 124, 192, 231, 236 f., 245, 264
Sippori 129, 196, 254
Sivkhi Lake, see also Hula Lake
Smith, Eli 8, 22, 171, 236–238, 240, 242, 245
Socoh 208
Sodom 111, 140, 166, 195, 204, 208, 215
Sokolow, Nahum 271–274, 276, 290
Solomon 71, 73, 117, 126, 131, 167, 178, 186, 207, 223 f., 243, 251, 262, 276
Sons of Noah 142
Spitz, Jonas 250 f.
St. Victor Abbey 20
Star of David 112, 262, 269, 271, 278, 289

Succoth   41, 49, 82, 88, 124, 187, 194, 196
Suez, Gulf of   232, 264, 267, 281
Survey of Western Palestine   247, 281

Ta'anakh   36
Tabernacle   21, 23, 28, 30, 50, 57, 60, 62, 65, 93, 117, 120, 146, 210, 257
Tadmor   169
Talmud   1f., 8, 13f., 28, 32, 38, 69, 72f., 76, 78, 87, 89, 118, 141, 145, 163, 166f., 171, 176, 196, 212, 215, 251f., 266, 285, 289
Tamar   178
Tantura   217, 257, 267
Tarnegola the Upper   255
Tarnogrod   174
Tayibe   237
Temple   2, 13, 21, 65, 73, 76, 82, 85, 93, 96f., 117–119, 135, 150f., 177f., 186, 192, 198, 202, 212, 232, 235, 243, 251, 254, 257–260, 274, 276, 278, 289f.
Teomim, Joshua Feivel ben Israel   174–180, 184, 190–192, 202, 236
Terah   208
Thiers, Adolphe   259
Tiberias   87, 126, 131f., 140, 166, 176, 196, 204f., 207f., 215, 220, 260
Tigris, River   168f., 175f., 194, 199, 203, 257, 267
Tilemanno Stella   125
Timnah   190, 251
Timnath-serah   280
Tirzah   127, 226, 280
Titus   131, 243, 260
Tor Amnon   7, 194, 266
Torai Amnon   28
Trans-Jordan   4, 217, 262
Tribal allotments   VII, XV, XVIII, XIX, XXI, XXII, XXIV, 1, 13, 19, 26, 83, 86, 87, 103, 111, 115, 121, 124, 127, 129, 131, 133, 137, 139, 142, 147, 149, 151–153, 155, 163, 164, 166, 170, 173–203, 224, 226, 227, 231, 243, 245, 248, 251, 252, 254, 257, 260, 266, 269, 274, 284–287.
Tripoli   238
Turek   209–212, 218
Tyre   36, 87, 117, 180, 195, 199, 203, 206, 208, 210, 215, 227

Urban II   17
Usha   254

Valley of Rephaim   190
Velde, Carel Willem Meredith van de   247
Venice   26, 38f., 42, 45, 47, 70, 81, 85, 87, 92, 95, 100
Vesconti, Pietro   103, 125
Vienna   121, 125, 140, 211, 213, 218, 251
Vilna Gaon, Elijah ben Shlomo Zalman   156, 186–190, 192, 195, 202, 203, 236, 271, 286, 287
Visscher, family of map-makers   117, 125, 147, 168f.
Vital, Hayyim   262

Wadowice   252
Warsaw   53, 55, 71, 73, 137f., 153, 156, 158f., 161, 264, 266, 272
Waters of Merom   166, 231, 268
Well of Miriam   60, 62, 65, 87, 147
Western Wall   207, 262
Wilderness of Kedemoth   187, 195
Wilderness of the people   30, 41, 49, 94, 147, 195
Wit, Fredrick de   117
Wood, Dennis   XVIII
Woodward, David   100, 102
Wronki   71
Würzburg   230

Yabok River   166
Yam Suf , see also Red Sea   81, 164, 166
Yarmouk River   194, 203
Yatir   280
Yavneh   36, 251
Yehudiah   217
Yesud haMa'ala, 217, 260, 266
Yiddish   65, 93, 129, 142, 210, 222, 226, 247, 257, 259f., 264, 267
Yir'on   280
Yoffe, Mordecai   47–50, 52–55, 65f., 78
Yozpa, Dov Baer,   152–159, 171, 286f.

Zach, Franz Xaver von   227
Zaddik, Yaacov ben Avraham   65, 103, 106, 108–112, 114, 119f., 129, 170, 195, 285, 287
Zamosc   125f.
Zanoah   208
Zarephath   215
Zebulun   26, 65, 103, 124, 129–131, 178, 180, 182, 190, 195, 210

Zechariah, Tomb of   86, 95, 195, 216, 226, 236
Zedad   6 f., 23, 41, 49, 51, 83, 178, 187, 194, 238, 290
Zered Brook   82 f., 187, 229
Zholkva   174, 182, 184, 222
Ziegler, Jacob   173
Zikhron Ya'akov   217, 260, 266, 269, 280
Zilpah   274, 276
Zin   4, 24, 28, 41, 73, 82, 84, 195
Zion   49 f., 66, 85 f., 96, 137, 181 f., 186, 192, 210, 215, 217, 249, 259, 265, 267, 276
Ziphron   6 f., 23, 41, 51, 83, 187, 208, 290
Ziskind, Alexander   137, 178–181, 190, 202
Zoan   159, 167
Zoar   251, 254
Zunz, Leopold   127, 221, 223, 226, 243
Zurich   80 f., 97, 121 f.

www.ingramcontent.com/pod-product-compliance
Lightning Source LLC
Chambersburg PA
CBHW060241240426
43673CB00048B/1938